Microsurgical Models in Rats for Transplantation Research

Edited by A. Thiede, E. Deltz,
R. Engemann, and H. Hamelmann

With 145 Figures and 76 Tables

Springer-Verlag
Berlin Heidelberg New York Tokyo

Editors:

Arnulf Thiede, Prof. Dr. med.
Eberhard Deltz, Dr. med.
Rainer Engemann, Dr. med.
Horst Hamelmann, Prof. Dr. med.

Chirurgische Universitätsklinik,
Abteilung Allgemeine Chirurgie,
Hospitalstrasse 40, D-2300 Kiel 1
Federal Republic of Germany

ISBN-13: 978-3-642-64899-1 e-ISBN-13: 978-3-642-61657-0
DOI: 10.1007/978-3-642-61657-0

Typesetting: Brühlsche Universitätsdruckerei, Giessen
Offsetprinting: Saladruck, Berlin · Bookbinding: Lüderitz & Bauer, Berlin
2124/3020-543210

Contributors

Abri, O.
Department of Surgery (Charité), Humboldt-University, 1020 Berlin, GDR

Ascherl, R.
Institute for Experimental Surgery, Technical University of Munich, Munich, FRG

Baumeister, R.G.
Department of Surgery, University of Munich, Großhadern, FRG

Baumgartner, D.
Department of Surgery A, University Hospital Zurich, Switzerland

Bernhard, A.
Department of Cardiovascular Surgery, University of Kiel, FRG

Bismuth, H.
Groupe de Recherche de Chirurgie Hépatique, INSERM U 17,
Hôpital Paul Brousse, Villejuif, France

Blümel, G.
Institute for Experimental Surgery, Technical University of Munich, Munich, FRG

Bockhorn, H.
Department of Surgery, University of Tübingen, FRG

Bonner, J.
Department of Surgery, University of Tübingen, FRG

Brendel, W.
Institute for Surgical Research, University of Munich, Großhadern, FRG

Brölsch, C.E.
Department of Surgery, University of Chicago, Chicago, Illinois, USA

Brunaud, M.D.
Groupe de Recherche de Chirurgie Hépatique, INSERM U 17,
Hôpital Paul Brousse, Villejuif, France

Brüsch, U.
Clinic for Abdominal and Transplantation Surgery,
Medizinische Hochschule Hannover, Hannover, FRG

Charra, M.
Groupe de Recherche de Chirurgie Hépatique, INSERM U 17,
Hôpital Paul Brousse, Villejuif, France

Chisholm, P.M.
Immunology Section, Chelsea College, University of London,
London SW3 6LX, England

Cohen, Z.
Department of Surgery and Medicine, Toronto General Hospital,
University of Toronto, Toronto, Ontario, Canada

Coles, M.I.
Harefield Hospital, Harefield, Middlesex, England

Cox, J.H.
Immunology Section, Chelsea College, University of London,
London SW3 6LX, England

Crougneau, S.
Groupe de Recherche de Chirurgie Hépatique, INSERM U 17,
Hôpital Paul Brousse, Villejuif, France

Deltz, E.
Department of General Surgery, University of Kiel, FRG

Eigler, F.W.
Department of General Surgery, University of Essen, FRG

Engemann, R.
Department of General Surgery, University of Kiel, FRG

Filler, G.
Clinic for Abdominal and Transplantation Surgery,
Medizinische Hochschule Hannover, FRG

Fox, M.
Urology/Transplant Laboratories, Royal Hallamshire Hospital,
Sheffield, England

Gabka, Ch.
Insitute for Surgical Research, University of Munich, Großhadern, FRG

Gaudecker, B. von
Department of Anatomy, University of Kiel, FRG

Geißdörfer, K.
Institute for Experimental Surgery, Technical University of Munich, FRG

Gigou, M.
Groupe de Recherche de Chirurgie Hépatique, INSERM U 17,
Hôpital Paul Brousse, Villejuif, France

Gnutzmann, K.H.
Department of Nephrology, University of Kiel, FRG

Gokel, J.M.
Department of Pathology, University of Munich, FRG

Goldmann, U.
Insitute for Surgical Research, University of Munich, Großhadern, FRG

Gracie, J.A.
Harefield Hospital, Harefield, Middlesex, England

Gray, D.W.R.
Nuffield Department of Surgery, University of Oxford, John Radcliffe Hospital, Headington, Oxford, England

Greenberg, G.R.
Department of Surgery and Medicine, Toronto General Hospital, University of Toronto, Toronto, Ontario, Canada

Griffin, R.J.
Harefield Hospital, Harefield, Middlesex, England

Groth, C.G.
Department of Transplantation Surgery and the Department of Medicine, Karolinska Institute, Huddinge Hospital, Stockholm, Sweden

Gugenheim, J.
Groupe de Recherche de Chirurgie Hépatique, INSERM U 17, Hôpital Paul Brousse, Villejuif, France

Gunnarsson, R.
Department of Transplantation Surgery and the Department of Medicine, Karolinska Institute, Huddinge Hospital, Stockholm, Sweden

Günther, E.
Department of Immunogenetics, University of Göttingen, FRG

Gutsche, H.-U.
Department of Nephrology, University of Kiel, FRG

Hahn, H.-J.
Central Institute for Diabetes "G. Katsch", Karlsberg, GDR

Hamelmann, H.
Department of General Surgery, University of Kiel, FRG

Hammer, C.
Insitute for Surgical Research, University of Munich, Großhadern, FRG

Hardy, M.A.
Department of Surgery, Columbia University College of Physicians and Surgeons, New York, NY, USA

Hepp, H.K.
Department of Gynecology and Obstetrics, University of Saarland, Homburg, FRG

Herold, A.
Urology/Transplant Laboratories and Department of Pathology Royal Hallamshire Hospital, Sheffield, England

Hopt, U.T.
Department of Surgery, University of Tübingen, FRG

Houssin, D.
Groupe de Recherche de Chirurgie Hépatique, INSERM U 17, Hôpital Paul Brousse, Villejuif, France

Iga, Ch.
Department of Surgery, Columbia University College of Physicians and Surgeons, New York, NY, USA

Ingram, N.P.
Urology/Transplant Laboratories and Department of Pathology Royal Hallamshire Hospital, Sheffield, England

Jakubowski, H.D.
Department of Surgery, University of Essen, FRG

Jessen, H.
Department of Nephrology, University of Kiel, FRG

Kamada, N.
Department of Surgery, University of Cambridge, Addenbrooke's Hospital, Cambridge CB2 2QQ, UK

Klempnauer, J.
Clinic for Abdominal and Transplantation Surgery, Medizinische Hochschule Hannover, Hannover, FRG

Konertz, W.
Department of Cardiovascular Surgery, University of Kiel, FRG

Korell, M.
Department of Gynecology and Obstetrics, University of Saarland, Homburg, FRG

Kuromoto, N.
Department of Surgery, Columbia University College of Physicians and Surgeons, New York, NY, USA

Langer, B.
Department of Surgery and Medicine, Toronto General Hospital, University of Toronto, Toronto, Ontario, Canada

Lau, H.
Department of Surgery, Columbia University College of Physicians and Surgeons, New York, NY, USA

Lee, Sun
San Diego Microsurgical Institute and Training Center, La Jolla, CA, USA

Leimenstoll, G.
Department of Nephrology, University of Kiel, FRG

Liebich, H.
Department of Pathology, University of Munich, FRG

Liebich, H.G.
Institute for Surgical Research, University of Munich, Großhadern, FRG

Lindsey, N.J.
Urology/Transplant Laboratories and Department of Pathology Royal Hallamshire Hospital, Sheffield, England

Lippert, H.
Department of Surgery (Charité), Humboldt-University, Berlin, GDR

Lorenz, D.
Department of Surgery (Charité), Humboldt-University, Berlin, GDR

Lundgren, G.
Department of Transplantation Surgery and Department of Medicine, Karolinska Institute, Huddinge Hospital, Stockholm, Sweden

Lüneburg, M.
Department of Nephrology, University of Kiel, FRG

McShane, P.
Nuffield Department of Surgery, University of Oxford, John Radcliffe Hospital, Headington, Oxford, England

Martin, E.
Department of Pathology, Hopital Antoine Beclere, Clamart, France

Meister, R.
Department of Surgery, University of Erlangen-Nürnberg, Erlangen, FRG

Morris, P.J.
Nuffield Department of Surgery, University of Oxford, John Radcliffe Hospital, Headington, Oxford, England

Müller, G.H.
Department of Surgery, University of Tübingen, FRG

Müller-Hermelink, H.K.
Department of Pathology, University of Würzburg, FRG

Müller-Ruchholtz, W.
Department of Immunology, University of Kiel, Kiel, FRG

Niedermayer, W.
Department of Nephrology, University of Kiel, FRG

Nolan, M.S.
Urology/Transplant Laboratories and Department of Pathology Royal Hallamshire Hospital, Sheffield, England

Nordgren, S.
Department of Surgery and Medicine, Toronto General Hospital, University of Toronto, Toronto, Ontario, Canada

Noss, U.
Department of Gynecology and Obstetrics, University of Saarland, Homburg, FRG

Oluwole, S.F.
Department of Surgery, Columbia University College of Physicians and Surgeons, New York, USA

Östman, J.
Department of Transplantation Surgery and Department of Medicine, Karolinska Institute, Huddinge Hospital, Stockholm, Sweden

Paul, K.
Department of Surgery, University of Erlangen-Nürnberg, Erlangen, FRG

Randzio, G.
Department of Radiology, University of Kiel, FRG

Reece-Smith, H.
Nuffield Department of Surgery, University of Oxford, John Radcliffe Hospital, Headington, Oxford, England

Reemtsma, K.
Department of Surgery, Columbia University College of Physicians and Surgeons, New York, USA

Rose, M.L.
Harefield Hospital, Harefield, Middlesex, England

Salter, D.N.
Urology/Transplant Laboratories and Department of Pathology Royal Hallamshire Hospital, Sheffield, England

Satake, K.
Department of Surgery, Columbia University College of Physicians and Surgeons, New York, USA

Schang, T.
Department of General Surgery, University of Kiel, FRG

Schareck, W.
Department of Surgery, University Hospital, Tübingen, FRG

Scheidel, P.H.
Department of Gynecology and Obstetrics, University of Saarland, Homburg, FRG

Scherer, M.A.
Institute for Experimental Surgery, Technical University of Munich, FRG

Schmid, T.
Department of General and Abdominal Surgery, University of Saarland, Homburg, FRG

Schmid, Ursula
Department of General and Abdominal Surgery, University of Saarland, Homburg, FRG

Schoel, R.
Institute of Pathology, University of Kiel, FRG

Seifert, J.
Institute for Surgical Research, University of Munich, Großhadern, FRG

Semik, M.
Department of Cardiovascular Surgery, University of Kiel, FRG

Shinomiya, T,
Department of Surgery, University of Cambridge, Addenbrooke's Hospital, Cambridge, England

Steiniger, B.
Clinic for Abdominal and Transplantation Surgery,
Medizinische Hochschule Hannover, Hannover, FRG

Strahlendorf, F.
Institute of Pathology, University of Kiel, FRG

Strowitzki, T.
Department of Gynecology and Obstetrics, University of Saarland, Homburg, FRG

Thiede, A.
Department of General Surgery, University of Kiel, FRG

Timmermann, W.
Department of General Surgery, University of Kiel, FRG

Tydén, G.
Department of Transplantation Surgery and Department of Medicine, Karolinska Institute, Huddinge Hospital, Stockholm, Sweden

Ulrichs, K.
Department of Immunology, University of Kiel, FRG

Wiebecke, B.
Institute of Pathology, University of Munich, FRG

Wiedemann, R.A.
Department of Gynecology and Obstetrics, University of Saarland, Homburg, FRG

Wolff, H.
Department of Surgery (Charité), Humboldt-University, Berlin, GDR

Wonigeit, K.
Clinic for Abdominal and Transplantation Surgery,
Medizinische Hochschule Hannover, Hannover, FRG

Wottge, H.-U.
Department of Immunology, University of Kiel, FRG

Wunderlich, A.
Department of Surgery, University of Tübingen, FRG

Yacoub, M.H.
Harefield Hospital, Harefield, Middlesex, England

Yankah, A.C.
Department of Cardiovascular Surgery, University of Kiel, FRG

Zander, E.
Central Institute for Diabetes "G. Katsch", Karlsburg, GDR

Zelder, O.
Department of Surgery, University of Marburg, FRG

Zimmermann, F.A.
Department of General and Abdominal Surgery, University of Saarland,
Homburg, FRG

Preface

Preclinical experimental transplantation research that is based on microsurgical models in rats fulfills two indispensable conditions for modern organ transplantation research: Almost all organ grafts can be performed on the rat with an amount of technical effort that is still justifiable. Thus transplantation models that are analogous to human organ transplantation can be developed, tested, and evaluated. This fulfills a necessary condition from the standpoint of surgery. With the species rat, we have a great variety of genetically different inbred strains. From the immunological point of view this is an indispensable prerequisite for the investigation of preclinical transplantation models that can be expected to produce controllable, reproducible results. In vivo experimental results can be supplemented by and correlated to in vitro tests. Lately these experimental results are being greatly expanded and more precisely defined by the application of immunohistological methods that have been established recently in Kiel.

In this book we hope to present a cross section of the microsurgical models in use today and of current immunological and immunohistological models. Furthermore, we wish to record the present state of microsurgical organ transplantation research and to show its relationship to the current state and development of clinical organ transplantation. A special aspect of our Kiel research group is the long-term, well-functioning, interdisciplinary cooperation between surgery, immunology, and pathology. Through this cooperation we attempt to provide an atmosphere in which theoretical and practical viewpoints can mutually influence each other.

The first part of this book presents the fundamentals, such as microsurgical models (A) and immunological principles (B). The experimental and immunological background of transplantation research is discussed. The second part continues with results of experimental transplantation research. The topics are: vessel and heart (C), small bowel (D), pancreas and intestines (E), liver (F), and miscellaneous organs (G). In the third part (H) experimental results are illuminated from the clinical point of view, and we attempt to appraise the value of these experimental results for the clinical questions related to organ transplantation. To stress particularly the clinical relevance, experienced clinical transplantation surgeons with both experience and an interest in experimentation present their clinical data and discuss and elucidate the experimental results.

Altogether we wish to answer the following three questions:

1. Which model is appropriate for which immunological question?
2. Which experimental rat transplantation model is appropriate for solving which clinical problem related to organ transplantation, and which model is of no significance for clinical transplantation?
3. With what model does a young physician have a chance to contribute to a solution of a clinical problem related to organ transplantation?

We hope this book will stimulate young physicians who are doing experimental work and are interested in transplantation. We would like to motivate them to use a suitable transplantation model and the immunogenetic conditions found in the rat in their experimental transplantation research.

The Editors

Contents

B. Immunological Principles

C. Results of Vessel and Heart Transplantation

D. Results of Small-Intestine Transplantation, Graft Versus Host Reaction, and Host Versus Graft Reaction

E. Pancreas Transplantation Versus Islet Transplantation

F. Results of Liver Transplantation

History and Development of Experimental Microsurgery in Rats

Sun Lee *

Shortly after the end of World War II, most American universities were equipped with dog laboratory facilities where vascular surgeries were tested to develop open heart surgery and kidney transplantation techniques. The University of Pittsburgh was no exception. When I completed general surgical residency at St. Francis Hospital in Pittsburgh in 1955, I was invited to join the surgical research team conducted by Dr. Bernard Fisher, Surgical Research Division of the Department of Surgery, University of Pittsburgh. At that time, surgical physiological observation of hypothermia, portacaval shunt, and vascular reconstruction was highlighted, employing the dog and skin graft experiment in the rat. At that period, rats were seldom used in surgical laboratories despite a long history of rat usage in allied biological research.

While we were actively engaged in kidney transplantation research, we had many occasions to use neonatal puppies for such purposes. Using these small animals, vascular anastomosis were practiced, in addition to acquiring thymectomy procedure and other micromanipulations including thyroid and parathyroid transplantations in dogs. In 1957, Mr. Singerman, Davis & Geck, American Cyanamid Company, brought my attention to a so-called microneedle swedged with 7–0 silk which was braided and siliconized to make it smooth. Dr. Fisher urged me to try a portacaval end-to-side procedure in the rat, and I made an extensive 6-month search for suitable microsurgical instruments. "During medical school I had tried to perform portacaval shunt in the rat but it was too difficult, the vessels tore, and I failed." Dr. Fisher was very eager to have me pursue this procedure. I tried various shapes and sizes of women's hairpins on the rat blood vessels but they were soon discarded, being either too loose or too sharp for such delicate blood vessels. By 1958, a miniature Satinsky clamp (Fig. 1) was designed with the help of Lawton Instrument Company instrument mechanics and the blades of the clamps were covered with rubber tubing (Davol Co.) which was air passage tubing for an endotracheal tubal cuff. After many attempts and failures due to malalignment of the two blood vessels, blood loss from the needle holes and gaps, constriction of the stoma, and anesthesia problems, end-to-side porta-

* The author's name was changed from Sil Heung Lee to Sun Lee in 1961 at the time of United States naturalization

Microsurgical Models in Rats
for Transplantation Research
Eds. A. Thiede et al.

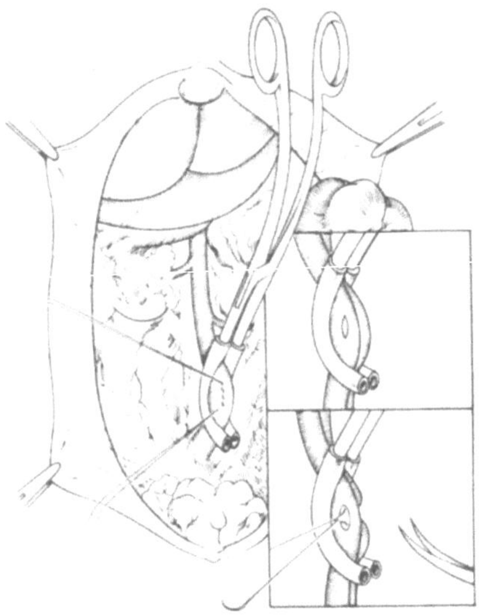

Fig. 1. A miniature Satinsky clamp applied to the aorto-vena cava in the rat

caval shunt was finally perfected in the rat in 1958. Dr. Fisher revised the 1959 edition of *Blood Vessel Surgery and Its Applications* written by Dr. Guthrie (Fig. 2), and in it he stated: "Do you know what you have done just now? You have opened an avenue to conduct allied physiological research, transplantation investigation, and it may well replace dog surgery in the years to come." He further stated: "We fully expect that vascular surgery will be applied to smaller and smaller vessels, for such is possible and will be utilized by those who are willing to master the technical skill. In our laboratory, end-to-side portacaval shunt is being performed by Dr. S. H. Lee on 300 gm rats with routine success by the same suture technique used in humans (7–0 silk), and the vascular anastomoses remain patent indefinitely. Such technical gymnastics might well be clinically utilizable. Particularly would this be so if organ transplantation became feasible. Of course, the smaller the blood vessel operated upon, the narrower becomes the permissible technical error that is comparable with success."

By the end of 1958, a paper entitled "Portacaval shunt in the rat" was sent to *Surgery* (Dr. Owen Wangensteen, chief editor), and a reply from Dr. Wangensteen about 6 months after submission stated: "Thank you for submitting the very interesting paper entitled, Portacaval shunt in the rat', but we cannot accept it for publication in *Surgery* at this time. We assume that such surgery has been performed by other(s) and we would like for you to write the following authors. If they deny such performance of the surgery in the rat, then you may claim your priority and it will be accepted for publication in *Surgery*." Enclosed were about ten names with addresses. Almost a year was spent in corresponding with these authors since many had moved. None claimed they had carried out suture anastomosis in the union of the inferior vena cava to the portal vein in the rat. We then resubmitted to *Surgery* and the paper appeared in 1961 (Lee and Fisher 1961 b). In the meantime, techniques of side-to-side portacaval shunt, arteriovenous fistula, liver arterialization, and kidney transplantation in the rat were cre-

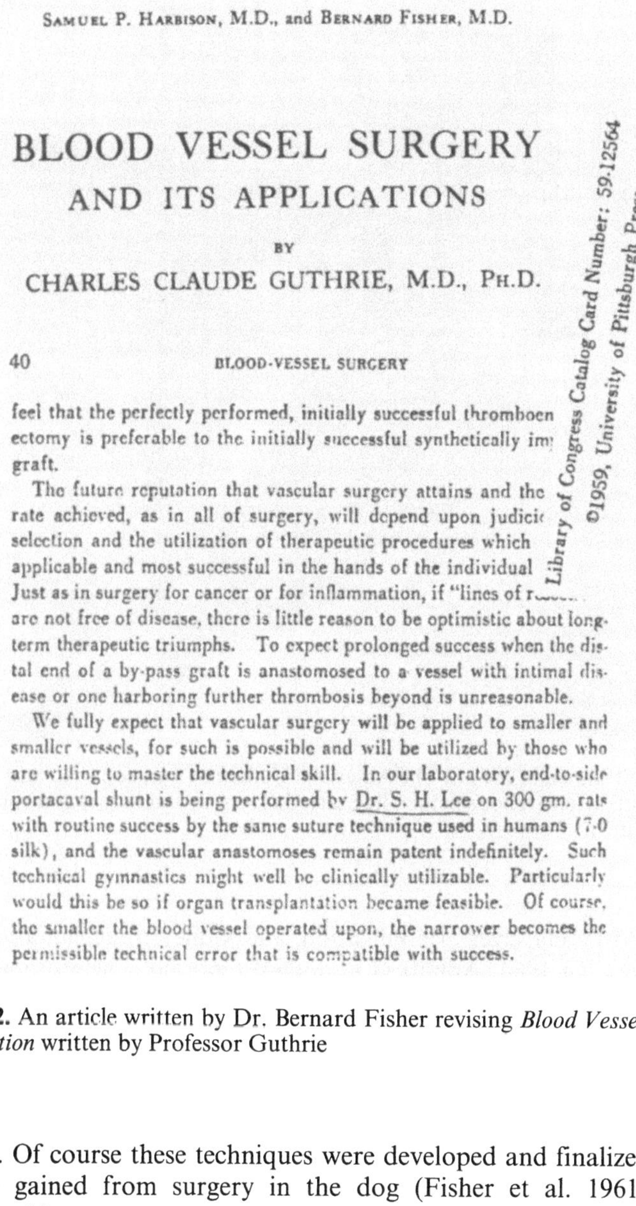
SAMUEL P. HARBISON, M.D., and BERNARD FISHER, M.D.

BLOOD VESSEL SURGERY
AND ITS APPLICATIONS
BY
CHARLES CLAUDE GUTHRIE, M.D., PH.D.

Library of Congress Catalog Card Number: 59:12564
©1959, University of Pittsburgh Press

40 BLOOD-VESSEL SURGERY

feel that the perfectly performed, initially successful thromboen ectomy is preferable to the initially successful synthetically im graft.

The future reputation that vascular surgery attains and the rate achieved, as in all of surgery, will depend upon judici selection and the utilization of therapeutic procedures which applicable and most successful in the hands of the individual Just as in surgery for cancer or for inflammation, if "lines of r are not free of disease, there is little reason to be optimistic about long-term therapeutic triumphs. To expect prolonged success when the distal end of a by-pass graft is anastomosed to a vessel with intimal disease or one harboring further thrombosis beyond is unreasonable.

We fully expect that vascular surgery will be applied to smaller and smaller vessels, for such is possible and will be utilized by those who are willing to master the technical skill. In our laboratory, end-to-side portacaval shunt is being performed by Dr. S. H. Lee on 300 gm. rats with routine success by the same suture technique used in humans (7-0 silk), and the vascular anastomoses remain patent indefinitely. Such technical gymnastics might well be clinically utilizable. Particularly would this be so if organ transplantation became feasible. Of course, the smaller the blood vessel operated upon, the narrower becomes the permissible technical error that is compatible with success.

Fig. 2. An article written by Dr. Bernard Fisher revising *Blood Vessel Surgery and Its Application* written by Professor Guthrie

ated. Of course these techniques were developed and finalized from the knowledge gained from surgery in the dog (Fisher et al. 1961; Lee and Fisher 1961 a, b).

While we were busy with fundamental liver physiology using the portacaval shunt model in the rat, the Cincinnati group headed by Miller (Miller et al. 1962) reported a model of rat kidney transplantation using the neck blood vessels. Because it is a nonphysiological model, this technique has not been popular among immunological investigators. In the meantime, we reinforced the technical feasibility of rat blood vessel surgery at the 1961 Annual American College of Surgeons Congress in the scientific exhibit section (Fig. 3), which included portacaval

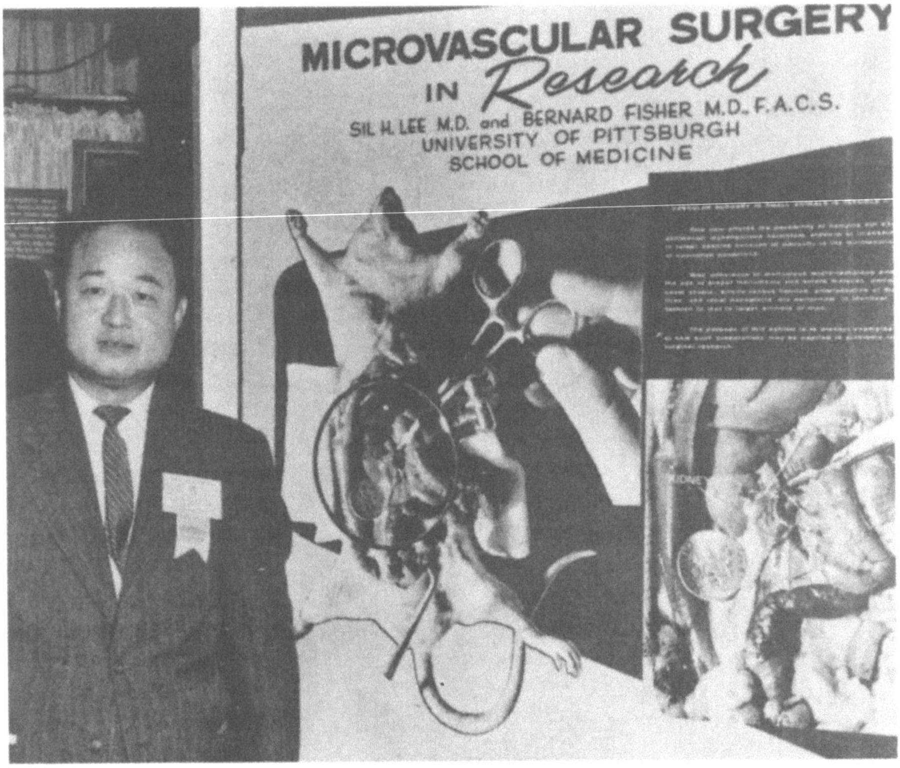

Fig. 3. Scientific exhibit shown in 1961 at The American College of Surgeon's Annual Congress: portacaval shunt, arteriovenous fistula, liver arterialization, and kidney transplantation in the rat were shown

shunt, arteriovenous fistula, liver arterialization, and kidney transplantation in the abdominal cavity. In 1964, Abbott et al. (1964) reported a heterotopic rat heart transplantation in an end-to-end anastomosis between the donor aorta and pulmonary artery and the recipient's abdominal aorta and inferior vena cava. In 1963 Dr. Sakai from Japan joined our research group and we geared many projects to demand at that time using dogs and rats. Since I was engaged in a clinical renal transplantation project, I was very uncomfortable due to my lack of transplantation immunology background. Therefore, I accepted an invitation to join Dr. Frank Dixon's institution in La Jolla, California, in 1964 for a period of 3 years. At that time I used my technical skill to investigate allied transplantation immunology and to acquire transplantation immunology background. At Scripps Clinic and Research Foundation, I have had a good opportunity to refine renal transplantation technique in the rat so that we were able to investigate the fate of both Masugi and Heyman's nephritis. Dr. Emil Unanue, an able assistant of Dr. Dixon's, and I studied intensively on the Masugi nephritis model in the rat to observe the mode of transference of autoantibody from diseased to normal in the renal transplantation animals (Unanue et al. 1965, 1966). The end-to-end suture technique of Abbott was soon modified by Tomita (1966) in an end-to-side fashion. Tomita's suture technique was somewhat cumbersome because of up-and-down two raw stitching on each side of the vascular edges and was soon further modified. Of course, when Tomita was suturing these vessels, only 7–0 silk sutures were available.

Two major events occurred early in 1964 in La Jolla. The first event was that Dr. Sakai was no longer with Dr. Fisher and wished to learn kidney transplantation in the rat. He spent several weeks in La Jolla to master the transplantation technique. Since he was familiar with the portacaval shunt procedure in the rat, rat blood vessel anastomosis was soon not a difficult problem for him in the renal transplantation procedure. Second, I was invited to the Harvard University Peter Brigham Hospital twice to teach scholars of Professor Merrill, where I met Dr. Rowinski and others. I taught several surgeons for 1 week and then returned to La Jolla. Within a few weeks, Dr. Merrill called me to Boston to teach another group of doctors, where I had the pleasure of meeting a young scholar, Dr. Guttmann. According to Dr. Merrill, Dr. Rowinski was the only surgeon who could perform rat renal transplantation and he had left for Poland. Dr. Guttmann, who was an internist, wanted to learn the delicate renal transplantation technique. He was highly motivated. After a 1-week course at Harvard, we exchanged letters and telephone calls, and I sent several hand drawings to him. As we at Scripps (Drs. Feldman, Edgington, and fellows) published several scientific papers on the use of rat renal transplantation (Edgington et al. 1969; Feldman et al. 1968; Lee 1967; Unanue et al. 1965, 1966), Dr. Guttmann and his associates started to report their results on rat renal transplantation in *Transplantation* and other leading journals (Guttmann and Lindquist 1969; Guttmann et al. 1967, 1969 a, b). In the meantime, Dr. Sakai again paid a lengthy visit to our laboratory prior to moving to Europe.

In 1966, at the Annual Congress of the American College of Surgeons, we reported the feasibility of transplanting partial liver (Lee and Edgington 1966) a heterotopic fashion with hepatoarterial revascularization alone. This was followed by a report of immunopathological sequences of partial liver grafts supplied by both hepatic arterial and portal venous afferent blood (Lee and Edgington 1968) and that of liver grafts in the rat supplied by hepatic arterial blood alone. We had a steady flow of visitors to our laboratories as Dr. Dixon advocated my work whenever he was a speaker at various centers. Dr. Coburn came to consult us on various aspects of rat liver and spleen transplantations.

Professor Cortesini in Rome, advised by Dr. Dixon, invited me to conduct the First Italian Microtransplantation Course at the University of Rome in late 1967, and I was able to meet more than 20 Italian scholars. Among these, Mazzoni et al. (1968), Pettirosi et al. (1968), Ancona et al. (1968), and a few others (Bui-Mong-Hung et al. 1969; Lauschke and Hermann 1968) were able to perform kidney and heart transplantation in rats. The following year I met Dr. Mikaeloff from Leon while attending the Liver Symposium; I visited Leon after the symposium to conduct a small course on rat kidney and liver transplantation. Dr. Mikaeloff and his associates thereafter published several papers pertaining to rat liver transplantation (Mikaeloff et al. 1968, 1969).

In 1968, I joined Dr. Marshall Orloff in the Department of Surgery of the newly established School of Medicine at the University of California, San Diego. Here, emphasis was on surgical techniques and surgical physiology as a means of perfecting procedures in humans. While kidneys and livers were being operated upon daily at the temporary facilities at Mercy Hospital in San Diego, I focused upon the liver regeneration problems using the partial liver transplant models in the rat. We reported the rat spleen transplantation technique (Lee and Orloff 1969), as did Coburn in 1969. We also reported on rat heart-lung transplantation in 1970. Dr. Koopmans (graduate student of Professor Deutch), Dr. Smallwood, and I spent almost every evening at the Scripps Clinic Immunology Laboratory to perfect heart-lung (Lee et al. 1970), testicle (Lee et al. 1971), and stomach transplantations (Lee et al. 1972a) in the rat. Dr. Koopmans and I later spent more time together to perfect pancreas transplantation (Lee et al. 1972b).

Once a week, from midsummer of 1969, Dr. Yoko Mullen visited Scripps Clinic Immunology Laboratory, where I was involved with rat kidney transplantation. She became a very competent microtransplant surgeon, contributed very valuable publications, and became a right-hand assistant to Dr. Hildemann. Dr. Takeuchi from Tokyo University, a veterinarian fellow at the Veterans Administration Hospital in Livermore, California, visited me several times to master the technique of bilateral kidney transplantation in the rat; later he reported the results of antigen load in the organ transplantation at the Second International Microsurgical Congress held in La Jolla. In the meantime, many important contributions were made by various scholars; while Fox et al. (1971) reported renal transplantation results in genetically hypertensive rats, Reemtsma et al. (1968) reported the partial rat pancreas transplantation in 1960 and 1968. In 1971, Monchick and Russell reported small-intestine transplantation in the rat. Further contributions on rat small-intestine transplantation were made by Kort et al. (1973) and Thiede et al. (1976).

In order to assemble transplantation microsurgeons, Professor Van Bekkum, in Holland, organized the First International Microsurgical Congress at his Radiological Institute in 1969. Kidney, liver, and other organ transplantation techniques were discussed. During the workshop, rabbit kidney was transplanted to rat, and a mouse kidney transplantation was attempted.

In 1971, the second congress of the International Microsurgical scholars was held in La Jolla, California. Dr. Chandler was one of the best organizers at that meeting. Among several papers, two outstanding contributions are worth mentioning: the first one was presented by Cho et al. (1972), who compared free versus immediately revascularized skin grafts in rats, and the second was a comparison of single and double kidney grafts in the rat by Takeuchi (1972).

In 1973, the third congress was held in Rome-Valomontone under the direction of Professors Mazzoni and Cortesini. Both clinical and experimental microsurgical scientific papers were discussed. While we presented techniques of liver and pancreas transplantations in rats, many researchers reported both physiological and immunological sequences following heart and kidney transplantations in the rat.

The fourth congress held in Mexico City under Dr. Chavez-Peon was very meaningful and one to the remembered because of the official designation of the

Society to be known as The International Microsurgical Society, Dr. Jacobson, the pioneer of microsurgery, was present at the meeting and many valuable scientific papers were read and discussed. Both ourselves and others reported a pancreas transplant model in diabetic rats to treat diabetes, and some institutions had also tried to transplant the whole liver in an orthotopic fashion. Clinical replantation cases among the world's microsurgeons had tremendously increased.

The German meeting in 1977 was very well organized under the direction of Professor Lie in Bonn. It was the fifth such congress and was divided into two parts, clinical and experimental microsurgical sections, because of an increased number of participants. There, fertility and infertility problems were highlighted and rabbit and rat models were used to study the competent reconstruction of fallopian tubes by Boeckx et al. (1976), Urue-Buscarons et al. (1979), and others (Lee 1974). Dr. Biemer reported many cases of plastic reconstruction in Germany. At this time, orthotopic liver transplants in the rat were carried out by many centers including Kiel, Bonn, and France, in addition to our group, and the question was raised as to the hepatic arterial reconstruction of the newly transplanted rat liver.

In 1980, because of geographical reasons, not many scholars gathered at the Sixth International Microsurgical Congress in Sydney under Dr. Owen. It was the first time Chinese delegates participated in the congress. They presented various problems in dealing with industrial accidents including microsurgical application, and the hundreds of cases of toe transfer to replace thumb were indeed worth mentioning from Peking and Shanghai.

The Seventh International Microsurgical Congress was held in Leon in 1982, organized by Dr. Dubernard. A proposal to merge two microsurgical journals, the International Journal of Microsurgery and the Journal of Microsurgery, into one was made because of the difficult financial situation of the publishers.

The value of microsurgery has been fully understood by many surgical subspecialists because of accomplishments in both clinical and research microsurgery. It has been noted that microsurgery no longer remains in the laboratory as a curious part of medicine, but is used to treat and ease the agony, pain, and psychologically detrimental aspects, such as disfigured faces. The presence of Dr. Terzis was greatly appreciated because of her discussion of various problems in dealing with neural reconstruction in both animals and man. Because of a lack of clinical practice materials for beginners, it has been urged that a suitable animal model (rat) is used for training, since rats are hardy and not much public sentiment is attached to them.

While liver heterotopic and orthotopic transplants have been perfected by us and the Holland group, heart and kidney transplants in the rat have been widely employed in immunological studies. Before changing from the renal transplant model to the heart transplant model, Guttmann, Fabre and Morris (1973) undertook vigorous investigation on the immunological consequences of using the renal transplant model beginning in 1973. There are many men to be remembered as pioneers, such as Drs. Kort and Thiede, who extensively studied small-intestinal transplantation, and Pettirosi et al. (1973), who observed the lymphatic drainage from freshly transplanted small intestines.

Over 20 years have passed since the first publication of rat blood vessel surgery was reported by us, and many contributors have studied both surgical physiology and transplantation immunology in rat organ transplantation. Among them, we should not forget these names in addition to the ones mentioned previously: Drs. Ono and Tilney, who carried out extensive work on rat heart surgery; Hess, Thiede, and Lie, who followed our techniques of either heterotopic or orthotopic liver transplantations; Morris, Sakai, Marquet, and others, who investigated kidney transplantations; and Nozawa, Weil, and the Minnesota group (Drs. Najarian, Sutherland, and others), who have studied pancreas transplantation in diabetic rats. Drs. Altwein and Brekke played a major role in testicular transplantation and endocrinological investigations along with ourselves. Of course, there are many unsung heroes I could not enumerate in these limited pages. During these years, many associates have shared difficult times with me and, more frequently, joyous times. Dr. Rosen was very practical and knowledgeable in transplantation physiology; Drs. Edgington, Tung, Feldman, and Wolf were very helpful in the interpretation of histological sections. Drs. Chandler and Skivolocki from the U.S.; Drs. Nozawa, Ueda, and Niwa from Japan; Drs. Broelsch and Altwein (testicle transplantation studies) from Germany; Drs. Storck and Borjosson from Sweden; Drs. Flamont and Domerque from France; Dr. Li from China; Drs. deMacedo and Bacchi from Brazil; Drs. Gonzales, Jaureguizar, and Fernandez-Cruz from Spain; and our residents all aided in the development of various transplantation techniques in the rat. Finally, I am highly indebted to three outstanding men: Drs. Bernard Fisher, Frank Dixon, and Marshall Orloff, who "pulled and pushed" to develop these techniques.

References

Abbott CP, Lindsey ES, Creech O Jr, DeWitt CW (1964) A technique for heart transplantation in the rat. Arch Surg 89:645

Ancona E, Perachia A, Cortesini A, Frasson P, Omodei S (1968) Homotransplantation of heart in the rat. Prog Transplant Congr 2:136

Boeckx W, Brosens I, Vasquez G (1976) Microsurgical reanastomoses of the fallopian tube after sterilization. Proc Int Microsurg Soc 2:49

Bui-Mong-Hung, Licopolous P, Vigano M, Leandri J (1969) Homotransplantation hétérotopique du coeur chez le rat; étude des stades intiaux du reject et influence de la conservation. Third Congress of the European Society of Surgery, Munich

Cho SI, Marcus FS, Kountz SL (1972) A new model for study of allograft rejection in the rat: use of skin with an intact vascular pedicle. I. Effect of vascularity on allograft survival. Transplantation 13:486

Coburn RJ (1969) Spleen transplantation in the rat. Transplantation 8:86

Edgington TS, Lee S, Dixon FJ (1969) Persistence of the autoimmune pathogenetic process in experimental autologous complex nephritis. J Immunol 108:528

Fabre JW, Morris PJ (1973) Dose response studies in passive enhancement of kidney grafts in rats. Transplantation 15:397

Feldman JD, Pick E, Lee S, Silvers WK, Wilson DE (1968) Renal transplantation in rats. II. Tolerant recipients. Am J Pathol 52:687

Fisher B, Fisher ER, Lee SH (1961) The effect of alteration of liver blood upon experimental hepatic metastasis. Surg Gynecol Obstet 122:11

Fox U, di Francisco GF, Longoni F (1971) Nota su di un tipo di anastomosi uretero-vesicale net trapianto di rene nel ratto. Minerva Urol 29:240

Guttmann RD, Lindquist RR (1969) Renal transplantation in the inbred rat. XI. Reduction of allograft immunogenicity by cytotoxic drug pretreatment of donors. Transplantation 8:490
Guttmann RD, Carpenter CB, Lindquist RR, Merrill JP (1967) Renal transplantation in the inbred rat. III. A study of heterologous antithymic sera. J Exp Med 126:1099
Guttmann RD, Lindquist RR, Merrill JP (1969a) Renal transplantation in the inbred rat. VIII. Quantitation of cell proliferation during unmodified rejection. Transplantation 8:19
Guttmann RD, Lindquist RR, Ockner SA (1969b) Renal transplantation in the inbred rat. IX. Hematopoietic origin of an immunologic stimulus of rejection. Transplantation 8:472
Kort WJ, Westbroek DL, McDicken I, Lameijer LDF (1973) Orthotopic total small bowel transplantation in the rat. Eur Surg Res 5:81
Lauschke H, Hermann G (1968) Eine Technik zur heterotopen Nierentransplantation bei Ratten. Z Gesamte Exp Med 147:29
Lee S (1967) An improved technique of renal transplantation in the rat. Surgery 61:771
Lee S (1974) Experiences with organ transplantation in the rat. IX. Implantation vs transplantation of the ovary. J Res Inst Med Sci Korea 6:112
Lee S, Edgington TS (1966) Liver transplantation in the rat. Surg Forum 17:220
Lee S, Edgington TS (1968) Heterotopic liver transplantation utilizing inbred rat strain. I. Characterization of allogeneic graft rejection and the effects of biliary obstruction and portal vein circulation on liver regeneration. Am J Pathol 52:649
Lee SH, Fisher B (1961b) Portacaval shunt in the rat. Surgery 50:668–672
Lee SH, Fisher B (1961a) Microsurgery in research. Scientific exhibit (portacaval shunt, liver arterialization, arteriovenous fistula and renal transplantation in the rat). Annual Congress of American College of Surgeons
Lee S, Orloff MJ (1969) A technique for splenic transplantation in the rat. Surgery 65:436
Lee S, Willoughby WF, Smallwood CJ, Dawson A, Orloff MJ (1970) Heterotopic heart and lung transplantation in the rat. Am J Pathol 59:279
Lee S, Tung KSK, Orloff MJ (1971) Testicular transplantation in the rat. Transplant Proc 3:586
Lee S, Edgington TS, Smallwood CJ, Thompson AG, Chandler JG, Rosen H, Orloff MJ (1972a) Secretory function and relative resistance of the gastric mucosa of the rat to allogeneic rejection. Transplantation 14:641
Lee S, Tung KSK, Koopmans H, Chandler JG, Orloff MJ (1972b) Pancreaticoduodenal transplantation in the rat. Transplantation 13:421
Mazzoni G, Valli A, Demofonti A, Francesconi M (1968) Transplantation of rat kidney. International College of Surgeons, Italian Section, Florence
Mikaeloff P, Levrat R, Nesmoz P, Rassat JP, Philippe M, Dubernard LM, Ramesau A, Bel A (1968) Transplantation hépatique chez le rat. Presse Med 76:1521
Mikaeloff P, Levrat R, Nesmoz P, Rassat JP, Philippe M, Dubernard LM, Bel A (1969) Heterotopic liver transplantation in the rat. Value, technique, results of about 70 cases. Lyon Chir 68:133
Miller BF, Gonzales E, Wilchins LJ, Nathan P (1962) Kidney transplantation in the rat. Nature 194:310
Monchick G, Russel PS (1971) Transplantation of small bowel in the rat. Technical and immunological action of homologous antiserum. Transplant Proc 3:697
Pettirossi O, Ronzoni G, Maisano A (1968) Kidney transplantation in the rat and results in assessment of antirejection treatment. Chir Patol Sper 16:291
Pettirossi O, di Pietrantono M, Cherubini M (1973) Intestinal transplantation in the rat with cannulation and external drainage of the transplanted intestinal lymphatic duct. Eur Surg Res [Suppl] 125
Reemtsma K, Gialdo N, Depp DA, Eichwald EJ (1968) Islet cell transplantation. Ann Surg 168:438
Takeuchi A (1972) Single vs double kidney allografts: effect of antigen overload. Second Congress of the International Microsurgery Society, La Jolla

Thiede A, Deltz E, Lee S (1976) Allogene Dünndarmübertragung bei Inzuchtratten. Transplantationstechnik und morphologische Befunde. Langenbecks Arch Chir Suppl 29–32
Tomita F (1966) Heart transplantation in the rat. Sapporo Med J 30:165
Unanue ER, Lee S, Dixon FJ, Feldman JD (1965) Experimental glomerulonephritis. VII. The absence of an autoimmune antikidney response in nephrotoxic serum nephritis. J Exp Med 122:565
Unanue ER, Dixon FJ, Lee S (1966) Experimental glomerulonephritis. VIII. The in vitro fixation of heterologous nephrotoxic antibodies to, and their exchange among tissues of the rat. Int Arch Allergy Applied Immunol 29:140
Urue-Buscarons J, Ribans-Barba C, Mestres-Estratus J, Fargas-Moreno F-J (1979) Effects of microsurgical techniques on the rat uterus. Proc Int Microsurg Soc 3:180

Reflections on the Requirements and Perspectives of a Microsurgical Research Group

A. Thiede, E. Deltz, and *R. Engemann*

To establish a research group that will work on microsurgical techniques one needs not only the necessary rooms and instruments, but also young scientists with the motivation to work. Of equal importance, however, is a mentor who will stimulate the team to long-lasting and productive cooperation. For 10 years we have considered Sun Lee, who is known as the father of rat microsurgery, as our mentor (Fisher and Lee 1965; Lee and Fisher 1961; Lee 1967; Lee and Orloff 1979).

The definition of microsurgery as surgery with auxiliary optical equipment, such as magnifying glasses or operating microscopes, clearly classifies microsurgery not as a special discipline but as an auxiliary technique for various surgical disciplines. Microsurgery can be a link between various medical disciplines and, acting as such a link, it has been the basis of long-term interdisciplinary development of a scientific research program at the University of Kiel, finally leading to the establishment of a clinical organ transplantation group.

Microsurgery for experimental transplantation research is dependent on the following requirements:

1. Auxiliary optical equipment
2. Microsurgical instruments
3. Small animals (such as rats)
4. Particularly fine atraumatic needle-suture combinations

The optical equipment is described in Table 1. We distinguish between magnifying glasses and operating microscopes, which are compared in this table. To some extent they compete with each other in microsurgery, but on the whole they should be considered to be complementary. Magnification is, of course, much better with the operating microscope (up to 30-fold), but it therefore has a smaller depth of focus and field of view. Illumination of the area of operation is superior with the operating microscope, due to the built-in lamp; hence additional lamps are unnecessary. Vascular anastomoses can be performed down to a minimum of 1 mm in diameter with magnifying glasses, whereas with the operating microscope it is possible to perform anastomoses, e.g., in the femoral region of the rat, with a diameter of 0.3 mm. Film and photographic documentation is also possible with the operating microscope. Minimal training is required to use magnifying

Microsurgical Models in Rats
for Transplantation Research
Eds. A. Thiede et al.

Table 1. Comparison of magnifying glasses and the operating microscope

	Magnifying glasses	Operating microscope
Magnification	×6	×30
Depth of focus	Good	Moderate
Field of view	Relatively large	Relatively small
Illumination of the operating field	Available	Available, built-in
Vascular anastomoses	Down to a minimum of 1 mm in diameter	Down to a minimum of 0.3 mm in diameter
Film and photographic documentation	–	Possible
Training requirements	Low	High
Handling	Easy	Moderate
Cost	$ 30–$2000	$ 2000–30,000

Table 2. Advantages of inbred rat strains for transplantation research

1. Constant genetic difference within an experimental group
2. Syngeneic control groups for checking the transplantation technique
3. Defined variation of the genetic difference between experimental groups
4. Breeding of F_1-hybrid combinations for separating, analyzing, and manipulating isolated graft versus host and host versus graft

Table 3. Inbred rat strains (available in Kiel, SFB 111)

Strain	MHC	F_1 hybrids	MHC
AS	$RT1^l$		
AS2	$RT1^f$		
AVN	$RT1^a$		
BN	$RT1^n$		
CAP	$RT1^c$		
DA	$RT1^{av1}$		
F344	$RT1^{lv1}$	$(CAP \times F344)F_1$	$RT1^{c/lv1}$
LEW	$RT1^l$	$(BN \times LEW)F_1$	$RT1^{n/l}$
PVG	$RT1^c$		
L.1N	$RT1^n$		

glasses, but very much more is needed for the operating microscope. Handling of magnifying glasses is easy, whereas that of the operating microscope is more difficult. In 1984 costs range from $ 30 to $ 2000 for magnifying glasses with individually adapted lenses, whereas operating microscopes cost anything from $ 2000 to $ 30000, depending on the type and equipment. The special surgical instruments and the particularly fine atraumatic needle-suture combinations have been described in detail elsewhere (Silber 1979; Engemann et al. 1982).

The choice of the species rat for experimental microsurgery has the advantage that inbred strains are available (see Table 2). Outbred rats are used only for the training program. To study transplantation immunology we use inbred rats. The following rats are bred in Kiel (Table 3): AS, AS2, AVN, BN, CAP, DA, F344, LEW, PVG, and LEW 1N. The F_1 hybrids we have used so far are CAP × F344 and BN × LEW. Both were bred for special questions related to small-bowel transplantation. Additionally we have used congenic and intra-major histocompatibility complex (MHC) recombinant congenic strains in recent years. The inbred strains enable different immunogenetic combinations to be made. In a syngeneic combination, both the MHC and non-MHC must be compatible, for example LEW-LEW. In allogeneic combinations (not using recombinant congenic strains) one can distinguish between the following possibilities:

1. MHC compatible, non-MHC incompatible (e.g., AS-LEW)
2. MHC incompatible, non-MHC compatible (e.g., L.1A-LEW)
3. MHC and non-MHC incompatible (e.g., BN-LEW, DA-LEW)

Besides the immunogenetic barrier between donor and recipient the role of regulatory *Ir* genes (Butcher and Howard 1982) has recently become more and more important, since the strength of an immune response to an allograft may also depend on the direction of grafting (e.g., L.1A→L.1U liver grafts rejected, L.1U→L.1A liver grafts accepted; Engemann 1984, unpublished data). Summarizing, one can say that the use of inbred strains offers us a number of special possibilities from the view of transplantation biology (Thiede 1977).

The following part deals with the program we offer the young scientist who starts to work with experimental microsurgery. First he (or she) is instructed on how a normal healthy rat looks. Then we emphasize some special symptoms of pneumonia, peritonitis, etc. This is important for a reliable follow-up after major surgery in the rat, especially if the animals have been treated with immunosuppressive drugs.

Surgery in the rat as well as in humans can be performed only with excellent anesthesia. We favor ether anesthesia for rat microsurgery, because it is easy to learn for the beginner and even major surgery such as pancreaticoduodenal transplantation or liver transplantation can be performed without anesthesiological problems. Of course, there are other forms of anesthesia, e.g., with barbiturate, i.p. or a combined ether-neuroleptic anesthesia. This may be of value for special surgical procedures.

However, the main object is to introduce the beginner step-by-step to vascular and organ transplantation techniques. In our opinion there is no point in confronting the junior microsurgeon, who is usually a student, with a difficult transplantation technique. He must be allowed to gain experience in simple vascular suture techniques, which he must be able to perform automatically. Only then is it worthwhile teaching him difficult and complex organ transplantation techniques and perhaps in the last stage introducing him to experimental liver transplantation. The introduction of the training program begins with preparation and suturing of the aorta (diameter 1.2 mm); the simplest operating technique is the end-to-end anastomosis of the aorta. The suture of the aorta can be performed as an interrupted suture, or for the advanced surgeon as a running suture. In addition to the simple aortic suture, we have been concerned with the transplantation of segments of the aorta for years, to be able to study questions related to

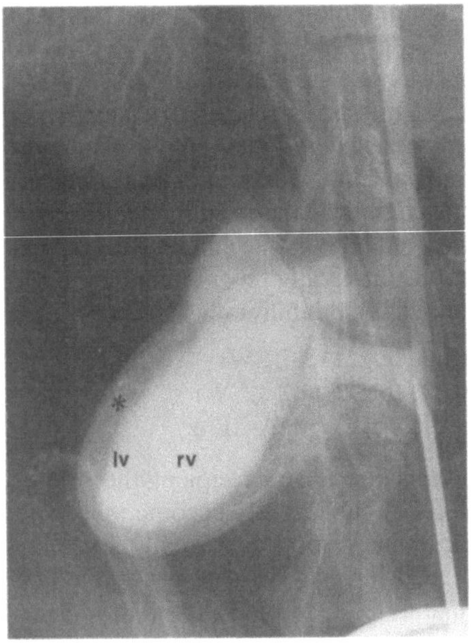

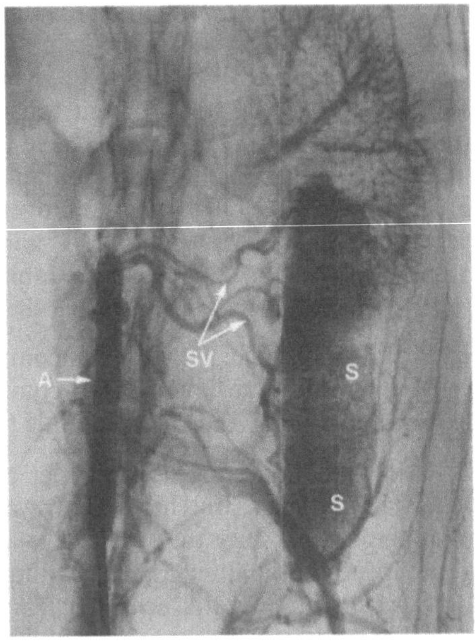

Fig. 1 **Fig. 2**

Fig. 1. Abdominal heart transplant. *lv*, left ventricle; *rv*, right ventricle; *, heart wall with coronary arteries

Fig. 2. Aorta (*A*), splenic vessels (*SV*), and transplanted spleen (*S*) demonstrated by microradiography

the immunogenicity of blood vessels (Thiede et al. 1981). The next step in the training program includes an anastomosis of a vascular graft of the femoral artery (Timmermann and Thiede 1981). The diameter of the vessel is 0.3–0.4 mm, and the suture materials is 11-0. The sutures are almost invisible to the naked eye. Thus the operating microscope is a necessity here. With sufficient training these anastomoses can also be performed with a continuous suture. With training it is possible to achieve a long-term patency rate of 80%.

Venous sutures are slightly more difficult because when the vein is clamped and not filled with fluid it collapses. The edges of the vessel are transparent and barely recognizable. With sufficient training, however, if the vessel is flooded repeatedly, anastomoses can be performed as a running suture without stenosis. The end-to-side anastomosis is essential for rat organ transplantation. It is a more difficult vascular anastomosis, but it is necessary for the following transplantation models: abdominal heart (Fig. 1), lung, spleen (Fig. 2), vascularized skin flap, kidney (Fig. 3), small intestine, pancreas (Fig. 4), testicle, and liver transplantation. Three different examples are outlined here.

A simple operating technique is used in spleen transplantation. Only two end-to-side anastomoses must be performed, one in the arterial system and one in the venous system. The problem in spleen transplantation lies in the preparation of the small splenic vessels, which are interspersed with pancreatic tissue. An angiogram of the transplanted spleen after 6 weeks shows the perfused organ and its

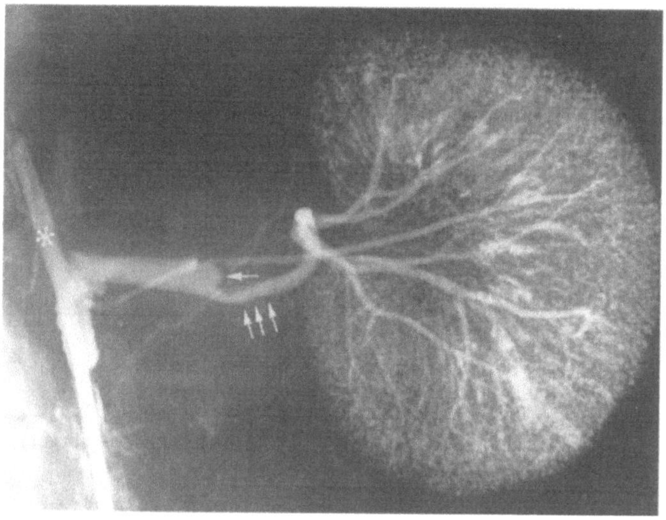

Fig. 3. Microradiographic control of a transplanted kidney. *, aorta. Individual glomeruli are visible. ←, aortic segment of the transplant artery; ↑↑↑, original transplant artery

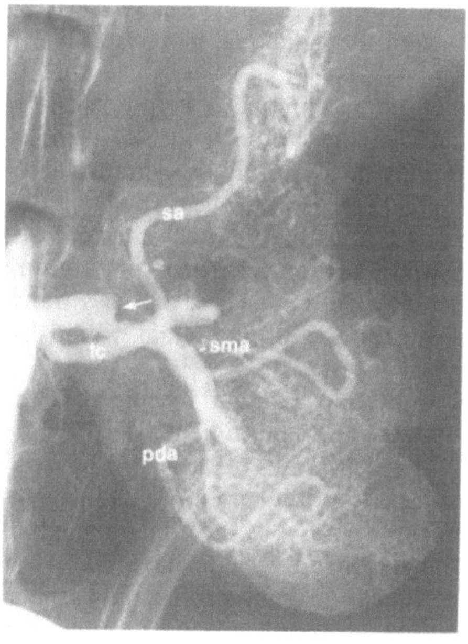

Fig. 4. Pancreaticoduodenal transplant; microradiographic demonstration of truncus celiacus (*tc*), arteria mesenterica superior (*sma*), arteria lienalis (*sa*), and arteria pancreaticoduodenalis (*pda*)

vessels (Fig. 2). Kidney transplantation is a standard operation procedure with high preclinical value for testing immunosuppressive agents. Two end-to-side anastomoses must be performed to ensure the blood circulation. The ureter anastomosis can be performed by implanting the ureter into the bladder; a segment of the donor bladder can be connected to the recipient's bladder, or the ureter can

Table 4. Avantages of experimental microsurgery for a surgical resident in specialized clinical training

1.	Further training of microsurgical skills
2.	Experimental surgical work without absolute dependence on technical assistance
3.	Continuation of an experimental program alongside a full-time clinical job
4.	Close relationship between clinical programs and experimental program because one person is confronted with both fields

Table 5. Surgical disciplines in which microsurgery is performed

1. Cardiovascular surgery	7. Ophthalmology
2. Gynecology	8. Orthopedics
3. General surgery (including gastroenterological surgery)	9. Otolaryngology
4. Hand surgery	10. Traumatology
5. Neurosurgery	11. Transplantation surgery
6. Plastic surgery	12. Urology

be anastomosed by a splint, which is, in our hands, the most advantageous technique. An angiogram of an 8-week-old kidney transplant shows the vessels and even the fine structures of the renal cortex with the individual glomeruli (Fig. 3, syngeneic control).

The most difficult abdominal transplantation technique is orthotopic liver transplantation. Elsewhere in this book we discuss the various models of auxiliary and orthotopic liver transplantation. Especially in our group, liver transplantation is performed with arterialization. Five anastomoses must be carried out: two caval anastomoses, a portal vein anastomosis, and the bile duct anastomosis, which is performed over a splint. Finally, arterialization is carried out to ensure the blood supply to the liver hilus and the choledochus in particular. Liver transplantation is the most challenging part of our training program and will be discussed in detail later.

A special nonsuture technique for vessel connection is the cuff technique, which has been used in our group for heterotopic heart transplantation in the neck. This technique has the advantage that end-to-end anastomoses of vessels with incongruent lumina can be performed. It can be used for all other end-to-end anastomoses as well. The splint techniques for the anastomoses of the ureter in the kidney or the bile duct in liver transplantation have already been mentioned.

Finally, we would like to reflect on some educational and professional aspects of the concept and structure of a microsurgical research group like ours. The young physician who enters our team naturally asks himself what advantages a microsurgical training program will have for him. In addition to the items listed in Table 4, we believe that learning experimental microsurgery early can be of great value for later clinical work, provided a young physician or a student already knows that he will be entering a surgical discipline. He has the chance to apply his microsurgical knowledge in a number of fields, which means that he has an advantage over others of his age who have not received this training. (The surgical specialities in which microsurgery is performed are listed in Table 5.) Naturally, a training program of this kind is not only important for the prospective surgeon, it also gives the leader of such a research group the chance to test whether a student or young physician is suited for surgical work. The following aspects can be considered:

Skill. The success rate of microsurgical operations depends to a high degree on the accuracy with which the animals and the technique are handled. Our experience shows that students who do not work extremely precisely from the beginning will have a high failure rate. According to our experience as clinicians, everyone who has performed excellent work in microsurgery has later proved successful in clinical surgery.

Perseverance and Patience. For difficult organ transplantation, in addition to skill and techniques, perseverance and patience are required. The ability to carry out these operations in the rat and to struggle against one's own frustration trains the ongoing surgeon for those clinical situations in which patience and perservance are often at least as important as skill.

Character Suited for Team Work. Meanwhile we are a relatively large team with a considerable amount of mutual dependence. Thus, we have a good opportunity to test whether a new member is able to cooperate and to fit into the team. The students and younger physicians who have succeeded in our microsurgical training program have later proved to be fair colleagues in the clinic.

Scientific Interest. Working in such a group, it soon shows up clearly whether a person has scientific interest or not. This may be helpful in the decision whether a more practical or a more scientific field of medicine offers the most advantages and is most suitable for an individual student.

Finally we want to emphasize that a microsurgical research group offers a lot of possibilities for interdisciplinary cooperation at the university, e.g., with immunologists, immunopathologists, physiologists, and clinicians. Especially close cooperation with the latter may lead to further development of preclinical transplantation models for analyzing clinical transplantation problems.

References

Butcher GW, Howard JC (1982) Genetic control of transplant rejection. Transplantation 34:161–166

Engemann R, Deltz E, Thiede A (1982) Nahtmaterialien und Nahttechniken in der experimentellen Mikrochirurgie. In: Thiede A, Hamelmann H (eds) Nahtmaterialien und Nahttechniken. Springer, Berlin Heidelberg New York, pp 90–106

Fisher B, Lee S (1965) Microvascular surgical technique in research, with special reference to renal transplantation in the rat. Surgery 58:904–914

Lee S (1967) A improved technique of renal transplantation in the rat. Surgery 61:771–773

Lee S, Fisher B (1961) Portocaval shunt in the rat. Surgery 50:668–672

Lee S, Orloff MJ (1979) Microvascular surgery in the rat. In: Silber SJ (ed) Microsurgery. Williams and Wilkins, Baltimore, pp 93–129

Silber SJ (1979) Microsurgical technique. In: Silber SJ (ed) Microsurgery. Williams and Wilkins, Baltimore, pp 1–29
Thiede A (1977) Gefäßtransplantation. Die Bedeutung immunologischer Reaktionsmechanismen. Untersuchungen an standardisierten Ratteninzuchtstammkombinationen. Schattauer, Stuttgart. (Ergebnisse der Angiologie, vol 15)
Thiede A, Engemann R, Körner H, Deltz E, Müller-Ruchholtz W (1981) Vein replacement with fresh vital veins. A comparison of transplantation in RT1-different rat strain combinations. Ann Surg 193:283–287
Timmermann W, Thiede A (1981) Histologische und funktionelle Untersuchungen an Femoralarterientransplantaten bei Ratten. Vasa 10:141–147

Techniques of Microsurgical Vascular Anastomoses in Experimental Microsurgery

R. Engemann

After the introduction of microvascular suture techniques in animals as small as rats by Fisher and Lee in 1961 (Fisher et al. 1962; Lee and Fisher 1961), nearly all organ transplantation techniques were developed within the following 15 years, most of them pioneered by Lee [for an overview see Thiede (1977)]. In addition to a wealth of ideas and inexhaustible patience on the part of the respective surgeons, however, the following factors had to coincide before microsurgical techniques for experimental rat surgery could be established worldwide:

1. The availability of magnifying glasses with up to 6-fold magnification and of operating microscopes with up to 30-fold magnification
2. The development of suitable and precise instruments made, e.g., of titanium
3. The industrial production of thin monofile sutures
4. The breeding of laboratory rats, especially inbred rat strains

The suture material we use for the anastomoses in our laboratory is usually nylon (Ethilon) or polypropylene (Prolene), although it is also possible to use absorbable sutures (Thiede et al. 1979) in experimental microvascular surgery. For ligation of side branches we use 5-0 or 6-0 silk. The size of the sutures for the anastomoses in our organ transplantation program ranges from 7-0 to 11-0 United States Pharmacopeia (USP), which corresponds to a thread diameter of 45–15 μm. In an experimental laboratory specializing in organ transplantation, suture material may represent the second largest item in the budget, following the animals. Since it is comfortable and even essential to use short sutures if one is working with a microscope (so one can manage to suture and knot within one's small area of vision), double-armed sutures can easily be divided into two parts of appropriate length (8–10 cm).

Most of the needles are 3-8 curved needles with a diameter of 0.14–0.07 mm. Though the ratio between needle and thread diameter is too large, sutures with metalized tips, as invented by O'Brien et al. (1970), are not used in experimental microsurgery. The disparity between suture and needle sometimes leads to bleeding from stitching holes, especially in arterial anastomoses. To avoid or minimize this, one should take care to pull the needle through the vessel wall in the direction of its curvature, to keep the holes as small as possible.

Microsurgical Models in Rats
for Transplantation Research
Eds. A. Thiede et al.

The knots are usually made instrumentally, and three knots are sufficient (one tied in the opposite direction from the others). The advantages of the running suture techniques, such as shorter anastomosing time and better primary hemostasis, are opposed by the disadvantage of a stenosis, if the knots are pulled too tight. This is particularly true of anastomoses of small caliber. As a matter of principle one should control the tightness of the knots optically. Whereas in replantation surgery single-stitch techniques are preferable to a running suture technique (Biemer and Schmidt-Tintemann 1982), in experimental surgery we prefer the running suture technique. Especially in the transplant models the shorter anastomosing time is of value in keeping the warm ischemic time as short as possible.

The following microsurgical techniques for vessel anastomoses have been established for rats weighing 250–300 g. Anesthesia is performed with ether. It is induced in a glass container and maintained by placing a syringe that is stuffed with gauze soaked with ether over the rat's nose. The depth of the anesthesia is easily controlled by moving the syringe away from or closer to the nose.

End-to-End Anastomosis

For the beginner in experimental microsurgery the infrarenal abdominal aorta of the rat is the most suitable object to start with, and one should start with an end-to-end anastomosis. After the aorta has been stripped of the surrounding connective tissue and separated from the adjacent vena cava inferior, it is clamped with

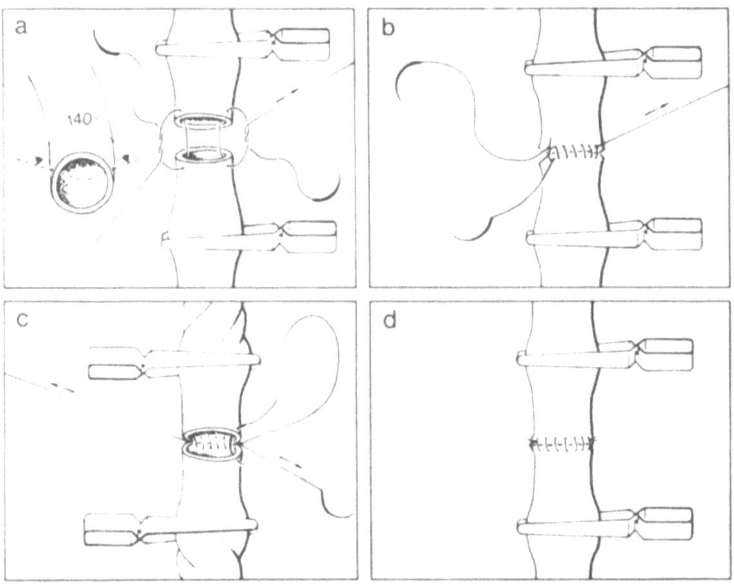

Fig. 1 a–d. End-to-end anastomosis; running suture technique. **a** Positioning and knotting the angle suture in "eccentric biangulation"; **b** suturing the anterior wall; **c** rotation by 180 ° and suturing the posterior wall; **d** completed anastomosis

microbulldog clips and divided. The vessel stumps are rinsed with saline solution to remove clots. Heparin is not necessary. The angle sutures are placed in an eccentric biangulation (Cobbett 1967), as can be seen in Fig. 1 a. This holds the opposite side a little distance away, so that it cannot be grasped by mistake. The suture itself is made in a running technique (Fig. 1 b), as mentioned above. Coming to the angle suture, the thread is tied, the vessel is rotated 180° (Fig. 1 c), and the posterior wall is completed (Fig. 1 d). To reestablish the blood flow, the distal clamp is released first, to allow preclotting in the stitching holes and to minimize blood loss during that period. After 30–60 s the proximal clamp is released. Persistent bleeding is managed by gentle compression with gauze or a cotton swab. Additional single stitches should and can be avoided by training, since in rat organ transplantation the blood loss might be harmful to the rat. Training for accuracy and meticulousness is important for the young surgeon and a prerequisite for successful (micro) surgery.

If the anastomosis of a rat aorta with an inside diameter of 0.9–1.1 mm can be accomplished with the aid of magnifying glasses or even with the naked eye, for anastomoses of smaller vessels, such as the femoral artery, a microscope is mandatory. Suitable magnification is between 10-fold and 16-fold, sometimes up to 30-fold. The suture material is 10-0 or 11-0 nylon. The patency rate, e.g., for syngeneic femoral artery grafts in rats, is 80% with a running suture technique (Timmermann and Thiede 1981).

End-to-end anastomoses of veins for the purpose of vein transplantation (Engemann et al. 1981; Thiede et al. 1981) or for organ transplantation are performed

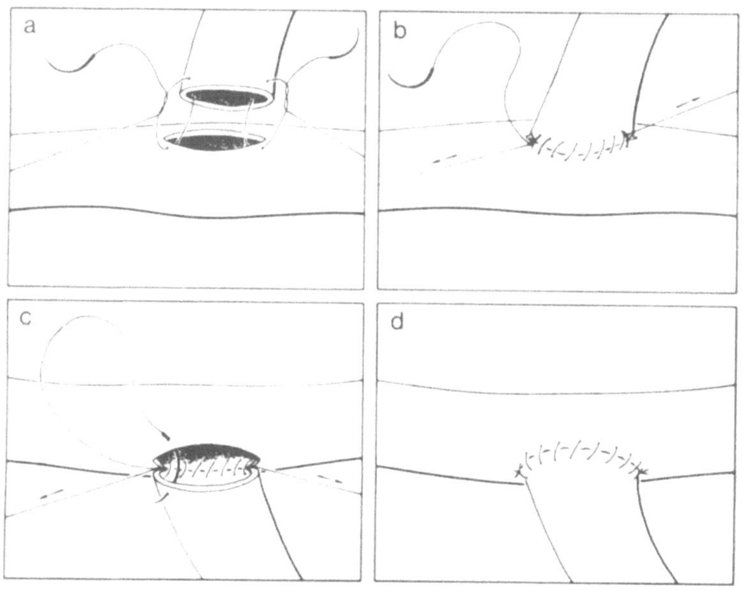

Fig. 2 a–d. End-to-side anastomosis; running suture technique. **a** Positioning and knotting the angle suture; **b** suturing the anterior wall; **c** suturing the posterior wall from the *outside* after folding back the vessel; **d** completed end-to-side anastomosis

in the same fashion. Veins have very thin walls and are so fragile that they must be handled very carefully to avoid trauma. Eccentric biangulation is especially useful for veins. Thus, e.g., for fixed vessels, the posterior suture row can easily be done from the inside. The resulting inverting anastomosis makes no difference in experimental rat vein surgery.

End-to-Side Technique

Using the basic suture techniques described above – eccentric biangulation, fixed-angle sutures, and the running suture technique – end-to-side anastomoses are easy to manage. When clamping the vessel that is to receive the end-to-side anastomosis, one should ensure that the vessel is *filled* between the clamps. Thus, when the venotomy or arteriotomy is performed with a beveled cannula or feather scissors, the posterior wall cannot be damaged. After the angle sutures have been knotted, the anterior wall is established as a running suture. The vessel is folded to the other side, and the suture of the posterior wall is also accomplished as a running suture (Fig. 2a–d). If, as in organ transplantation or in portocaval anastomosis, the joining vessel is fixed, the posterior row is again sutured from the inside, after the angle sutures have been positioned (Fig. 3a–d).

Cuff Technique

To avoid sutures in microvascular anastomoses, the cuff technique was described recently (Dunn 1976; Heron 1971; Kamada and Calne 1979). Details will be given later in this book by Timmermann. The principle is that one vessel – the smaller one, if possible – is everted over a polyethylene cuff, and then the other vessel stump – the bigger one, if possible – is pulled over the endothelialized cuff. Then the anastomosis is secured by a circular ligature or additional single stitches. The advantage is a short anastomosing time and optimal hemostasis. The disadvantage is the additional vessel length needed and the possibility of an indentation in the anastomosis.

Splint Technique

In addition to arterial and venous anastomoses, in liver and kidney, and to some extent in pancreas transplantation, microanastomoses of the bile duct and the ureter must be performed. We use the following modified technique according to Zimmermann et al. (1979) for bile duct anastomoses in rat liver transplantation: a transverse incision is made into the fixed donor bile duct, and a beveled polyethylene splint is inserted (Fig. 4a) and fixed with a ligature (Fig. 4b). The distal end of the splint, which is also beveled, is inserted into the recipient's bile duct, which is held open by a suture and the forceps. By crossing the ends of the stay suture, the splint is secured against sliding off. The anastomosis is fixed by two or three additional single stitches. The ureter anastomosis in rat kidney transplantation can be accomplished in the same manner.

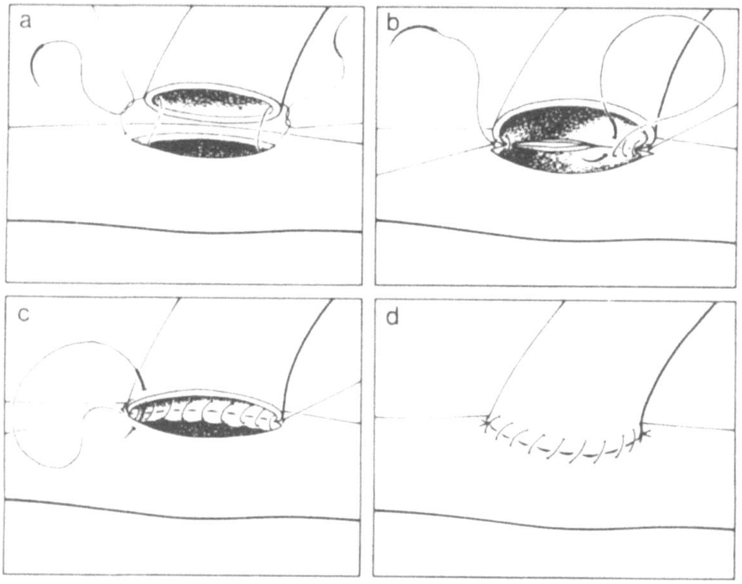

Fig. 3a–d. End-to-side anastomosis; running suture technique. **a** Positioning and knotting the angle sutures; **b** suturing the posterior wall from the *inside*; **c** suturing the anterior wall; **d** completed end-to-side anastomosis

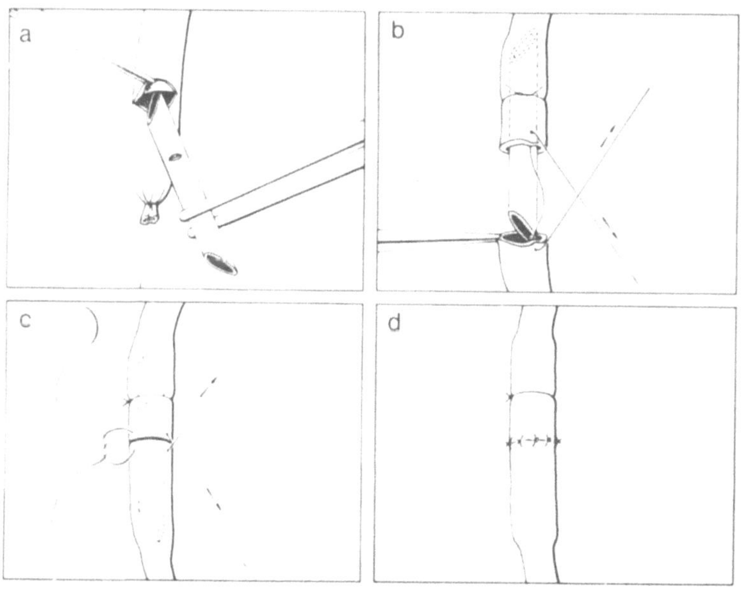

Fig. 4a–d. Splint technique (bile duct, ureter). **a** Inserting the polyethylene splint into the bile duct; **b** fixation of the splint with a tie over the mark; **c** connection of the ducts; **d** completed anastomosis

In conclusion, these suture and anastomosing techniques demonstrated here are a basic approach to microsurgery. Anyone with experience in microsurgery will develop his own variations of these techniques. With these basic techniques, however, some surgical skill, and a lot of patience it is possible to perform all the organ grafts in rats that will be described in the following chapters.

References

Biemer E, Schmidt-Tintemann U (1982) Anatomische und funktionelle Grundlagen für die Wahl von Nahtmaterialien und Nahttechniken in der klinischen Mikrochirurgie. In: Thiede A, Hamelmann H (eds) Moderne Nahtmaterialien und Nahttechniken in der Chirurgie. Springer, Berlin Heidelberg New York, p 400

Cobbett J (1967) Small vessel anastomosis. Br J Plast Surg 20:16–20

Dunn DC (1976) Orthotopic renal transplantation in the rabbit. Transplantation 22:427–433

Engemann R, Körner HH, Schang T, Thiede A (1981) Venentransplantation in RT1-different Stammkombinationen. Vasa 10:211–217

Fisher B, Lee S, Fisher ER, Saffer E (1962) Liver regeneration following portocaval shunt. Surgery 52:88–102

Heron J (1971) A technique for accessory cervical heart transplantation in rabbits and rats. Acta Pathol Microbiol Scand [A] 79:366–372

Kamada N, Calne RY (1979) Orthopic liver transplantation in the rat. Technique using cuff for portal-vein anastomosis and biliary drainage. Transplantation 28:47–50

Lee S, Fisher B (1961) Portocaval shunt in the rat. Surgery 50:668–672

O'Brien B McC, Henderson PN, Crock GW (1970) Metallized microsutures. Med J Aust 1:717–719

Thiede A (1977) Gefäßtransplantation. Die Bedeutung immunologischer Reaktionsmechanismen. Untersuchungen an standardisierten Ratteninzuchtstammkombinationen. Schattenhauer, Stuttgart (Ergebnisse der Angiologie, vol 15)

Thiede A, Lütjohann K, Beck C, Blunck F (1979) Suture reactions with application of non-absorbable and absorbable materials in microsurgery. Standardized comparable investigations in rat vessels. J Microsurg 1:216–222

Thiede A, Engemann R, Körner HH, Deltz E, Müller-Ruchholtz W (1981) Vein replacement with fresh vital veins. A comparison of transplantation in RT1 different rat strain combinations. Ann Surg 193:283–287

Timmermann W, Thiede A (1981) Histologische und funktionelle Untersuchungen an Femoralarterientransplantaten bei Ratten. Vasa 10:141–147

Zimmermann F, Butcher GW, Davies HJ, Brons G, Kamada N, Türel O (1979) Techniques for orthotopic liver transplantation in the rat and some studies of the immunologic responses to fully allogeneic liver grafts. Transplant Proc 11:571–577

Heart Transplantation in Rats

W. Konertz, M. Semik, and *A. Bernhard*

Cardiac transplantation in rats has been performed preferentially with models of empty beating hearts that were connected either to the neck vessels or to the abdominal aorta and vena cava. Whether this nonphysiological approach with altered coronary perfusion and proven functional impairments (Bernhard and Konertz 1983; Dittmer and Goss 1973) has any influence on the immunological behavior of the graft or the recipient is still unknown. On the other hand, orthotopic transplantation, which would be a very desirable procedure in rats, is troublesome to perform due to difficulties with extracorporeal circulation in such small animals (Konertz et al. 1979a). We introduced heterotopic thoracic transplantation as left ventricular bypass in rats (Konertz et al. 1979b, c, 1980b). This superior procedure gives immunological and functional results under standardized and repeatable conditions. This paper will describe technical aspects of heart transplantation in rats.

Heterotopic Abdominal Heart Transplantation

The heterotopic abdominal heart transplantation is a very often used model in transplantation research. This is mainly because of its technical simplicity and the ease of determining graft function by simple palpation. The operative technique consists of end-to-side anastomosis of the graft's ascending aorta to the infrarenal aorta of the recipient and the graft's pulmonary artery to the inferior vena cava of the host (Abbott et al. 1974; Konertz et al. 1980a).

Donor Operation

The donor is heparinized with 1000 IU heparin i.v. After longitudinal laparotomy the animal is bled off by transection of the great abdominal vessels. The abdominal incision is extended to a bilateral thoracotomy and the anterior chest wall is removed. After blunt dissection of the retrosternal fat, the aorta and pulmonary artery are transected in the transverse sinus in a very far cranial position. All cardiac veins are closed with a single 4-0 ligature. Now the graft is excised and brought into ice-cold Ringer's solution.

Microsurgical Models in Rats
for Transplantation Research
Eds. A. Thiede et al.

Recipient Operation

Under ether anesthesia a midline laparotomy is performed. The abdominal vessels in their infrarenal section are freed from retroperitoneal tissue. With a curved vascular clamp both vessels are clamped. A small longitudinal aortotomy of 4-5 mm length is performed, and the aorta of the graft is sutured to the infrarenal abdominal aorta of the host using standard microvascular techniques with 8-0 Prolene. After completion of the aortic anastomosis the vena cava is incised, the clot is carefully removed, and the pulmonary artery of the graft is connected to the inferior vena cava with 8-0 Prolene with a continuous suture. After the clamp is released hemostasis is achieved by gentle compression of the anastomoses. In most instances the heart starts beating spontaneously. The laparotomy is closed with a one-layer suture.

Comment

In the hands of a skilled microsurgeon the operation can be performed without the operation microscope, and the duration of the whole procedure should not exceed 30–40 min including ischemic time of a maximum of 20 min. Within these time limits hypothermia without chemical cardioplegia seems sufficient for myocardial preservation and after release of the aortic clamp the heart should start spontaneously with vigorous contractions.

Due to the nonphysiological mode of transplantation, functional studies should be carried out with caution (Bernhard and Konertz 1983). The heterotopic abdominal heart is prone to fibrosis and even ossification (Dittmer and Goss 1973). Functional studies in our laboratory showed functional impairment of the left ventricle as soon as 2 weeks postoperatively (Konertz et al. 1980 a). For immunological research, however, heterotopic abdominal heart transplantation is a very valuable model and used widely.

Cervical Transplantation

The technique of cervical transplantation in rats with the cuff technique is described by Timmermann et al. later in this volume (p. 31). This model is equally valuable as the heterotopic abdominal heart transplantation with the same advantages and disadvantages in the case of functional evaluation.

Heterotopic Thoracic Heart Transplantation

The above-mentioned drawbacks of heterotopic abdominal transplantation regarding functional evaluation of these grafts stimulated our interest in a model with a pumping heart, and we found heterotopic thoracic transplantation quite suitable.

Donor Operation

A bilateral thoracotomy is performed after induction of ether anesthesia and systemic heparinization with 1000 units of heparin given intravenously. After blunt

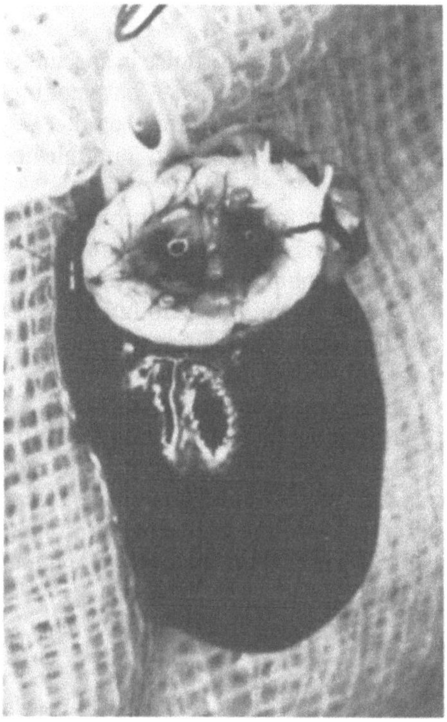

Fig. 1. Graft ready for transplantation. Within the Dacron ring the pulmonary artery (*left*) and the left atrium (*right*) are visible

preparation of the retrosternal fat the aorta and pulmonary artery were cut in the transverse sinus as far cranially as possible. Cardioplegic arrest is achieved with cold Ringer's solution. All great veins are ligated with 4-0 silk. The hearts are rapidly excised and immersed in ice-cold Ringer's solution. Then a 3-mm Dacron ring is sutured onto the left atrium and the pulmonary artery is also connected to this ostium. Figure 1 shows a graft ready for transplantation.

Recipient Operation

Anesthesia is introduced and maintained with ether. After orotracheal intubation and under artifical respiration the chest is entered via a left thoracotomy in the fourth intercostal space. The left lung is gently pushed caudally. Sometimes the hemiazygos vein is ligated and transected. The descending aorta is dissected and controlled with two vessel snares. A longitudinal aortotomy is made and an end-to-side anastomosis between the graft's ascending aorta and the descending aorta of the recipient with 8-0 Prolene is performed using standard microvascular techniques. The graft's aorta is cross-clamped and blood flow through the descending aorta of the recipient is restored. The left atrium of the recipient is partially clamped and an atriotomy is performed. Anastomosis between the Dacron ring and the left atrium is carried out using a continuous suture of 8-0 Prolene. After meticulous hemostasis all clamps are removed. With the restoration of coronary perfusion the graft starts beating spontaneously. If the experimental protocol prescribes a loaded heart, the descending aorta proximal to the anastomosis is closed with a hemoclip after appropriate reperfusion.

Comment

Heterotopic thoracic transplantation has gained clinical importance in special indications (Barnard 1976; Losman 1977; Losman and Barnard 1977; Yacoub et al. 1983). Experimental work was based on studies in dogs (Marcial et al. 1972; Moore and Ross 1976; Suros and Woods 1973), pigs (Konertz, unpublished data), and primates (Losman 1978).

In our laboratory we were able to scale down the model of Moore and Ross (1976) ro rat size, which is accomplished by draining the coronary sinus return into the recipient's left atrium with a common pulmonary artery-atrio-atrial anastomosis. Artificial respiration during the whole procedure is mandatory. After appropriate training a microsurgically skilled person can perform the operation with 60–80 min. There is no special postoperative care necessary. Under these conditions we have a survival of 95% with 90% good functioning grafts (Fig. 2). The special value of this model is the fact that functional evaluation of the transplanted heart becomes possible using inbred animals. This allows manipulation of the genetic difference. Syngeneic transplantation with indefinite graft survival is very valuable for investigation of late alterations of the graft after prolonged extracorporeal preservation (Konertz et al. 1982, 1983). Transplantation in major

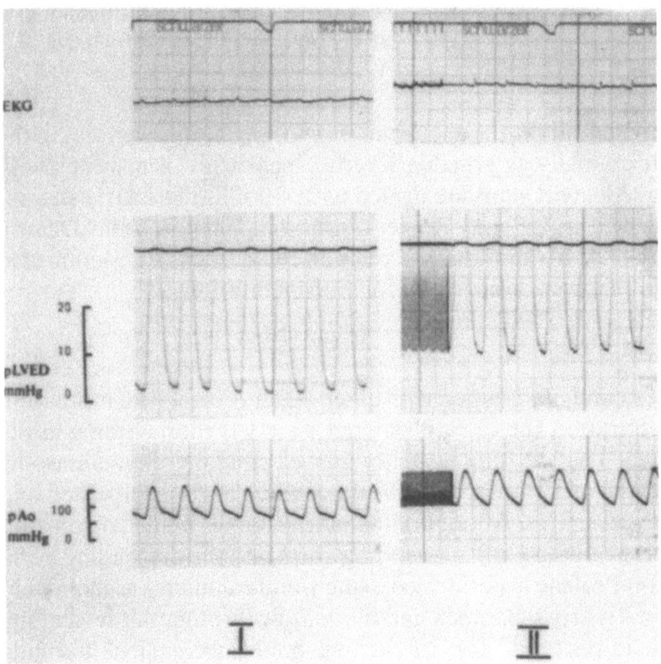

Fig. 2. Original registration after heterotopic thoracic heart transplantation in the rat. The aortic pressure is measured distally via the femoral artery after proximal ligation of the descending aorta. *I*, baseline measurements; *II*, volume loading with 25 ml/kg 5% dextrose; *EKG*, electrocardiogram; *pLVED*, left ventricular end-diastolic pressure; *pAo*, aortic pressure

histocompatibility complex (MHC)–incompatible strain combinations allows repeatable functional and immunological data sampling for evaluation of monitoring methods or drug trials.

Clinical Heart Transplantation

The growing number of centers beginning heart transplantation seems due to increased survival and rehabilitation following this procedure. The introduction of cyclosporin A has led to more than 80% survival of heart recipients at 1 year and has substantially reduced the costs. Currently applied techniques are the orthotopic transplantation, the heterotopic transplantation as left- or biventricular bypass and the en bloc heart-lung transplantation. These different approaches allow special procedures for every individual patient with end-stage cardiac and/or pulmonary disease. The main problems to be solved in the field of heart transplantation lie in organ preservation. With currently applied techniques the heart and lung allow ischemic times of only a few hours. Microsurgical experimental models of heart transplantation in inbred animals stimulate research with unique effectiveness at a low cost.

References

Abbott CP, Lindsay ES, Creech O, et al. (1974) Technique for heart transplantation in the rat. Arch Surg 89:645–652

Barnard CN (1976) Heterotopic versus orthotopic heart transplantation. Transplant Proc 8:15–19

Bernhard A, Konertz W (1983) Experimental heart transplantation. J Thorac Cardiovasc Surg 86:314

Dittmer JE, Goss RJ (1973) Size changes of auxillary adult heart grafts in rats. Cardiology 58:355–363

Konertz W, Regensburger D, Semik M, et al. (1979a) Heart transplantation in rats (Abstr). Transplant Proc 11:593

Konertz W, Semik M, Bernhard A (1979b) Early postoperative pump performance of thoracic heterotopic cardiac grafts in rats (Abstr). Transplant Proc 11:592

Konertz W, Semik M, Bernhard A (1979c) Early postoperative pump performance of thoracic cardiac grafts in rats (Abstr). J Mol Cell Cardiol 11:32

Konertz W, Thiede A, Bernhard A (1980a) Heterotopic heart transplantation in rats – an improved technique of functional evaluation. Excerpta Med Int Congr Ser 465:359–361

Konertz W, Semik M, Bernhard A (1980b) Left ventricular bypass in inbred rats – a new experimental model in microsurgery. Operative technique and hemodynamic evaluation. Thorac Cardiovasc Surg 28:277–279

Konertz W, Zaworka P, Johanson W, et al. (1982) 24 hours heart preservation with modified Eurocollins Solution. In: Belenger J (ed) Life support systems. Proceedings of the 9th annual meeting of ESAO. Saunders, London, pp 151–152

Konertz W, Zaworka P, Johanson W, et al. (1983) 24 hours heart preservation. Heart Transplant 2:278–280

Losman J (1977) Review of the cape town experience with heterotopic cardiac transplant. Cardiovasc Dis 4:243–255

Losman J (1978) Cardiac surgical research and heart transplantation in *Papio ursinus*. J S Afr Vet Assoc 49:211–215

Losman J, Barnard CN (1977) Hemodynamic evaluation of left ventricular bypass with a homologous cardiac graft. J Thorac Cardiovasc Surg 74:695–708
Marcial BH, Armelin E, Stolf N, et al. (1972) Assisted circulation with an auxiliary heart transplant: an experimental study. J Thorac Cardiovasc Surg 63:696–704
Moore CH, Ross DN (1976) Experimental auxiliary heart transplantation for left ventricular assistance. Transplant Proc 8:41–44
Suros J, Woods JE (1973) Synchronized intrathoracic auxiliary heart transplantation in parallel. J Thorac Cardiovasc Surg 65:415–424
Yacoub MH, Rose ML, Fraser AK, et al. (1983) T-Cell phenotype characterization in human cardiac allografts. Heart Transplant 2:259–265

Accessory Heart Transplantation in the Neck: A Simple Model in the Rat Using the Cuff Technique for Vascular Anastomoses

W. Timmermann

Introduction

Heterotopic heart transplantation represents the easiest model of a vascularized organ graft in the rat. Nevertheless, two end-to-side anastomoses in microsurgical technique have to be performed in the "conventional" abdominal heart transplantation.

An alternative is to use a nonsuture technique for the vascular anastomoses, the so-called cuff technique. It can only be applied for end-to-end anastomoses, so the heart graft has to be placed in the neck and the cervical vessels have to be used for blood supply. The idea of connecting blood vessels in a nonsuture techniques was first published by Nitze in 1897. He suggested slipping the two vessels over each other by means of an apparatus manufactured from ivory and holding the vessels in position by a circular ligature. In 1900, Payr took up the idea and suggested several methods of connecting vessels without suturing them by using various modifications of magnesium tubes. These tubes looked lift cuffs, hence the name cuff technique.

Cuff techniques have only rarely, if ever, been used clinically, but there have been a number of publications about their application with laboratory animals in recent years.

The model of heterotopic cervical heart transplantation is, for example, very useful for evaluating the effect of immunosuppressive protocols, because the function of the heart can be observed either by inspection or palpation in non-anesthetized animals. An example of such a study is given below.

Accessory Heart Transplantation in the Rat

The heart graft is harvested following the technique of Ono and Lindsey (1969). The ascending aorta and main pulmonary artery are transsected 3–4 mm from their origin from the heart; all other vessels to the heart are ligated and cut. The heart is perfused with cold saline solution and stored at 4 °C.

The recipient operation is started by a longitudinal incision on the right ventral side of the animal's neck. Then the common carotid artery and external jugular

Microsurgical Models in Rats
for Transplantation Research
Eds. A. Thiede et al.

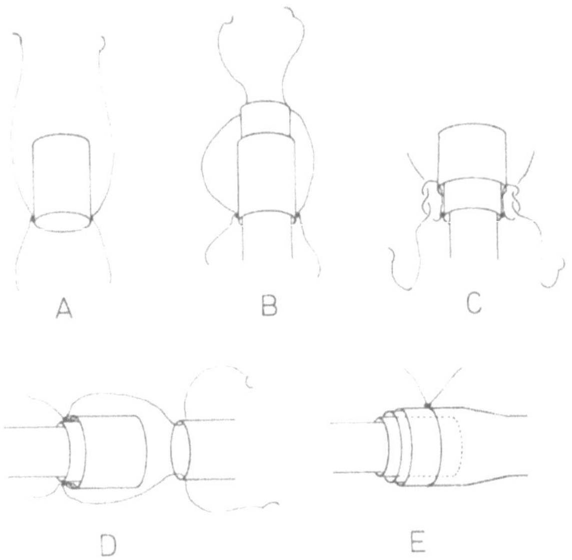

Fig. 1A–E. Connection of two vessels in a nonsuture technique by means of a plastic tube (cuff technique)

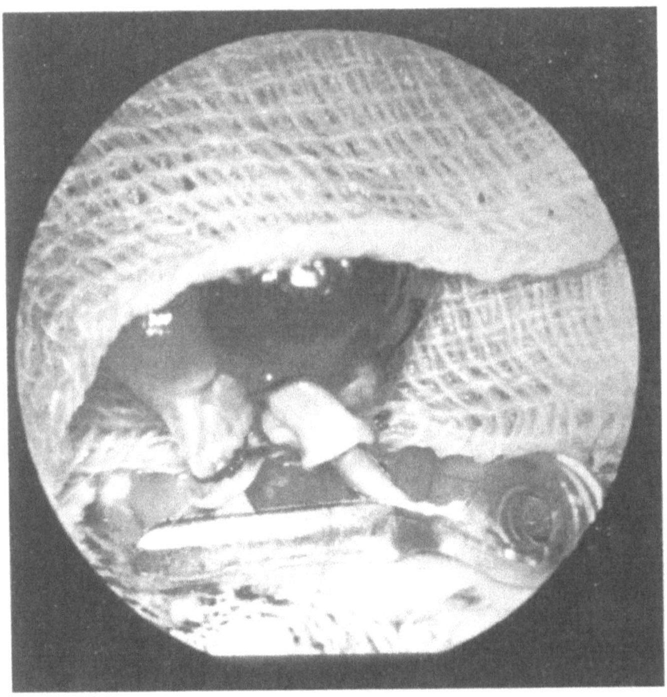

Fig. 2. Heterotopic transplantation of a rat heart in the neck. Graft in the *top*, visible completed anastomosis between the pulmonary artery and external jugular vein (*left*), and carotid artery with cuff before sliding into the stump of the aorta (*right*)

vein are exposed, clamped, ligated, and cut as far distally as possible to obtain long vascular stumps. The cuffs are made from plastic tubes or cannulas, which should be slightly larger in diameter than the corresponding vessel. For this operation, 3- to 4-mm pieces of Argyle Medicut cannulas are used, size G 16 (diameter: 1.7 mm) for the venous cuff and G 20 (diameter: 1.2 mm) for the arterial cuff. Two atraumatic sutures, 7-0 or 8-0 nylon, are placed in one end of the cuff and fixed by a knot, leaving the end with the needle long (Fig. 1 A). Then the cuff is placed over the corresponding vessel and after placing one or both sutures in the vessel (Fig. 1 B), it is possible by gently pulling on the sutures to evert the vessel over the cuff (Fig. 1 C).

Then the graft is brought in to position. To complete the anastomoses, the vessels of the graft have to be pulled over the cuffs. This is made possible by placing one or both sutures from the corresponding cuff through the vessel of the graft (Fig. 1 D) and pulling on the needles until the vessels slip over the cuff. A circular ligature completes the anastomosis (Fig. 1 E). In this way, the carotid artery is anastomosed to the stump of the ascending aorta and the external jugular vein to the pulmonary artery as shown in Fig. 2. The heart is finally placed subcutaneously in the right side of the neck.

Example of Application of Accessory Heart Transplantation

The evaluation of the potency of immunosuppressive drugs is a very common problem in transplantation research. Transplantation between inbred strains of rats makes possible an indefinite number of repetitions of the same experiment so that the effect of any immunosuppressive can be observed without interferences resulting from varying histoincompatabilities between donor and recipient.

The effect of two drugs, cyclosporin A (Sandimmun) and ASTA 5122 (Cytimun) on the survival of vascularized organ grafts were evaluated (Timmermann et al. 1980). We chose the accessory heart transplantation as the easiest model in the rat and transplanted the heart over a major histocompatibility barrier. The strain used, the groups and numbers of animals, and the treatment are listed in

Table 1. Groups, strain combinations, numbers, and treatments of animals which received accessory heart grafts

Group	Strain combination	Histocompatibility	Treatment	*n*/group
C1	DA ($RT1^{av1}$) → DA ($RT1^{av1}$)	Syngeneic	–	11
C2	PVG ($RT1^{c}$) → DA ($RT1^{av1}$)	Allogeneic	–	8
A	PVG ($RT1^{c}$) → DA ($RT1^{av1}$)	Allogeneic	Cyclosporin A 15 mg/kg/d 14 days p.o.	11
B	PVG ($RT1^{c}$) → DA ($RT1^{av1}$)	Allogeneic	ASTA 5122 5 mg/kg/d 14 days p.o.	9

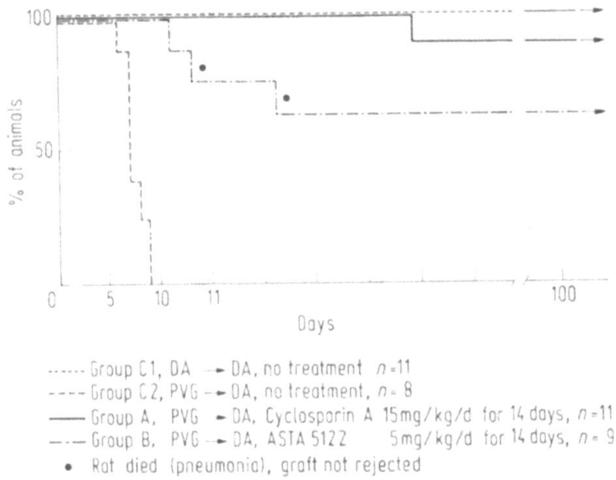

Fig. 3. Duration of function of accessory heart transplants: untreated controls (groups C1 and C2) and treated allogeneic transplants (groups A and B)

Table 1. Recipients of allografts received no treatment (group C2), cyclosporin A 15 mg/kg for 14 days orally (group A), or ASTA 5122 5 mg/kg for 14 days orally (group B).

The duration of function of the heart grafts is shown in Fig. 3. Both drugs can induce long-term tolerance for heart grafts, cyclosporin A in 10 out of 11 animals, and ASTA 5122 in 6 out of 9 animals; two animals died from pneumonia with functioning grafts.

Untreated allografts are rejected very quickly; isografts survive without complications for over 100 days.

These data demonstrate that accessory heart transplantation is very useful as a simple and reliable model for transplantation research. The cuff technique used for the anastomoses shows very good long-term results, as demonstrated by the syngeneic control group (C1), and the effect of the immunosuppressive treatment is clearly visible (groups A and B).

Cuff Technique in Other Models

The cuff technique has been used in the rabbit for accessory heart transplantation by Heron (1971) and for kidney transplantation by Dunn (1976). Dunn reported a series of 300 transplantation; in the last 100 there were no vascular complications; in the first 200 the rate of thrombosis was 1%.

For transplantation of the rat liver, Kamada and Calne (1979) published a technique with the portal vein anastomosis performed with the cuff technique. In 124 operations, there was no vascular complication. Limmer and Calne (1981) presented a series of 300 rat liver transplants with two anastomoses with the cuff technique; he did not notice bleeding or early thrombosis.

Nolan et al. (1982) described a technique for pancreatic transplantation in the rat with both venous and arterial anastomoses performed with the cuff technique. He reported no arterial thrombosis in 52 operations.

All authors describe the cuff technique as a quick and easy procedure.

References

Dunn DC (1976) Orthotopic renal transplantation in the rabbit. Transplantation 22:427–433

Heron I (1971) A technique for accessory cervical heart transplantation in rabbits and rats. Acta Pathol Microbiol Scand [A] 79:366–372

Kamada N, Calne RY (1979) Orthotopic liver transplantation in the rat. Transplantation 28:47–50

Limmer J, Calne RY (1981) A simplified technique for orthotopic liver transplantation in the rat using a cuff technique for portal vein and infrahepatic vena cava anastomoses. Eur Surg Res 13:236–242

Nitze (1897) Discussion remark. Zentralbl Chir 39:1042

Nolan MS, Lindsey NJ, Savas CP (1982) Pancreatic transplantation in the rat. Transplantation 33:327–329

Ono K, Lindsey ES (1969) Improved technique of heart transplantation in rats. J Thorac Cardiovasc Surg 57:225–229

Payr E (1900) Beiträge zur Technik der Blutgefäß- und Nervennaht nebst Mitteilungen über die Verwendung eines resorbierbaren Metalles in der Chirurgie. Arch Klin Chir:67–93

Timmermann W, Calne RY, White DJG, White D, Thiede A (1980) Die Wirkung von Cyclosporin A und Cytimun auf die Überlebenszeit von RT-1 allogenen Rattenherztransplantaten. Langenbecks Arch Chir 353:139–142

Microvascular Pulmonary Transplantation in Rats

W. Konertz

The increasing number of heart lung transplantations being undertaken (Reitz 1982; Reitz et al. 1982) makes in an in-depth understanding of rejection phenomena occurring in lung allografts necessary. For experimental studies with this purpose, however, little investigation has been carried out in inbred animals under defined immunogenetic conditions. This is mainly attributable to the lack of simple orthotopic transplantation techniques in the rat. The aim of this paper is to present procedures of pulmonary transplantation in rats.

Orthotopic Left Lung Transplantation

The first attempt to perform orthotopic lung transplantation in rats was made by Asimacopulous et al. in 1971. They used very large rats and had less than 50% survival. An improved technique using microvascular techniques including the use of an operation microscope was introduced by Mark et al. in 1978. Their technique was modified in our laboratory.

Donor Operation

Donors were adult Lewis rats 200–250 g wheiging and obtained from commercial sources. After systemic heparinization with 1000 IU heparin i.v. a bilateral paramedian thoracotomy was performed with the rat under ether anesthesia. The left hilar structures were dissected free and the lung flushed with ice-cold Ringer's solution. During pneumonectomy it was important to preserve a button of left atrial tissue on the pulmonary vein. Figure 1 shows the excised graft ready for transplantation.

Recipient Operation

Anesthesia was introduced and maintained with ether. After orotracheal intubation the animals were placed on a volume-constant respirator set on 50 cycles/min and maximum inspiration pressure of 26 mm HG with a positive end expiratory pressure (PEEP) of 6 mm HG. A left lateral thoracotomy was performed in the fourth intercostal space. The hilar vessels and bronchus were freed and the left

Microsurgical Models in Rats
for Transplantation Research
Eds. A. Thiede et al.

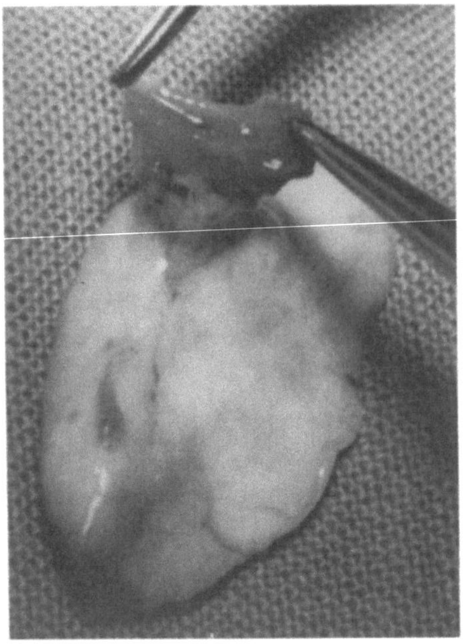

Fig. 1. Graft ready for transplantation. Note the left atrial tissue on the pulmonary vein

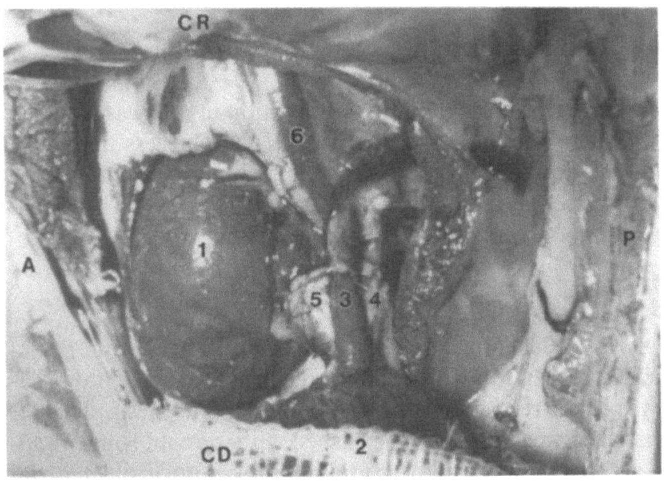

Fig. 2. Lung hilus after transplantation. *A*, anterior; *P*, posterior; *CR*, cranial; *CD*, caudal; *1*, heart; *2*, retracted left lung graft; *3*, pulmonary artery; *4*, bronchus; *5*, pulmonary vein; *6*, left superior vena cava

pulmonary artery and left main stem bronchus were clamped with a large heavy clamp to obtain an operative field quiet and free from respiratory movements. The pulmonary veins were ligated and a left pneumonectomy was performed, leaving long arterial and bronchial stumps. The pulmonary artery anastomosis was then performed between three stay sutures in a running fashion with 9-0 monofilament material. The bronchi were anastomosed end-to-end with a 7-0 continuous Prolene suture. The pulmonary artery was then clamped with a small bulldog clamp and the large clamp removed. The pericardium was entered dorsal to the phrenic nerve and the left atrial appendage clamped. After incision the appendage was connected to the button on the pulmonary vein with a continuous 8-0 Prolene suture. The atrial clamp was removed first followed by removal of the bulldog clamp on the artery.

Figure 2 shows the lung after transplantation. After meticulous hemostasis and thorough flushing with warm normal saline the graft was carefully located in the thoracic cavity and fully inflated. The chest was closed and the animal allowed to breath spontaneously. Usually at the end of the operation the rats were extubated.

Comment

All the techniques using the operation microscope in thoracic procedures in the rat are very complex and time consuming; this might be responsible for a survival of only 65% of the animals in the series of Mark and Wildevuur (1981). We looked for points of improvement of his technique and found the very delicate venous anastomosis the most troublesome point, leading to complications in 50% of the surviving animals in the series of Mark and Wildevuur (1981). We modified the technique (Konertz et al. 1980a) by avoiding a direct end-to-end anastomosis of the pulmonary veins. Instead we connected the pulmonary vein of the graft directly to the left atrial appendage. This maneuver shortened the duration of the operation considerably, made the use of the operation microscope unnecessary, and led to satisfying results.

In our series no attempt to increase ischemic tolerance of the graft was undertaken. Therefore the ischemic time had to be in the order of 30 min. An increase

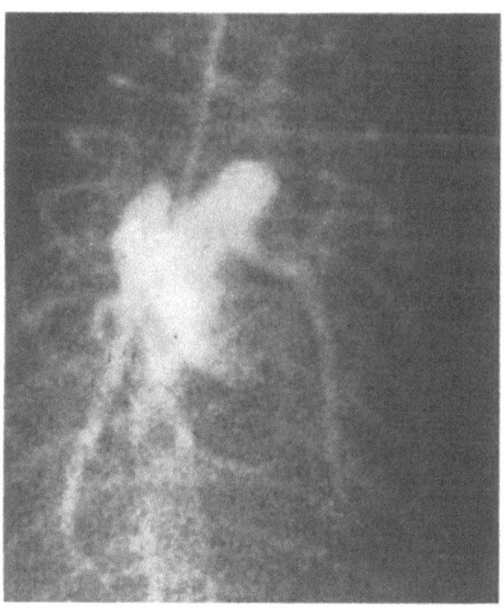

Fig. 3. Angiography on postoperative day 14 in syngeneic lung transplantation

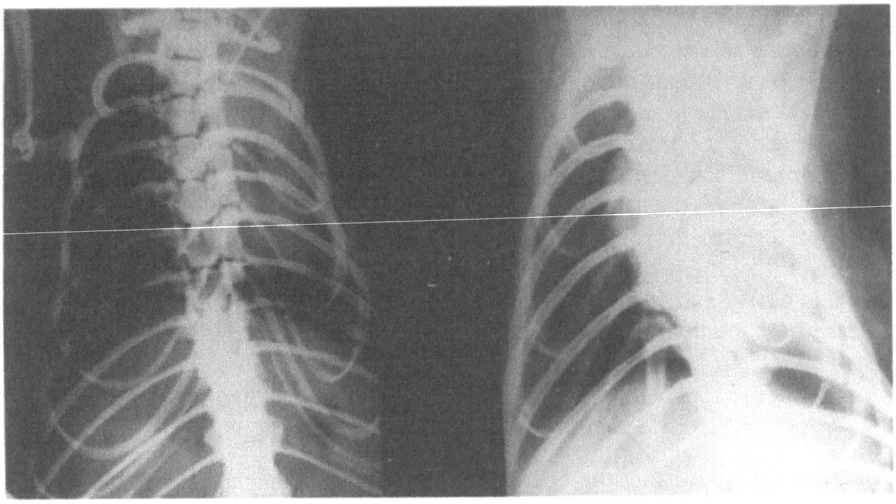

Fig. 4. Chest X-ray on postoperative day 8 in syngeneic (*left*) and allogeneic (BN-LEW) transplantations (*right*). The opacification of the left lung seems due to acute rejection

in ischemic time to over 45 min often resulted in pulmonary edema postoperatively.

The lack of an easy and repeatable follow-up method in such lung-allografted rats is really cumbersome. Cohen et al. (1983) used gas analysis after contralateral pneumonectomy. Performing this procedure on postoperative day 3 led to 100% mortality. Histological examination of the graft also results in the loss of the animals. Mark et al. (1981) used perfusion scintigraphy and chest X-rays, while we performed angiography (Fig. 3) and daily chest X-rays (Fig. 4). Scintigraphy and angiography demonstrate only the perfusion of the lungs while cest X-rays show the aeration. Fine-needle aspiration and cytological examination of the aspirate (Konertz et al. 1980a) seem promising methods for producing further information on immunological processes occurring in the transplanted lung. In conclusion there is no single global monitoring technique that delivers functional and immunological data of lung allografts.

Heterotopic Heart Lung Transplantation

This technique was introduced by Lee et al. in 1970. The heart and both lungs are transplanted to the abdomen of the recipient, and vascular connections are made by connecting the aorta of the graft and the infrarenal aorta of the recipient, with all other vessels on the graft ligated. Due to the nonphysiological functional status of both organs this technique has found no widespread application. However, a recent publication from Womble et al. (1983) used heart lung transplants in rats for evaluation of immunosuppressive properties of propranolol.

Thoracic Heart Lung Transplantation

This technique is a combination of orthotopic lung transplantation (Konertz et al. 1980a) and heterotopic thoracic heart transplantation (Konertz et al. 1980b).

Donor Operation

After systemic heparinization a bilateral thoracotomy is performed. The right hilar structures are ligated with a single 4-0 ligature. Both superior caval veins are also ligated and the heart and the left lung excised leaving long aortic, tracheal, and inferior caval stumps. The graft is immersed in ice-cold Ringer's solution until transplantation.

Recipient Operation

Under ether anesthesia a left lateral thoracotomy in the fifth intercostal space is performed. The left lung is gently pushed caudally. Sometimes the hemiazygos vein has to be ligated and transsected. The descending aorta is dissected and controlled with two vessel snares. A longitudinal aortotomy is made and an end-to-side anastomosis between the graft's aortic arch and the descending aorta of the recipient with 8-0 Prolene is performed. The vessel snares are released and blood flow to the lower part of the body of the recipient as well as to the donor heart is established. Now the left pneumonectomy in the recipient is performed and a tracheal connection with the graft is made by a running 7-0 Prolene suture. The inferior caval connections are then made preferentially with the cuff technique. The heart and lung are located in the thoracic cavity with special care to avoid kinking of the vascular supply of the graft. At the end of the operation the animal is extubated. Figure 5 shows how the graft is connected in the recipient's cardiovascular system.

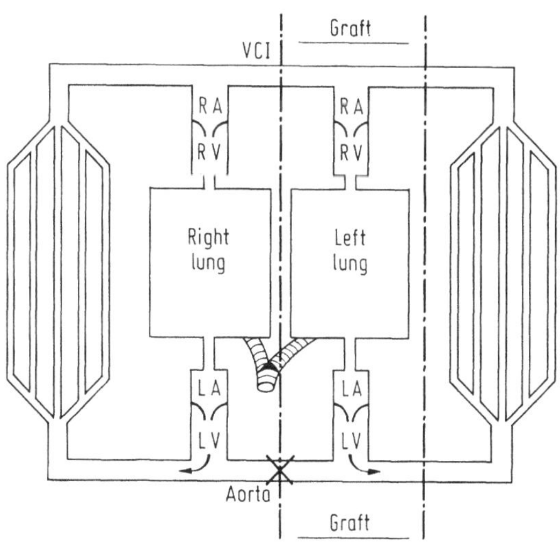

Fig. 5. Flow chart after thoracic heart lung transplantation

Comment

This functional model of heart lung transplantation in rats makes evaluation of graft functional very easy. By pressure measurements at the tail or femoral artery the function of the heart can be evaluated, while arterial or capillary blood gas analyses from the lower part of the body represent the functional properties of the transplanted left lung, because this graft supplies the lower part of the body. However, this model is still under experimental investigation and the number of procedures performed is still limited.

References

Asimacopoulos PJ, Molokiha FAS, Peck CAS, et al. (1971) Lung transplantation in the rat. Transplant Proc 3:583

Cohen RG, Sievert FC, Jamison SW, et al. (1983) Functional evaluation of syngeneic lung transplants in rats. Heart Transplant 2:171–174

Konertz W, Bürsch J, Semik M, et al. (1980a) Techniques of pulmonary transplantation and postoperative monitoring in inbred rats. Eur Surg Res [Suppl 1] 12:79–80

Konertz W, Semik M, Bernhard A (1980b) Left ventricular bypass in inbred rats. A new experimental model in microsurgery operative technique and hemodynamic evaluation. Thorac Cardiovasc Surg 28:277–279

Lee S, Willoughby WF, Smallwood CJ, et al. (1970) Heterotopic heart and lung transplantation in the rat. Am J Pathol 59:279–283

Mark KW, Wildevuur CRH (1981) Lung transplantation in the rat. I. Technique and survival. Ann Thorac Surg 34:74–80

Mark KW, Hess F, Wildevuur CRH (1978) Microsurgical aspects of unilateral lung transplantation in the rat. Excerpta Med Int Congr Ser 465

Mark KW, Piers DA, Wildevuur CRH (1981) Lung transplantation in the rat. II. Lung perfusion scintigraphy in normal and left lung transplanted rats. Ann Thorac Surg 34: 81–88

Reitz BA (1982) Heart lung transplantation: review. Heart Transplant 1:291–298

Reitz BA, Wallwork JL, Hunt SA, et al. (1982) Heart lung transplantation – successful therapy for patients with pulmonary vascular disease. N Engl J Med 306:557–564

Womble JR, Larson DF, Copeland JG, et al. (1983) Propranolol. An effective immunosuppressive agent in rat heart lung transplantation. Transplantation 35:417–420

Spleen and Vascularized Skin Flap Transplantation

G. H. Müller and *A. Wunderlich*

A rat model for spleen grafting has already been described in 1969 by Lee and Orloff and by Coburn. They published a method using an aortic and a portal vein segment for end-to-side anastomoses to the aorta and vena cava of the recipient. In Tübingen, we prefer a technique which allows end-to-end anastomoses and avoids cross-clamping of the aorta and vena cava.

The topographical position of the spleen is as follows: The arterial supply is provided by the celiac axis via the splenic artery, and the venous outflow by the splenic vein, which drains into the portal vein. A major problem in spleen grafting can occur during dissection of the donor, which has to be carried out with great care.

After making a midline incision, the intestines are retracted to the right, and the stomach to the diaphragm. The fine ligaments between the spleen, pancreas, stomach, and intestines are dissected. The vascular branches from and to the stomach are ligated and divided, as are the branches to the mesenteric artery. The portal vein is then tied off below the splenic vein. The arterial supply provided by the celiac axis is identified, gastric and hepatic arteries are tied off, and the graft is removed. No attempt is made to divide the pancreatic tissue from the spleen completely. Depending on the size of the recipient's renal artery, we use the celiac artery itself, the artery on a small aortic patch, or an aortic segment for

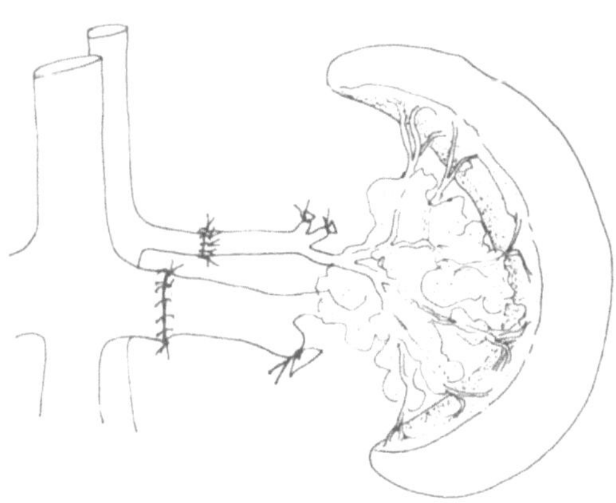

Fig. 1. Splenic transplantation: site of implantation. Celiac artery end-to-end to renal artery and portal vein end-to-end to renal vein

Microsurgical Models in Rats
for Transplantation Research
Eds. A. Thiede et al.

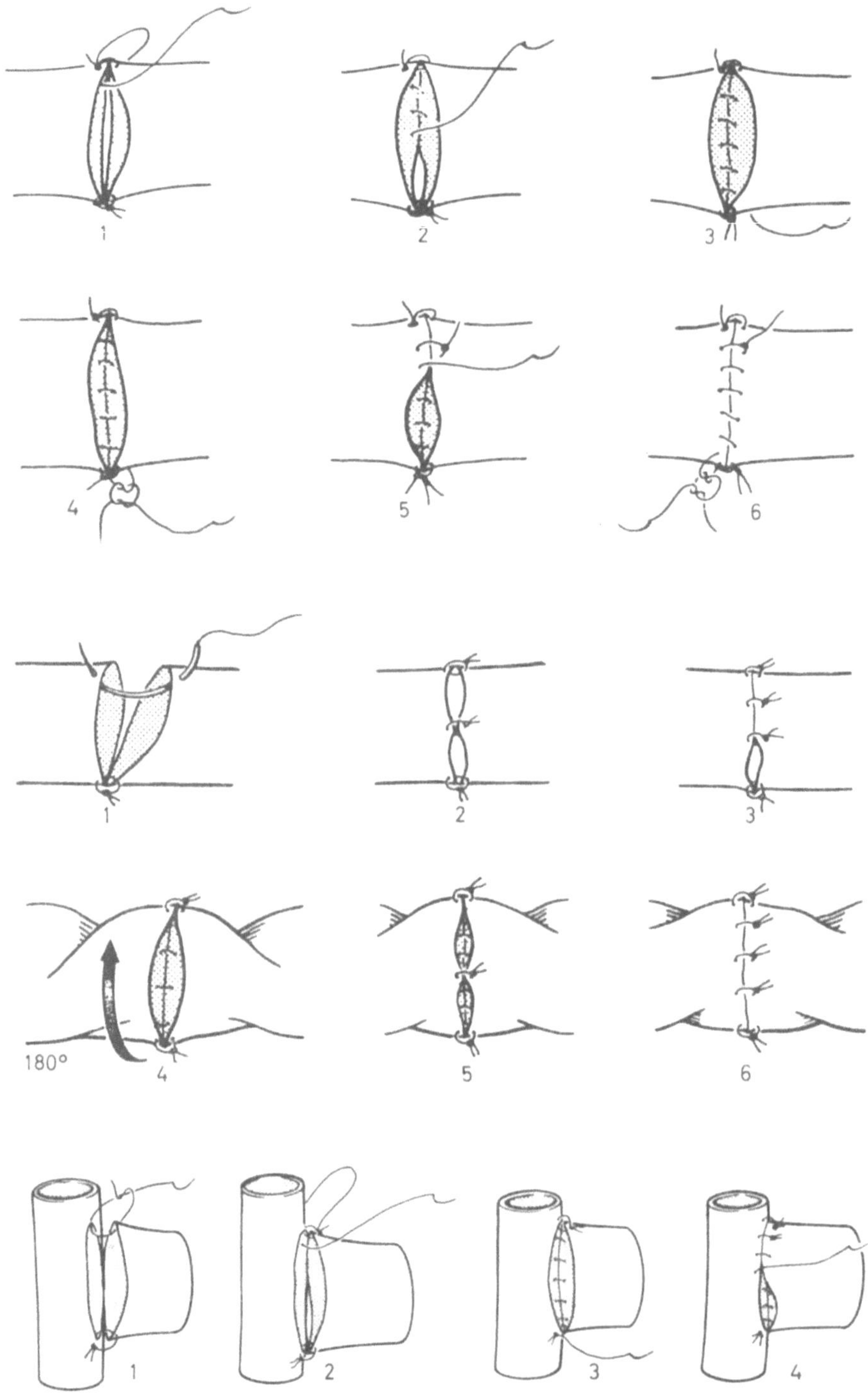

Fig. 2. Microvascular techniques used: end-to-end, single stitches; end-to-end, continuous sutures; end-to-side, continuous sutures

an end-to-end anastomosis (Fig. 1). Only single stitches are used for the artery. The microsurgical technique we use for the venous end-to-end anastomoses is different: We prefer continuous stitches for all venous anastomoses (Fig. 2).

Technical problems arise more often than in other transplant models, such as in the case of the heart and kidneys. In our tests, approximately one out of five had to be rejected as a result of thrombosis, or were lost due to other technical failures.

Spleen grafts are interesting for various reasons: Depending on the genetic barrier between donor and recipient, spleen transplants may lead to a host-versus-graft rejection, to a graft-versus-host reaction, or to a status which is still being investigated and which may lead to a stable donor-specific unresponsiveness. Our experiments performed with DA and Lewis rats showed perfect isografts after 3 months but a strong rejection of all spleen allografts. In this combination, even cyclosporin A was not able to prevent rejection.

Some recent studies using cyclosporin A (CyA) have led to the development of another transplant model in the rat: Recipients of a liver allograft that were treated with CyA showed a complete unresponsiveness to secondary vasularized grafts, but constantly rejected donor-specific skin grafts. To investigate this result more fully, we performed both conventional skin grafts and vascularized skin grafts.

A groin flap, with femoral vein and artery intact, was taken from a heparinized donor (Fig. 3). Care was taken to avoid injury to the fine superficial epigastric vessels. The left common carotid artery and the jugular vein of the recipient were then exposed. The carotid was tied off and prepared for an end-to-end anastomosis with the femoral artery of the skin flap. The jugular vein was cross-clamped for an end-to-side anastomosis. Standardized techniques were used with single stitches for the artery. For the veins, continuous sutures were again preferred, as is shown in detail (Fig. 2).

The following results were observed: DA skin in Lewis rats survived only 7 days without immunosuppression. In contrast to this, vascularized skin from DA donors survived significantly longer. By days 6–8, an edema developed and the graft doubled in size.

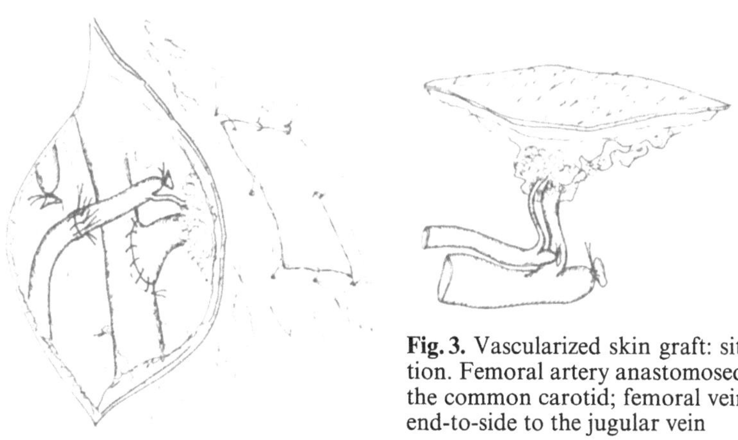

Fig. 3. Vascularized skin graft: site of implantation. Femoral artery anastomosed end-to-end to the common carotid; femoral vein anastomosed end-to-side to the jugular vein

Between days 8 and 12, the grafts lost hair and started to shrink. Vascular thrombosis occurred and in the following 30 days the graft was reabsorbed and replaced by healthy skin.

By day 7, the day of rejection for conventional skin grafts, there were heavy cell infiltrations consisting of mononuclear cells in vascularized free flaps. Only a few macrophages were observed at this time; the skin looked well perfused, in contrast to the rejected nonvascularized skin allografts.

Thus, rejection in vascularized skin grafts appears to be more like rejection in vascularized kidney or heart grafts.

Transplant models are necessary for examination of the rejection mechanism. This model for vascularized grafts may help to explain some interesting immunological aspects of skin grafting. Biopsies are easy to perform and, in contrast to life-supporting organ grafts, the recipient survives after rejection. This model may help to work out the difference between vascularized and nonvascularized grafts and the difference in their rejection mechanism. This model is also of interest for practice in microsurgery: an arterial or venous anastomosis can be carried out without a laparotomy, and the two basic techniques (end-to-end and end-to-side anastomoses) can both the demonstrated, again using two different techniques: interrupted or continuous sutures. We therefore recommend this model for further studies in microsurgery.

References

Coburn RJ (1969) Spleen transplantation in the rat. Transplantation 8:86–88

Lee S, Orloff MJ (1969) A technique for splenic transplantation in the rat. Surgery 65/3:436–439

Renal Transplantation in the Rat

H. D. Jakubowski

Introduction

Renal transplantation in the rat is one of the most essential experimental models in transplantation research for several reasons: Many inbred rat strains are defined by their histocompatibility system and are therefore very suitable for immunological investigations (Festing and Staats 1973). Furthermore, in addition to the economic aspects, the fact that rats are available in large numbers and can be handled without problems is important. Experimental questions are thus normally answered in a short time.

The operative procedure described here represents the author's technique used since 1977 (Jakubowski et al. 1981). Apart from several personal modifications it is essentially the method described by Fisher and Lee (1965) and improved upon by Lee (1967) later on.

Operative Technique

Open ether anesthesia is routinely applied in all these experiments. Anastomoses of the vessels and ureter are sutured using an operating microscope with six fold magnification.

Donor Nephrectomy (Fig. 1)

The donor is anesthetized, and the abdomen is shaved and disinfected. The rat is then placed on its back with its tail toward the investigator. After opening the abdominal cavity through a long midline incision, the bowels are wrapped in a saline-moistened gauze pack, displaced by a retractor to the right, and the left kidney is exposed.

The retroperitoneum is simply opened with blunt fingertip dissection and removed from the great vessels and the left renal vein. Small vessels inosculating into the renal vein (e.g., suprarenal, spermatic, nontypical lumbar vein) are ligated with 5-0 silk and cut. Two holding sutures are placed through the fat tissue at the upper and lower poles of the kidney and fixed with a straight clamp, thus making possible simple handling of the graft without touching it with fingers or

Microsurgical Models in Rats
for Transplantation Research
Eds. A. Thiede et al.

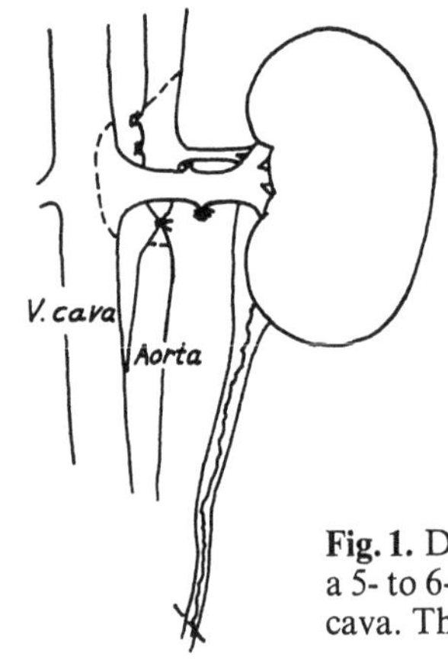

Fig. 1. Donor nephrectomy in the rat: The left kidney is removed with a 5- to 6-mm-long aortic cuff and a large vein patch of the inferior vena cava. The ureter is transected close to the bladder

instruments. Thereafter, the kidney is freed from its surrounding fat and connective tissue attachments and the ureter is mobilized up to the bladder. It is very important not to injure the periureteral adventitia, which contains the small nutritional vessels for the ureter. Otherwise ureter necrosis can subsequently occur.

The aorta has to be mobilized from about 1 cm below to 1 cm above the junction with the left renal artery. For this the superior mesenteric artery, right renal artery, and often a lumbar artery have to be ligated and cut. Finally a 3-0 silk ligature is applied loosely around the aorta 2–3 mm below the left renal artery. With this ligature the perfusion cannula is fixed and after graft perfusion the aorta is ligated.

For graft perfusion the aorta is clamped at least 6–8 mm above the left renal artery. Thereafter the infrarenal aorta is punctured with a 0.7-mm Butterfly cannula, which is connected to a 2-ml syringe containing chilled Ringer's solution. The prepared ligature is tied around the tip of the cannula and the kidney is flushed out. The graft pales totally within 2–3 min after an initial warm ischemia time of about 30 s. The cannula is then removed, the ligature is knotted, and the aorta is transected distally to the aortic clamp and the ligature. In this way the renal artery originates from a nearly 5- to 6-mm-long aortic cuff which can easily be anastomosed to the recipient's aorta. After elevating the graft by the holding sutures the renal vein is excised out of the vena cava together with a large vein patch of the cava. The ureter is finally transected as close to the bladder as possible, the graft then being placed in a beaker of cold Ringer's solution. This is stored in an iced water bath while the organ recipient is prepared. The nephrectomy results in the death of the donor rat.

Implantation of the Kidney (Fig. 2)

The recipient animal is anesthetized and prepared exactly like the donor. After opening the abdomen through a long midline incision the intestines are placed by a retractor to the left, thus exposing the right retroperitoneal region. The retroperitoneum is opened and removed from the inferior vena cava and aorta. Both vessels are then mobilized for a distance of about 2 cm distal to the renal vessels

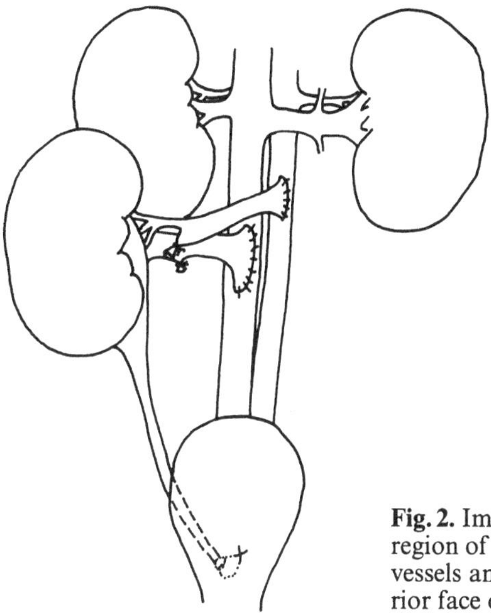

Fig. 2. Implantation of the graft into the right lumbar region of the recipient: end-to-side anastomosis of the vessels and implantation of the ureter into the posterior face of the bladder

and clamped with an artery forceps in a way which eliminates the blood circulation over a length of at least 1 cm in both. The vena cava is then incised on its right side for a length of 3 mm. A few millimeters proximal to this incision an elliptical opening is excised out of the ventral surface of the aorta. This should correspond in size to the diameter of the graft's aortic cuff.

For anastomosing the vessels and the ureter the rat is placed with its right side toward the investigator. At first the end-to-side anastomosis of the cava patch containing the renal vein is performed to the inferior vena cava. For this 8-0 or 9-0 monofilament suture material attached to a 6-mm curved atraumatic needle is used. After placing two stay sutures in the opposed ends of the anastomosis it can be sutured using a continuous technique. It is especially important for the vein anastomosis that the anastomatic area is kept moistened with saline, because this measure facilitates the handling of the extremely thin vascular walls. The arterial anastomosis is performed as oblique end-to-side anastomosis between the graft's aortic cuff and the recipient's aorta with the same suture material and in the same way as the vein anastomosis.

After completing the arterial anastomosis the aortic clamp is slowly removed. Normally profuse bleeding at the anastomoses stops soon after a tamponade with a gauze pad is used. If bleeding persists after several minutes, another suture must be placed at the bleeding point. If the anostomoses have been performed correctly, the kidney will become pink in color within a short time. If this does not happen, one must then assume that a technical mistake has been made. Such faults are very difficult to correct successfully.

The *ureteral anastomosis* is performed by inserting the end of the donor ureter into the bladder of the recipient. For this procedure the fundus of the bladder is clamped with atraumatic forceps and displaced toward the tail of the rat, exposing in this way the posterior face of the bladder. Now the posterior wall is opened in the neck area for a length of 2–3 mm. The tip of the ureter is freed from its surrounding tissue for a few millimeters. Then the ureter is loosely fixed in the bladder with on 8-0 atraumatic suture. This is accomplished by passing the suture through the anterior wall of the bladder at the level of the incision. The suture passes through the incision and into the open end of the ureter, exiting through the side wall of the ureter. The suture is passed back through the incision and anterior wall.

The suture is then drawn together, forcing the ureter into the incision. The suture is loosely knotted. It is important that the knotting does not force the open end of the ureter against the opposite wall, causing a closure. Finally the entrance of the ureter into the bladder is narrowed with a Z-suture so that the ureter remains mobile. This is necessary to avoid strictures of the ureter. The operation is finished with the closure of the abdominal cavity.

References

Festing M, Staats J (1973) Standardized nomenclature for inbred strains of rats. Transplantation 16:221

Fisher B, Lee S (1965) Microvascular surgical techniques in research, with special reference to renal transplantation in the rat. Surgery 58:904

Jakubowski HD, Niebel W, Dostal G (1981) Renal transplantation in the rat. Videofilm, Essen

Lee S (1967) An improved technique of renal transplantation in the rat. Surgery 61:771

Microsurgical Technique for Small-Intestine Transplantation

E. Deltz and *A. Thiede*

Since in small-intestine transplantation many problems remain unsolved (see Hardy et al., this volume, p. 135), different experimental settings for small-bowel grafting are necessary. In the rat model several operative procedures are possible.

Among microsurgical techniques for small-intestine transplantation it is important to distinguish between two different types of experimental design:

1. Auxiliary transplantation, in which the small intestine of the recipient is not removed and the transplant is grafted in addition to the recipient's organ
2. Orthotopic transplantation, in which the small intestine of the recipient is removed and replaced in orthotopic position by the transplanted organ.

Auxiliary Small-Intestine Transplantation

In principle, three types of auxiliary small-intestine transplantation are possible (Fig. 1). The difference between the various models of auxiliary heterotopic small-intestine transplantation lies in the manner in which the oral and aboral ends of the graft are handled. Both ends of the graft can be exteriorized as duodenostoma and ileostoma (Fig. 1 a), or only the oral end of the graft can be sutured to the

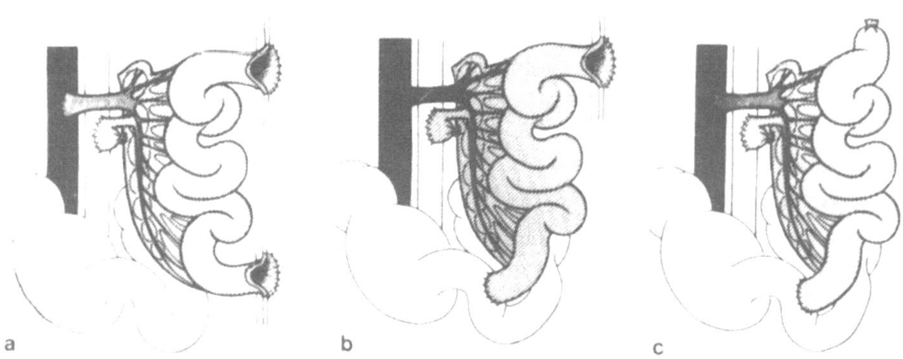

Fig. 1 a–c. Models of auxiliary heterotopic small-intestine transplantation

Microsurgical Models in Rats
for Transplantation Research
Eds. A. Thiede et al.

abdominal wall as a duodenostoma, while the aboral end of the graft is anastomosed to the terminal ileum of the recipient (Fig. 1 b). Finally, it is possible to close off the oral end and to anastomose the aboral end to the intestine of the recipient (Fig. 1 c). This is the technique used by the microsurgical group in Kiel. It is a modification of the methods used by Monchik and Russell (1971) in their pioneering experiments in the field of small-intestine transplantation. This technique was also used by Petirossi (1971), who was able to systematize the operative technique based on extensive anatomical studies.

This microsurgical method was taken up by us in cooperation with Sun Lee (Thiede and Deltz 1978; Deltz and Thiede 1984), thus beginning the systematic analysis of the immunological phenomena related to small-intestine transplantation.

Operative Technique

The operation is performed under ether anesthesia. A dissecting microscope with tenfold magnification is used as optical aid for some steps of the operative procedure.

Donor Operation

After shaving and disinfection the abdomen of the donor is opened by a midline incision. The operation starts with the separation of the small-bowel mesentery from the colon. The middle colic artery and the ileocolic artery and their branches are ligated and the mesocolon divided. Beginning from the porta hepatis, the portal vein and the superior mesenteric vein are divided after ligature of their tributaries and isolation from the pancreas.

The retroperitoneum is entered and the celiac trunk freed of surrounding tissue. The common hepatic artery, lienal artery, and a lumbar artery are ligated; then an aortic cuff with the superior mesenteric artery, which takes off from it, is ligated. This is followed by proximal and distal cross-clamping of the aorta, division of the portal vein close to where it enters the liver and division of the intestine at the duodenojejunal flexure and in the terminal ileum. The aortic cuff is punctured and the graft perfused with 0.9% NaCl solution at 4 °C until the graft appears white and the effluent clear. The aortic cuff is ligated distally and the graft removed by dissecting the aorta. The lumen of the intestinal graft is perfused with 20 cm^3 0.9% NaCl solution to remove remnants of the intestinal contents.

Transplantation

The abdomen of the recipient is opened by a midline incision. The small intestine of the recipient is retracted to the right, so that the abdominal aorta and the inferior vena cava are exposed. Both vessels are isolated along a stretch of 1 cm, clamped and opened longitudinally. To reduce the chance of thrombosis, the vessels must be rinsed free of blood.

The graft is brought into position and the end-to-side anastomosis of the aortic cuff of the graft to the abdominal aorta of the recipient is performed using a running suture. The posterior suture is performed from the lumen, the anterior suture from the exterior of the vessel. When the posterior suture is performed, it is important to make sure that there is absolutely no blood leakage, because after the anastomosis is completed no corrective stitches can be made. The next step is the end-to-side anastomosis of the portal vein to the inferior vena cava of the recipient. At first the anterior wall is sutured with 8-0 nonabsorbable material (Ethilon; Ethicon, Hamburg). In the vein-vein anastomosis, care must be taken to avoid stenosis, which would lead to stasis in the graft.

Next the clamps are removed and the blood stream released. Two milliliters of hydroxyethyl starch (Plasmagel) is injected into the tail vein as a substitute for blood loss. The oral end of the graft is closed with a circular ligature. The end-to-side anastomosis of the aboral end of the graft to the terminal ileum of the recipient is accomplished with a continuous one-layer suture with nonabsorbable material. The intestine of the recipient with the graft is returned to the abdomen, and the abdominal wall is closed with a continuous one-layer suture. The total operation time is about 120 min; the cold ischemic time is 30–40 min.

The most important complications are venous thrombosis in the region of the anastomosis, which leads to hemorrhagic infarctation of the graft and postoperative bleeding from the arterial anastomosis, which causes a protracted shock condition in the animal. Both complications lead to the death of the recipient within 24 h. With sufficient microsurgical practice and experience the operative mortality is 10%.

After the operation the animals receive a normal diet and water. Heparin and antibiotics are unnecessary. Briefly auxiliary small-bowel transplantation consists of the following steps.

1. Separation of the small bowel from the colon
2. Separation of the portal vein from the pancreas
3. Exposition of the small-bowel mesentery
4. Preparation of the aortic cuff with the upper mesenteric artery
5. Puncture of the aortic segment and perfusion of the graft
6. Removal of the graft
7. Aortic anastomosis
8. Portocaval anastomosis
9. Intestinal anastomosis

All auxiliary models of small-intestine transplantation have the disadvantage that they do not represent the normal anatomical situation. They are bypass models, in which the graft is shunted off and does not participate in the intestinal functions. Therefore, these models are not suited to absorption studies under physiological conditions. For such purposes orthotopic models must be considered. At the present time, however, the decisive questions that must be solved before clinical small-intestine transplantation can become established are immunological in nature. For investigation of immunological questions, however, auxiliary models are most suited. They enable long-term observation and evaluation of the immunological reactions in the graft and in the recipient organism, since the recipients survive even when the graft has sustained extensive damage, which in the orthotopic graft would lead to the early death of the recipient. In addition,

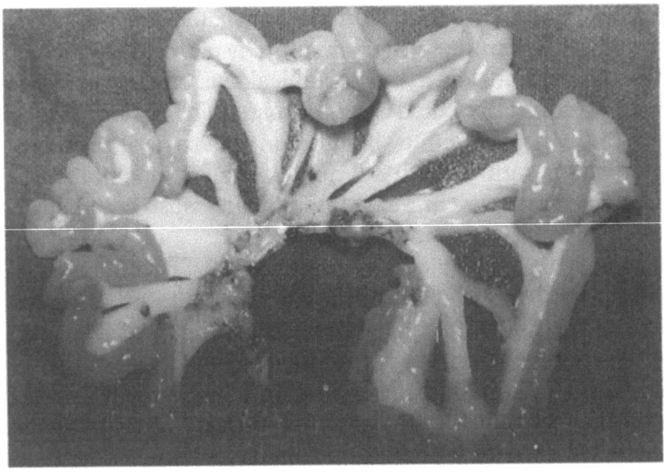

Fig. 2. Small-bowel graft with the mesenteric lymph nodes removed

in the bypass technique with one or two intestinal stomata it is possible to take biopsies which enable continuous evaluation of the rejection processes to be made.

Since the results of our investigations showed that the mesenteric lymph nodes of the grafts play an important role in the immunological reactions following small-bowel transplantation (Deltz et al., this volume p. 145), a technique for microsurgical removal of these lymph nodes was developed in order to elucidate their influence on immunological reactions and recipient survival.

The mesenteric lymph nodes can be removed by excision with fine scissors with the help of a dissection microscope. Hemostasis is achieved by electrocoagulation. Figure 2 shows a nonperfused small-bowel graft with the mesenteric lymph nodes removed.

As the advances in immunosuppression make it possible to control immune reactions better, it appears that it will be possible to perform clinical small-bowel transplantation in the near future (Hardy et al., this volume, p. 337; Stauffer 1983). Therefore, it is a prerequisite that investigations be done in a model of orthotopic small-bowel transplantation in which resorption studies can be performed in correlation with immunological reactions.

Because in the rat immediate replacement of the recipient's small bowel after resection is followed by an unacceptably high operative mortality we perform a two-step operative procedure, which results in a total replacement of the recipient's small bowel. A schematic diagram of the operative procedure is shown in Fig. 3.

Three weeks after heterotopic small-bowel grafting using the technique described above, when the lymph vessels of the graft have gained connection with the recipient's lymph vessels, the graft is inserted into the recipient's small bowel after its resection, thus reestablishing continuity of the alimentary tract. The recipient's small bowel is dissected beneath the duodenojejunal flexure and is anastomosed to the oral end of the graft after the ligated segment has been resected.

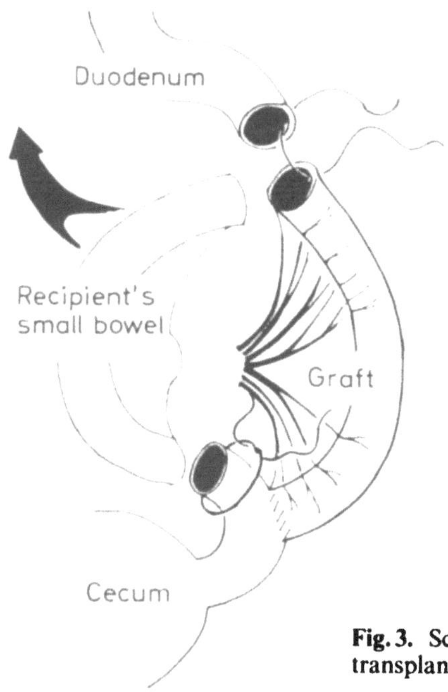

Fig. 3. Schematic diagram of orthotopic small-bowel transplantation (two-step technique)

The distal anastomosis is preserved and the recipient's terminal ileum ligated. The recipient's small bowel is then removed by stepwise ligation and dissection of the mesentery. By this technique an indefinite survival of the recipients can be achieved.

The application of these techniques in different experimental settings is the basis for investigation of the complex immunological problems following small-bowel transplantation. The elucidation of these phenomena is the basic prerequisite for clinical small-bowel transplantation.

References

Deltz E, Thiede A (1984) Experimental transplantation of small intestine. Microsurgical techniques and their applicability in research. In: Olszewski WL (ed) CRC, Boca Raton/Florida

Monchik GJ, Russell PS (1971) Transplantation of small bowel in the rat: technical and immunological considerations. Surgery 70:693–702

Petirossi O (1971) Un nuovo trapianto d'organo nel ratto: l'intestino. II. La technica operatoria. Ann Italiani Chir 47:181–191

Stauffer UG (1983) Die Dünndarm-Transplantation – Gegenwärtiger Stand und mögliche Zukunft. Z Kinderchir 38:286–289

Thiede A, Deltz E (1978) Morphological reaction in transplanted small intestines using immunogenetically defined rat strain combinations. Langenbecks Arch Chir 346:119–127

Various Techniques of Experimental Pancreas Transplantation in the Rat *

T. Schang, W. Timmermann, and *A. Thiede*

Pancreas transplantation as therapy for patients suffering from type I diabetes mellitus is far from being routine, especially in the state of terminal renal failure. In addition to immunological problems, the question of which is the most appropriate transplantation technique is a long way from being answered. Adequate handling of external secretion is a main subject of discussion. Today, transplantation of a duct-occluded pancreatic segment is the clinically preferred approach. The duct system is occluded by prolamine (Largiader 1981) or polymers like neoprene (Dubernard et al. 1983) for the purpose of immediate by stopping the exocrine secretion, followed by total atrophy of the exocrine tissue (Gebhardt and Gall 1980). Nevertheless, the islets of Langerhans continue their endocrine function. These techniques offer a considerable degree of protection against postoperative complications caused by pancreatic juice. Long-term results reported so far, however, give rise to the question of whether defects of the islets occur induced by the technically caused alterations of the gland.

On the other hand draining procedures may be able to avoid graft alterations, thus probably achieving a better rate of long-term islet function (Gliedman 1973; Calne and Groth 1983). Nevertheless, a greater risk of postoperative complications is present.

The technique of experimental pancreas transplantation in inbred rats was first presented by Lee et al. (1972). It offers the opportunity to study the technical problems in addition to the immunological question or the properties of new immunosuppressive agents. As a basic study for further research in kinetics of allograft rejection and methods of immunomodulation we have studied several transplantation techniques.

Materials and Methods

Three techniques served as models for secretion drainage procedures: pancreaticoduodenal graft, duodenal patch drained graft, and Roux-en-Y drained graft. Duct-ligated and duct-occluded segmental grafts represented nondraining techniques. Rejection was excluded by using a syngeneic LEW-LEW strain combina-

* This work was supported by the Deutsche Forschungsgemeinschaft SFB 111/B 12

Microsurgical Models in Rats
for Transplantation Research
Eds. A. Thiede et al.

tion. The animals used were aged 3 months and their body weight was 250–280 g. Recipients were made diabetic by 55 mg streptozocin/kg bodyweight i.v. 3 weeks prior to transplantation. Nonfasting serum glucose levels were above 20 mmol/liter in all diabetic animals. A nontransplantated control group (*n*, 10) showed no changes in diabetic state within 300 days. Endocrine function of pancreas grafts was estimated by repeated checks on serum glucose and body weight. Serum glucose levels above 11 mmol/liter were taken as hyperglycemia. The suture material was 8-0 Ethilon for vessel anastomoses and 6-0 Prolene for intestinal anastomoses; 5-0 or 6-0 silk was used for ligatures.

Techniques of Graft Harvesting

Pancreaticoduodenal Graft. After entering the abdominal cavity by a long midline incision the greater omentum is dissected and the omental bursa is opened. Vessels of the splenic hilum are ligated but not cut. The spleen is clamped; thus the pancreas can be moved without touching it. The middle colic vessels are ligated and cut at the lower pancreatic margin. Now the transverse colon is detached from the pancreatic head and the superior mesenteric vein is exposed. Then the duodenum is dissected from the posterior abdominal wall. The mesentery is divided at the duodenojejunal flexure up to the superior mesenteric vein. The superior mesenteric vessels and the duodenojejunal flexure are prepared for later dissection. Now preparation of the hook-shaped pancreatic processus at the pylorus follows by cutting the right gastroepiploic vessels. The common bile duct and hepatic artery are ligated and cut at the porta hepatis. Next the clamped stomach is turned upside down and the gastric vessels are ligated and cut. The pancreas is now turned to the animal's right side and preparation of the aorta between the diaphragma and left renal artery is performed. The right renal artery and some smaller tributaries of the aorta are cut. Finally, superior mesenteric vessels are cut at the lower pancreatic margin and graft perfusion is performed via the aorta by approximately 5 ml 4 °C cold 0.9% saline. An aortic segment is cut out, bearing the superior mesenteric artery and the celiac axis. Its proximal end is left open for arterial anastomosis. The portal vein is cut at the hepatic porta and used for venous anastomosis. The duodenum is flushed by cold saline to remove feces after both ends have been dissected. The oral end is closed by a simple ligature (Fig. 1).

Duodenal Patch Drained Graft. In addition to the procedure described above the pancreatic head is dissected from the duodenum carefully avoiding damage to its vascular arcades. A patch of the duodenal wall is excised around the duodenal papilla. The patch has to be supplied by two vascular branches (Fig. 2).

Roux-en-Y Drained Graft. In this technique the duodenum is totally dissected from the pancreatic head and the pancreatic duct is closed by a ligature. External secretion is led out retrogradely via the common bile duct, which is supported by a conical polyethylene stent made from a Medicut cannula. The surface of the stent is roughened by a knife. An additional ligature secures its position in the bile duct (Fig. 3).

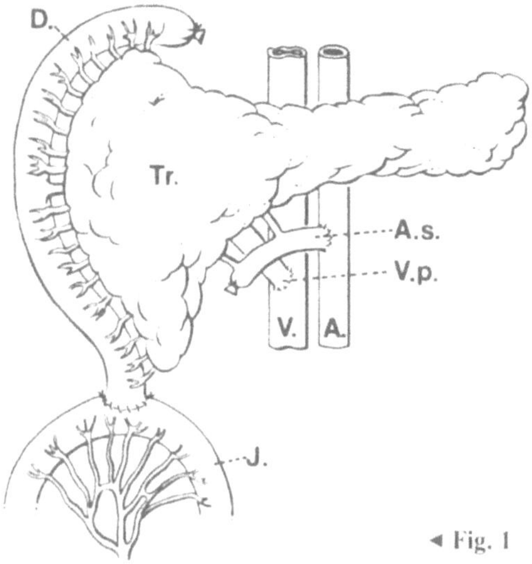

◄ Fig. 1

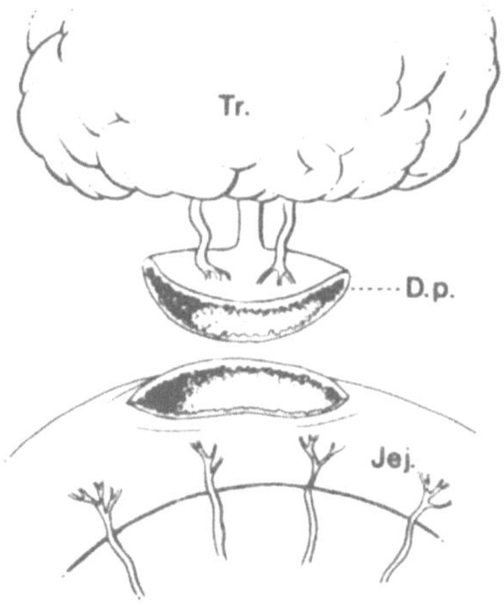

▲ Fig. 2

Fig. 1. Pancreaticoduodenal graft. *Tr.*, pancreas graft; *D.*, duodenum of donor; *A.s.*, aortic segment; *V.p.*, portal vein; *V.*, infrarenal vena cava of recipient; *A.*, aorta, *J.*, first jejunal loop of recipient

Fig. 2. Duodenal patch drained graft. *Tr.*, pancreas graft, *D.p.*, duodenal patch, *Jej.*, first jejunal loop of recipient

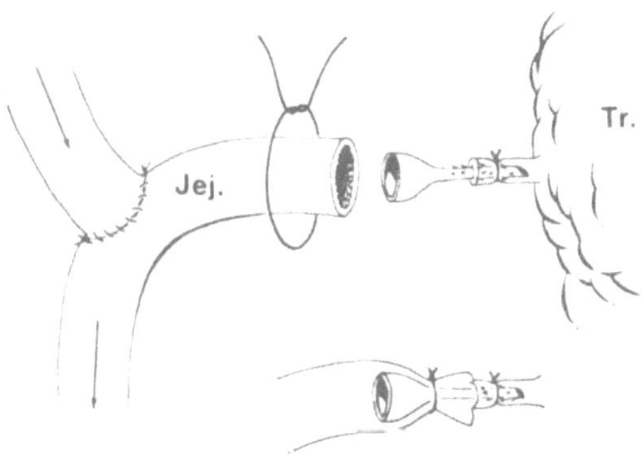

Fig. 3. *Roux-en-y* drained graft. *Tr.*, pancreas graft; *Jej.*, first jejunal loop of recipient

Duct-Ligated Segmental Graft. Preparation of the graft is similar to that described above except for the dissection of the pancreatic head from its body. The dissection line runs along the right side of the superior mesenteric vein and the portal vein. The pancreatic duct is carefully closed by a ligature in the procedure. The graft only needs an aortic segment bearing the celiac axis for blood supply (Fig. 4).

Duct-Occluded Segmental Graft. Occlusion of the pancreatic duct is performed by injection of Ethibloc prior to dissection of the pancreatic head. The bile duct is ligated as close as possible to the liver and at its orifice into the duodenum. A cannula is inserted into the duct system at the porta hepatis and approximately

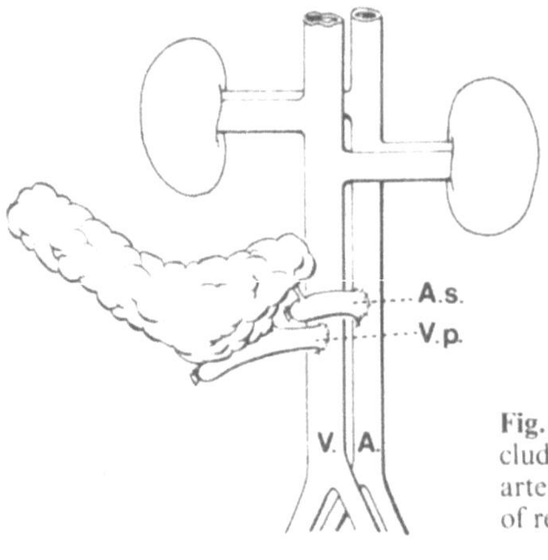

Fig. 4. Segmental graft (duct-ligated or duct-occluded). *A.s.*, aortic segment of donor with celiac artery; *V.p.*, portal vein; *V.*, infrarenal vena cava of recipient; *A.*, aorta

0.05 ml Ethibloc is injected. Injection is discontinued when extravasations are observed in the small pancreatic ducts.

Transplantation Techniques

Vessel anastomoses are always performed end-to-side by running an everting suture to the aorta and inferior vena cava below the renal vessels. Vessels are clamped for about 20 min. After removal of the vascular clamps, bleeding caused by stitch holes can be staunched by mild compression. For pancreaticoduodenal grafts, the intestinal anastomosis is performed to the first jejunal loop by running inverting sutures. The Roux-en-Y anastomosis is built subsequently to the vascular anastomoses. The stent is inserted into the open end of the Roux-en-Y anastomosis and secured by a circular ligature behind its funnel (Fig. 3). The duodenal patch is inserted into the first jejunal loop by single-stitch sutures – the patch is covered by the greater omentum (Fig. 2). All animals are given solid food and water immediately after transplantation. A total of 30 pancreaticoduodenal grafts, 8 patch drained grafts, 8 Roux-en-Y drained grafts, 10 duct-ligated, and 10 duct-occluded segmental grafts have been under long-term observation and subsequent histological study in order to compare the influence of these various techniques on the grafts.

Results

All results are summarized in Table 1. Secretion-drained grafts showed lasting endocrine function as demonstrated by serum glucose levels below 11 mmol/liter. One week after the operation all animals began to put on weight up to 480 g, whereas diabetic control animals lost weight down to 150 g. Histologically pancreaticoduodenal grafts possessed islets of normal shape. The majority of their cells were stained by Gomori's aldehyde fuchsin. Exocrine tissue showed no signs of fibrosis. These grafts could not be distinguished from a normal pancreas 300

Table 1. Results of experimental pancreas transplantation

Technique	*n*	Postoperative days	Preserved exocrine tissue	Lasting endocrine function
Pancreaticoduodenal graft	30	150–300	30	30
Roux-en-Y-drainage	8	30	5	8
Duodenal patch drainage	8	120	6	8
Duct-ligated segment	10	150	0	6
Duct-occluded segment	10	150	0	7

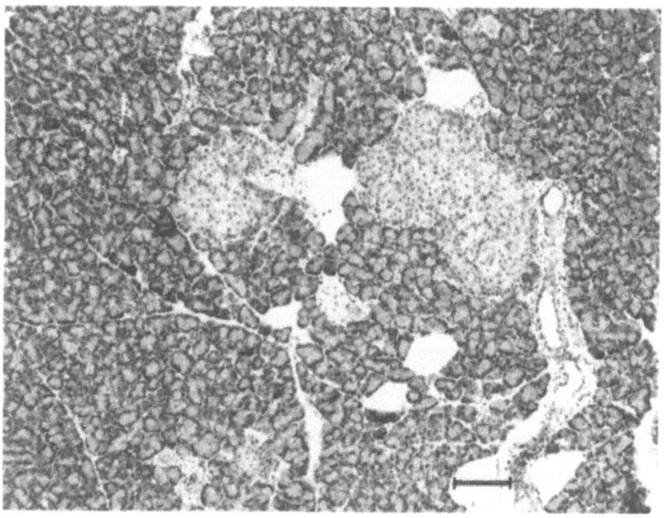

Fig. 5. Pancreaticoduodenal graft 300 days postoperatively. Goldner's trichrome stain; *bar*, 100 μm

days postoperatively (Fig. 5). Histologically, the transplanted duodenum presented a normal structure with long and slim intestinal villi. Six of the eight duodenal patch drained grafts looked like a normal pancreas after 120 days. However, two of the eight grafts showed cellular infiltrations consisting of macrophages and lymphocytes. Fibrous tissue spread out from the interlobular space. Some areas still contained exocrine tissue; some presented just proliferating ducts. Some islets looked normal; some appeared cap-deformed and fragmented by fibrous tissue. In these grafts the duodenal patch had become fibrotic, and so the drainage was ineffective.

Similar results were found in the Roux-en-Y drained grafts. Thirty days postoperatively, three of the eight grafts presented histological signs of ineffective drainage, caused by kinking of the stent, which had perforated the bile duct. The remaining five grafts in this group looked histologically normal 30 days postoperatively. Good results of endocrine function were found in most of the non-drained grafts. However, four of the ten duct-ligated grafts and three of the ten

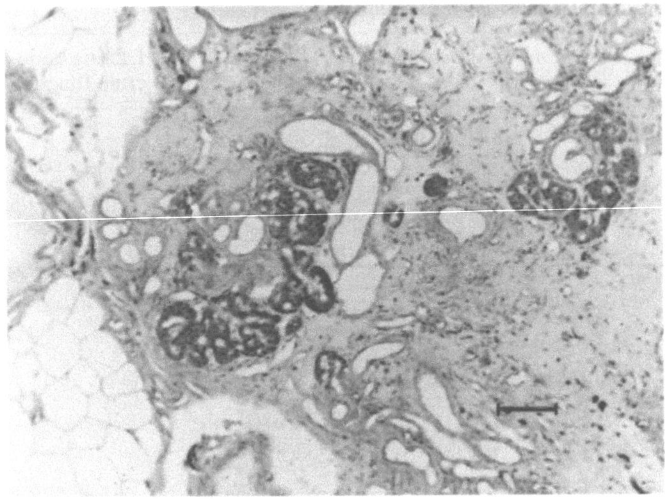

Fig. 6. Duct-ligated graft 150 days postoperatively. Goldner's trichrome stain; *bar*, 100 μm

duct-occluded grafts failed due to severe fibrosis (two in each group) or abscesses of the graft (one duct-ligated and two duct-occluded grafts) within 30 days after transplantation. Histologically well-functioning grafts in each of the non-drained groups looked very similar 150 days postoperatively. In normal pancreases or in drained grafts some very large islets were found constantly. All sections of non-drained grafts showed but relatively small islets. They were cap-like, deformed, and fragmented by fibrous tissue. Fibrous tissue between clusters of proliferating ducts was rich in fibroblasts, macrophages, and lymphocytes. Fatty tissue was also found in these grafts. Characteristically, islets were located in the periphery of duct clusters (Fig. 6).

Discussion

In conclusion, it may be said that all drained grafts presented good endocrine function. In contrast, 35% of the non-drained grafts lost endocrine function. Sixty-five percent of these grafts, however, showed excellent endocrine long-term function and other investigators reported even better results in similar models (Fairbrother et al. 1980; Klempnauer et al. 1983; Nolan et al. 1983), so all failures may be caused by avoidable technical faults. Nevertheless, a greater susceptibility to complications in these models should not be ignored.

A more important aspect is the occurrence of morphological alterations of the islets, even in well-functioning non-drained grafts. These alterations were also found in grafts presenting a relatively low grade of interstitial fibrosis. In a syngeneic rat transplantation model the functional reserve of the endocrine tissues seems to be sufficient to compensate the apparent damage of the islets and to keep the blood sugar in normal ranges. The question whether these changes in islet morphology may lead to an earlier loss of function in grafts under the condition of permanent immunosuppressive and several rejection crises has not been answered yet. In our experimental study the aim of secretion drainage, which should avoid technically caused alterations of the graft, was best achieved by the pancreaticoduodenal graft. Duodenal-jejunal anastomoses gave no rise to postoperative complications, whereas the duodenal patches and Roux-en-Y stint anastomoses were sources of failure. So we found that pancreaticoduodenal grafts would be the best-suited approach for further studies in performing allografts, because superposition of rejection and only technically caused alterations can be avoided.

References

Calne RY, Groth CG (1983) Segmental pancreas. Transplant Proc 15:1508–1510

Dubernard JM, Traeger J, Pozza G, Bosi E, Gelet A, Martin X, Kamel G, et al. (1983) Clinical experience with 31 pancreatic allografts in man. Transplant Proc 15:1318–1321

Fairbrother BJ, Boyle PF, Slater DN, George J, Nolan MS, Fox M (1980) Long-term results of transplantation of the duct-ligated pancreas in the rat. Transplant Proc [Suppl 2] 12:150–153

Gebhardt C, Gall FP (1980) Partielle Duodenopankreatektomie mit intraoperativer Pankreasschwanzverödung bei chronischer Pankreatitis. Langenbecks Arch Chir 353:57–62

Gliedman ML, Gold M, Whittaker J, et al. (1970) Clinical segmental pancreatic transplantation with ureter-pancreatic duct anastomosis for exocrine drainage. Surgery 74:171–180

Klempnauer J, Wonigeit K, Günther E, Pichlmayr R (1983) Pancreas whole organ transplantation in the rat. Evidence for a strong effect of non-MHC incompatibilities. Transplant Proc 15:1649–1651

Largiadèr F (1981) Pankreastransplantation. In: Pichlmayr R (ed) Transplantationschirurgie, p 775. Springer, Berlin Heidelberg New York (Allgemeine und spezielle Operationslehre, vol 3)

Lee S, Tung KSK, Koopmans H, Chandler JG, Orloff MJ (1972) Pancreaticoduodenal transplantation in the rat. Transplantation 13:421–425

Nolan MS, Lindsey NJ, Savas CP, Herold A, Beck S, Slater DN, Fox M (1983) Pancreatic transplantation in the rat. Long-term study following different methods of management of exocrine drainage. Transplantation 36:26–33

Experiences in Rat Pancreatic Islet-Cell Isolation and Transplantation [1]

K. Ulrichs, W. Timmermann, and *T. Schang*

Islet-cell transplantation appears to be a fast-growing new subject in the field of experimental research, as a look at the literature shows. There may be two main reasons for this expanding interest: (1) Isolated islets, provided they are clean, undamaged, and numerous, can cure artifically induced diabetes in rats and mice. This means that islet transplantation must be regarded as a potential therapy protocol for the treatment of diabetes in man as well. (2) Isolated islets provide the immunobiologist with an ideal microtissue to analyze and to manipulate in vitro, which is a chance second to none to overcome the problem of "tissue immunogenicity," regarded as one major barrier in both experimental and clinical transplantation. Whatever the basic interests are, one needs islets of high quality and sufficient quantity. This article aims to draw the reader's attention to only a few, but vital, technical points concerning isolation, cleaning, and transplantation of islets. Some of the details here may have already been analyzed thoroughly by others and written down elsewhere. These lines reflect the personal views, problems, and experiences of newcomers in the field of experimental islet transplantation.

In our laboratory we use the widely known isolation techniques of Lacy and co-workers (Lacy and Kostianovsky 1967; Lacy et al. 1972; Kemp et al. 1973; Finke et al. 1979) and of Lindall et al. (1969). We do not wish to repeat the details of the methods, which have already been described more precisely; but even though they are standard methods they can still cause serious difficulties in other laboratories. The techniques consist of four main steps: (a) mechanical separation of exocrine and endocrine tissue in situ by "blowing up" of pancreata with medium, (b) digestion of chopped pancreata with collagenase, (c) separation of islets from the digested material by Ficoll density centrifugation, and (d) cleaning of separated islets by handpicking. A separate point (e) concerns the transplantation of islets. Each procedure is of critical importance for the quality and quantity of islets.

ad (a). "Blowing up" of pancreata in situ via the common bile duct separates exocrine from endocrine tissue mechanically. This initial step is far more essential for

[1] This work was supported by the Deutsche Forschungsgemeinschaft, SFB 111/B12

Microsurgical Models in Rats
for Transplantation Research
Eds. A. Thiede et al.

obtaining sufficient and clean islets than is often believed. No problems usually occur in rat strains with large male individuals, i.e., Wistar-Furth or LEW rats. They possess large bile ducts which are easy to cannulate. For immunogenetic reasons it is sometimes necessary to change the donor strain; for simple logistic reasons it may be necessary to use the smaller female instead of male rats. Our experience is that in addition to Wistar-Furth, LEW: $RT1^{l}$ (male and female), CAP: $RT1^{c}$ (closely related to PVG, male and female), and DA: $RT1^{av1}$ (male and female) are excellent donor strains (2000 and more islets per 10 donors). F344: $RT1^{lv1}$ (male and female) and BN: $RT1^{n}$ (male) are both good donor strains (1200–1500 islets) but AS: $RT1^{l}$ (male) is a pretty poor one (less than 1000 islets). A genetically determined difference in islet-cell mass is already one severe handicap, but one should always remember that good or poor cannulating means good or poor "blowing up" and this means good or poor isolation results. Intensive technical training at this stage helped us to overcome frustration at much too early a stage.

ad (b). The removed pancreata, which are carefully cleaned of fat, blood clots, blood vessels, and macroscopically visible lymph nodes, are chopped and digested with collagenase. This enzymatic process remains the most sensitive part of the whole isolation procedure. Each investigator usually prefers his own supplier, type, and batch of collagenase. So far, we have also had some very good results with the often-used Worthington CLS-IV. The enzyme is usually kept at +4 °C in a dry place. We prefer to store it at −20 °C and feel that it retains its original activity over a much longer period. This may be of some importance whenever one wants to save a particularly good batch over periods of low laboratory activity. However, there are ways out of the dilemma once the collagenase activity starts to decline: (1) gradually increase the amount of enzyme per milliliter of chopped material (from about 5 to about 10 mg), (2) gradually increase the digestion temperature (from about 37° to about 39 °C), (3) gradually increase the digestion time (from about 7 to about 20 min), and (4) gradually increase the Ca^{2+} content of the Hanks' balanced salt solution (HBSS) preparation medium (from normally 185 to about 400 mg/liter; Ca^{2+} activates collagenase). To avoid islet malfunction, it is wise to change only one parameter at once and first assess its effect from the isolation result or islet function before going on to change the next one! In case a water bath with shaker is not available, shaking of the test tubes by hand is possible. This is a very tiring procedure but will eventually produce some marvelous islets. The end of the digestion process is usually indicated by the appearance of clear spots and streaks along the wall of the test tube. This is a truely reliable indicator system as long as it appears; however, there have been numerous isolations in the past in which neither spots nor streaks appeared and yet the digestion result was extremely good. So far, the reasons for this phenomenon are apparently unknown. In this special case one needs experience with the particular batch of collagenase to determine the optimal ending.

ad (c). Islets are separated from the digested material by Ficoll density centrifugation. Procedures performed with Ficoll should be precise and as brief as possible, because Ficoll is toxic to islets if they are exposed to it too long. We usually

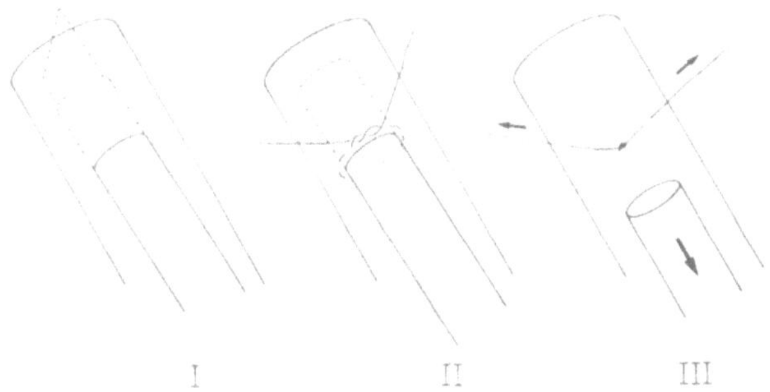

Fig. 1. *I* Plastic cannula with steel needle (Braunula V 18 G) in portal vein. *II* Steel needle removed and circular ligature placed. *III* Removal of plastic cannula and tying of suture after injection of islets

dialyze and lyophilize Ficoll-400 ourselves for the simple reason of cost. So far we have never had any real problems, apart from the fact that islets were not present where they should be: at the 11%–20.3% interphase! According to subsequent analysis it always turned out to be a collagenase digestion problem, as in those cases the total yield of islets was very poor. It was hardly ever a problem of the density gradient or the type and speed of the centrifuge as long as these operations were carried out and watched with the necessary skill and precaution.

ad (d). Separated islets are cleaned by handpicking under a dissecting microscope. A view into the petri dish before handpicking starts will give the investigator a quick impression of the quality and quantity of the islets, though at that stage neither can be changed any more. For immunogenic reasons, along with small blood vessels and debris handpicking must remove any small lymph node present. These vary remarkably from isolation to isolation, from rat strain to rat strain and from age to age. A valuable help for the beginner is the green filter light technique; islets are pink and lymph nodes are green. Picking too must be performed under sterile conditions whenever islets are not transplanted immediately; it can be very exhausting for untrained team members, but once they pick regularly the physical stress will very likely decline.

ad (e). Clean islets are transplanted into diabetic recipients. In the beginning we felt that a good yield of freshly isolated and clean islets was a tremendous achievement by itself. However, the transplantation procedure taught us a different lesson; when we lost three of our first five grafts (1200–1500 islets each) by major bleeding of the recipients – this also due to lack of experience – we decided to vary the standard transplantation technique which is performed via the portal vein: a minor open situs operation to prepare the portal vein is carried out under ether anesthesia. Around the injection hole of a plastic cannula (18 G, Braunula V) inserted into the vein, a finece circular suture is placed with 8-0 Ethilon under the operating microscope (Fig. 1). After injecting the suspended islets with 1–2 ml medium via the cannula into the portal vein, the cannula is carefully removed and the hole closed by gently tying the suture. This modified technique is very safe, i.e., even minor bleedings hardly occur and we never lost a single graft again. Furthermore, the complete transplantation procedure can be performed by one person, whereas the standard technique usually needs two, one to keep an eye on the injection needle on its way into the vein and the second person to resuspend and place the islets into the syringe on top of the needle. Admittedly, a slight disadvantage is the need for microsurgical training of the person in charge of the operation. In no way is this technical variation more time consuming than the standard technique, once the person in question has had microsurgical training.

References

Finke EH, Lacy PE, Ono J (1979) Use of reflected green light for specific identification of islets in vitro after collagenase isolation. Diabetes 28:612–613

Kemp CB, Knight MJ, Scharp DW, Ballinger WF, Lacy PE (1973) Effect of transplantation site on results of pancreatic islet isografts in diabetic rats. Diabetologia 9:486–491

Lacy PE, Kostianovsky M (1967) A method for the isolation of intact islets of Langerhans from the rat pancreas. Diabetes 16:35–39

Lacy PE, Walker MM, Fink CJ (1972) Perfusion of isolated rat islets in vitro. The participation of the microtubular system in the biphasic release of insulin. Diabetes 21:987–998

Lindall A, Steffes M, Sorenson R (1969) Immunoassayable insulin content of subcellular fractions of rat islets. Endocrinology 84:218–223

Technique for Orthotopic Rat Liver Transplantation*

R. Engemann

After the introduction of microvascular suture techniques by Lee and Fisher in 1961, it took more than a decade of microsurgical pioneer work before a technique for orthotopic rat liver transplantation could be presented (Lee et al. 1973). This technique, however, required a portosystemic shunt and was a difficult operation. To my knowledge, nobody, except its inventors, was able to reproduce the results. Even after the publication of a simplified technique by Lee et al. in 1975, several years went by before other groups started to publish their first re-

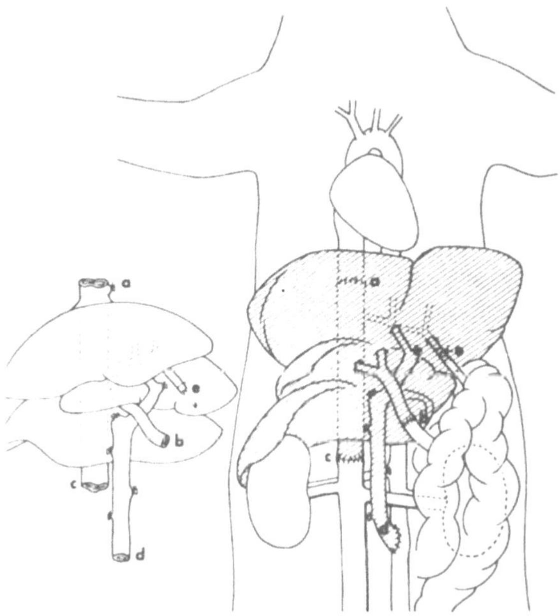

Fig. 1. Orthotopic rat liver transplantation with arterialization. *Inset*, transplant; *a*, inferior vena cava, suprahepatic; *b*, portal vein; *c*, inferior vena cava, infrahepatic; *d*, hepatic artery and abdominal aorta; *e*, bile duct

* This work was supported by the Deutsche Forschungsgemeinschaft SFB 111/B9

Microsurgical Models in Rats
for Transplantation Research
Eds. A. Thiede et al.

sults (Engemann et al. 1979; Hansen et al. 1979; Houssin et al. 1979; Kamada and Calne 1979; Limmer et al. 1980; Zimmermann et al. 1979). Heterotopic liver transplantation was first reported in 1966 by Lee and Edgington. In clinical liver transplantation it turned out, however, that the heterotopic model had the severe disadvantage of space problems (two livers in the abdominal cavity). In addition, an unexpected event occurred, the so-called interliver competition (i.e., the liver with the greatest physiological handicap – missing arterial blood supply, bile duct ligation, high venous outflow pressure – undergoes atrophy). Thus the orthotopic technique superseded the heterotopic one, although for special immunological questions it might still be an interesting model. I will refer only to the orthotopic technique. [For an overview of heterotopic liver transplantation see Hess et al. (1972)]. Since (for reasons given later) our technique of orthotopic rat liver transplantation includes the preparation of an aortoceliac segment to arterialize the graft with the hepatic artery, the harvesting and transplantation technique also includes the modes without arterialization (Fig. 1).

Donor Preparation

Under ether anesthesia we open the abdominal cavity via a long midline incision, which will be extended to the lateral abdomen below the thorax on both sides (avoid bleeding from the epigastric vessels!). We fold the two lateral abdominal walls to each side and fix them with pins. Then we wrap the small intestine up in moist gauze and fold it to the left side of the animal. The falciform ligament is then dissected, and the left infraphrenic vein is ligated and divided. This vein is constant, but varies in size and location of inflow in the different inbred strains. The next step is the double ligation and dissection of a small arterial branch arising from the left hepatic artery and supporting the lower esophagus. Then we dissect the inferior infrahepatic vena cava from the peritoneum and put a *single* ligation around the right suprarenal vessels. (They are divided when the graft is taken out; the ligature stays with the graft.) Then follows the sharp and blunt dissection of the posterior hepatic connective tissue.

In the preparation of the portal area, one must be very careful to avoid touching the hepatic artery, since vessels of this small caliber are very prone to develop spasms. The pyloric vein is double ligated with 5-0 silk and divided, and then the gastroduodenal artery, which lies close to the vein, is double ligated and divided. Then the small intestine, spleen, and stomach are folded to the right side of the animal to prepare the celiac axis. The aorta is cleaned of the connective tissue, which contains the main lymphatics of the upper abdominal organs, and the celiac artery and the superior mesenteric artery are isolated. The dissection of the celiac artery and the hepatic artery can be done with only three double ligations. The rest of the tissue can be divided without fear of bleeding. The next step is the blunt and sharp dissection of the abdominal aorta below the renal vessels. All branches must be ligated (from lumbar arteries only on ligation is mandatory). In addition, the left renal vein must be double ligated and divided. Finally, before perfusion starts, a Teflon catheter (Medicut 20, Sherwood), is inserted into the bile duct and secured by a tie. During the entire preparation phase one must carefully avoid any bleeding, since otherwise the animals gradually run into a volume deficiency situation, which might be increased because the venous return to the right heart chamber is diminished by cutting off parts of the circulatory system. (A reliable parameter for the circulation is the volume of the inferior vena cava).

Just before perfusion starts, the superior mesenteric artery is ligated; the inferior vena cava is double clamped below the liver and divided just above the distal clamp. Then perfusion starts via the portal vein by inserting a plastic catheter (Braunula, 2-G14). Saline solution at 4 °C is used without any addition. To facilitate the runoff of the perfusate, the diaphragm is divided and the intrathoracic inferior vena cava is cut. Up to this point the pale liver still shows reddish spots, indicating that the preparation of the hepatic artery is not harmful, and that the artery is still working. While the perfusion runs slowly via the portal vein, the aorta is clamped above the celiac axis, and the abdominal aorta with the celiac axis is perfused from the distal end with 2–3 ml cold saline solution. After ligation of the proximal end of the aorta, the liver is excised and stored in saline solution at 4 °C. The whole procedure takes about 50–60 min. When only a graft without the hepatic artery is being prepared, the hepatic artery must also be ligated when preparing the portal area. The preparation of such a graft takes about 10–15 min. For both techniques it is important to clean the vessel stumps as well as possible of fatty tissue, to avoid having to search for the collapsed vein lumina when the anastomoses are to be performed.

Recipient Operation

After a long midline incision under ether anesthesia, the recipient's own liver is removed according to the technique described above. There is, however, one important difference. In the recipient both the vessel stumps and the bile duct must be as long at is possible to leave them. The hepatic artery is double ligated and dissected. Before the final excision of the liver is made, the surgeon should take care that all items required are available on the table, to keep the following anhepatic time as short as possible. The inferior vena cava is cross-clamped as distally as possible between the liver and kidney, and then the portal vein is also cross-clamped, leaving a stump of approximately 0.8 cm in length. Finally, the inferior vena cava is clamped above the liver with a small Satinsky clamp with which an 0.5-cm edge of the diaphragm is also grasped. Then the liver is removed, the animal is rotated 180 °C, and the plastic tube filled with gauze soaked with ether is moved a little farther from the rat's nose to keep the anesthesia at as low a level as possible, since the animals have a very low cardiac output during the cross-clamping of the inferior vena cava and the portal vein. Now the graft is brought into its final position, and we place the angle sutures of the anastomosis of the inferior vena cava below the diaphragm, using 7-0 Prolene. The posterior vessel wall is sutured from the inside. Before completion of the front wall, the vessel (which runs behind the liver) must be emptied of air; because otherwise the first breath after declamping leads to air embolism of the lung. Two other hints: make sure that the stitches next to the angle sutures are tight, since they cannot be corrected. And, second, it is easier to suture the back wall of the inferior vena cava of the graft to the part of the cava passing through the diaphragm. This keeps the stitches from tearing out. Then the animal is again rotated 180 ° and the portal vein anastomosis is performed in the same fashion, using 8-0 or 9-0 nylon. Two angle sutures keep the anastomosis in place. Then first the back wall is completed as a running suture (five stitches), then the anterior wall (five stitches). Before the portal clamp is finally removed the portal vein is quickly compressed proximal to the liver. Thus after the clamp is released, first the air leaves through the anastomosis, and then the liver is immediately well perfused after the Satinsky clamp above the liver has also been removed.

The reestablishment of the portal blood flow returns the circulation of the animal to an acceptable condition, so that the third venous anastomosis can be carried out without hurry. This anastomosis of the inferior vena cava below the liver is also performed in the same manner: two angle sutures with 8-0 nylon, then a running suture of the posterior wall made from inside the vessel and then the anterior suture row. Here again we evacuate air from the vein by opening the distal clamp first, to allow the air to pass through the stitches. The circulation of the animal is now back to normal.

The last vessel anastomosis is the end-to-side anastomosis between the donor aorta, with the celiac artery and the hepatic artery, and the recipient's abdominal aorta below the kidney vessels. The aorta is separated from the surrounding connective tissue and clamped with two bulldog clamps, and then an aortotomy is performed with iris scissors. The donor aorta is gently rinsed with saline solution to remove any air bubbles and then sutured in a typical end-to-side anastomosis, using 8-0 nylon. Here we can perform both suture lines in a running fashion from the outside, since we can fold the vessel back and forth. In addition to the two bulldog clamps placed on the recipient's aorta, we have a third one on the donor's aorta just 5 mm proximal to the open end. After completion of the end-to-side anastomosis, first the distal clamp is removed and remaining air bubbles in the lumina are gently pressed through the suture lines, then the proximal clamp is released. When the anastomosis is satisfactory, the last clamp, the one occluding the donor aorta, is removed. Using this sequence we can avoid reclamping (which may lead to thrombosis, since we do not use heparin) or occlusion of the hepatic artery due to air embolism. In our hands the failure rate of arterialization is less than 5% (failures: twisting or air embolism). Successful arterialization is indicated by pulsating arteries at the porta hepatis and on the donor bile duct.

The last step is the bile duct anastomosis. Following the technique of Zimmermann et al. (1979), we connect the two ends of the bile duct with a Teflon splint (20 or 22 Medicut) and secure the splint by a circular tie and three single stitches with 8-0 nylon.

Finally, the abdominal cavity is checked for hemostasis. Then the guts are rinsed with warm saline solution and the midline incision is closed with a running suture using absorbable material (e.g., 2-0 Dexon). Then we give 1 ml saline solution i.v. through the penis vein and place another 2 ml saline solution s.c.; normally plasma expanders are not necessary; we use neither heparin nor antibiotics. Postoperatively the animals are allowed water and food ad libitum. For preparation of the graft, magnifying glasses and sometimes a microscope are very helpful.

The three venous anastomoses are carried out with magnifying glasses. As a safeguard at the end of such a complex operation, the arterial and bile duct anastomoses are performed with a microscope at × 10 magnification.

Let me now lay particular stress on some special problems of rat liver transplantation.

The Anhepatic Time. The anhepatic time – the time in which both the inferior vena cava and the portal vein are clamped – should be as short as possible; otherwise the animals enter an irreversible state of shock. Once you have started to gather experience in rat liver transplantation, you realize that the animals begin to survive when the anhepatic time is less than 25–30 min. This means that during this time you have to perform the vena cava-inferior suprahepatic anastomosis and the portal vein anastomosis. Our average time is 15 min; the fastest was 13 min. The cuff technique used by Kamada (Kamada and Calne 1979; Kamada et al. 1983) will shorten this time to an average of 12 min and probably means less stress. It may also be somewhat more reliable, especially for beginners. The total time for the recipient operation is about 40–45 min without arterialization and 60–70 min with arterialization. The operative and postoperative mortality is less than 10% for a trained microsurgeon.

The Bile Duct Anastomosis. The second important point is the bile duct anastomosis, which Calne called the Achilles' heel of liver transplantation, and this is true not only for human transplantation. Technically, one can perform an end-to-end anastomosis with single stitches or a running suture (Engemann et al. 1979), using 10-0 or 11-0 nylon or Dexon; however, this is time consuming and one needs a high-power microscope with × 15–30 magnification. The original method of Lee (Lee and Edgington 1966; Lee et al. 1973, 1975) was to pull the bile duct through into the duodenum and to wrap the duct with the omentum ma-

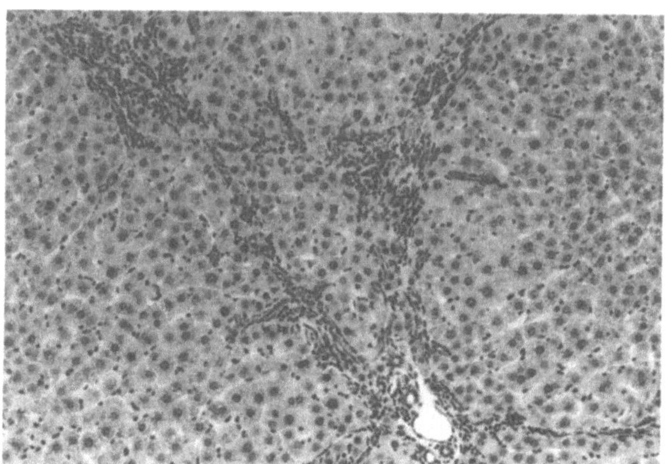

Fig. 2. Syngeneic liver graft without arterialization 48 days postoperatively; severe bile duct proliferation, grade IV according to Kamada and Calne (1979). H & E, × 100

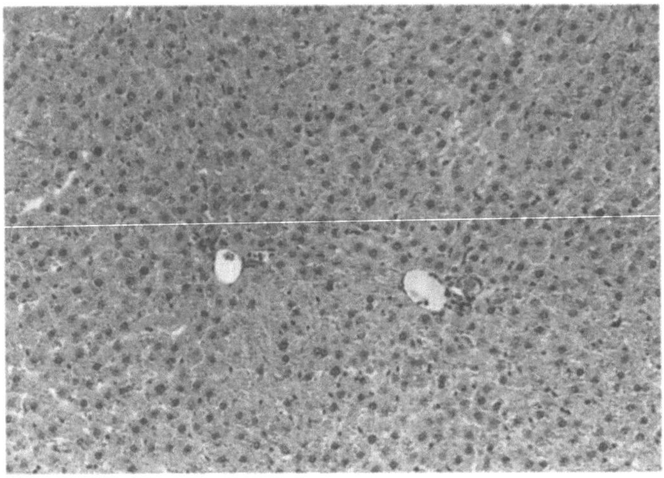

Fig. 3. Syngeneic liver graft with arterialization 48 days postoperatively; normal periportal area. H & E, × 100

jus. Common problems were biliary leakage, peritonitis, and central necrosis of the liver. This method was improved by Zimmermann et al. (1979), who used the splint technique. But even with this easy procedure we had a lot of problems in our laboratory, most of them because in the nonarterialized animals the donor bile duct has no arterial blood supply and is dependent on capillaries that either grow from the recipient's bile duct or are of gastroduodenal origin (Lie et al. 1983). One of the typical late results of bile duct occlusion or stenosis is bile duct proliferation, which can be seen to an extreme degree in Fig. 2. These undesirable side effects may only be a morphological disfigurement, but they may also reflect a severe disturbance of the liver's physiology. An unaffected liver graft, as Fig. 3 shows from an arterialized liver, might allow one to distinguish even very slight alterations better. The survival rates (more than 100 days) of our syngeneic groups have increased now to more than 90%. The long-term survival, however, depends not only on the skill of the respective surgeon, but also on the rate of lung infections of the rats, since after such a big operation animals are more susceptible than usual to lung infections.

References

Engemann R, Thiede A, Lee S (1979) Syngeneic orthotopic total replacement of the liver in rats. Post operative function control using angiography and transaminase courses. Excerpta Medica Int Congr Ser 465:402–405

Hansen HH, Kim Y, Lie TS (1979) Orthotopic liver transplantation in the rat – special reference to arterialization. Excerpta Medica Int Congr Ser 465:394–397

Hess F, Jerusalem C, Heyde N (1972) Advantages of auxiliary liver homotransplantation in the rat. Arch Surg 104:76–80

Houssin D, Gigou M, Franco D, Szekely AM, Bismith H (1979) Spontaneous long-term survival of liver allografts in inbred rats. Transplant Proc 11:567–570

Kamada N, Calne RY (1979) Orthotopic liver transplantation in the rat: technique using cuff for portal vein anastomosis and biliary drainage. Transplantation 28:49–50

Kamada N, Davies HS, Wight P, Culank L, Roser B (1983) Liver transplantation in the rat. Biochemical and histological evidence of complete tolerance induction in non-rejection strains. Transplantation 35:304–311
Lee S, Edgington TS (1966) Liver transplantation in the rat. Surg Forum 17:220–222
Lee S, Fisher B (1961) Portocaval shunt in the rat. Surgery 50:668–672
Lee S, Charters AC III, Chandler JG, Orloff MJ (1973) A technique for orthotopic liver transplantation in the rat. Transplantation 16:664–669
Lee S, Charters AC III, Orloff MJ (1975) Simplified technique for orthotopic liver transplantation in the rat. Am J Surg 130:38–40
Lie TS, Hansen HH, Niehaus KJ (1983) Bedeutung der Arterialisation des Transplantates bei Rattenlebertransplantation. Langenbecks Arch Chir 359:133–142
Limmer J, Herbertson BM, Calne RY (1980) Orthotopic rat liver transplantation using different combinations of four inbred strains. Eur Surg Res 12:343–348
Zimmermann F, Butcher GW, Davies HS, Brons G, Kamada N, Türel O (1979) Techniques for orthotopic liver transplantation in the rat and some studies of the immunologic responses to fully allogeneic liver grafts. Transplant Proc 11:571–577

Theoretical Aspects of Rat Organ Transplantation

W. Müller-Ruchholtz

This paper is not intended to survey transplantation in rats – not even theoretically; rather, it limits itself to some of the basic aspects, be they theoretical or practical. First of all, it may be pointed out that there is no particular field that merits the designation "rat transplantation immunology" since the basic immunobiological laws appear to apply to all mammals, from rat or mouse to man. It is this premise which mainly justifies the vivid interest of surgeons in experimental transplantation models.

As introduction three other general aspects may be added: (1) In basic terms, transplantation is *transfer experimentation*. It is mainly intended to transfer functioning cells successfully rather than hormones, metabolic products, etc., especially between genetically different individuals. (2) Transplantation research requires *inbred animals*. Syngeneic transplantation provides the only reliable means for establishing and controlling grafting techniques, which becomes increasingly important as more sophisticated organ transplantations come under study. In allogeneic transplantation comparability of the vast majority of experiments requires control of histoincompatibility, which cannot be fully provided by tissue typing for major histocompatibility complex (MHC)-determined antigen specificities. This statement is not commonplace, as long as many surgeons prefer transplantation studies in dogs, rabbits, or other species without inbred strains. (3) The *rat* is turning out to be increasingly *superior to the mouse*, as the importance of comparing the outcome of different organ grafts is being realized. For decades the mouse has been the favored species in transplantation research, and

Table 1. Immunobiology of tissue allotransplantation

Immunogenicity of the graft ↓	: Grade of histoincompatibility Antigen expression on cell surface Metabolic activities of the cell Privileged location
Reactivity of the recipient ↓	: Nonspecific immunosuppression ↓ Specific tolerance → Suppression type / Abortion type

Microsurgical Models in Rats
for Transplantation Research
Eds. A. Thiede et al.

there is no doubt that the hundreds of mouse inbred strains provide almost ideal means for analyzing basic immunobiological reactions. However, mice are far from ideal experimental animals when the question is asked to what extent the outcome of a cornea, trachea, liver, or pancreas graft differs from the rejection of a skin graft. Therefore, the growing number of increasingly well-defined inbred rat strains is of utmost importance in transplantation research.

The main immunobiological parameters which govern tissue allotransplantation are outlined in Table 1. Without going into details here it may be said that manipulative interventions may deal with either side of the "coin": reduction of the immunogenicity of the graft and/or reduction of the reactivity of the recipient. Some of these parameters will be gone into further below; others will be brought up in other papers of this volume.

Quite a few tissues and their transplantation aspects are outside the scope of a surgeon's interest. This is probably the reason why they were not included in this volume. However, some have provided very useful models for understanding basic immunobiological relationships and shall, therefore, be discussed briefly.

For historical reasons I would like to remind the older researchers of something which probably is almost unknown to the younger ones, namely parabiosis. Some 120 years ago the French physiologist Bert performed for the first time a "total transplantation," suturing one mouse to another side-to-side. This method was revived by the German surgeon Sauerbruch in 1908. The connection of the peritoneal walls was shown to facilitate capillary anastomoses. Rats were demonstrated to be similarly suitable. Also, a variety of hypotheses were forwarded in order to explain the so-called parabiosis intoxication, which mostly showed plethora/anemia syndromes and often killed at least one of the partners. It was only in the 1950s that it was discovered that this sort of "intoxication" was due to immunological transplantation rejection reactions which could easily be avoided by using animals of the same inbred strain.

Several years ago we became interested in the question of whether parabiosis could also be avoided, or altered, by replacing the natural capillary anastomoses in rats by cross-transfusion via arteriovenous shunts, allowing 2 ml of blood/min to pass from the carotid artery to the jugular vein by polyethylene cannulas for several hours or days (without additional means of blood flow regulation). We were able to show that the above mixture of host versus graft reactions and graft versus host reactions (GVHR) was avoidable only if one partner was treated for 2 days before and after a 6-h cross-transfusion with cytotoxic drugs such as cyclophosphamide or ifosfamide. The nontreated partner did not become sensitized; the treated partner became immunotolerant, as proven by selective acceptance of skin grafts of the transfusion partner's strain which were applied several weeks or months later (whereas third party skin grafts were rejected in a normal fashion).

In an extensive series of subsequent experiments the following could be shown: (1) This *tolerance-initiating effect of cytostatics* is dependent on the grade of *histoincompatibility* between the transfusion partners, allowing permanent tolerance only up to the MHC variant combination LEW ($RT1^{l}$)/F344 ($RT1^{lv}$). (2) This effect may vary strongly when *different drugs* are compared at equicytotoxic levels, e.g., the alkylating agent cyclophosphamide is clearly inferior to its derivative ifosfamide. (3) The tolerance-initiating potency of these substances is *independent of* their *immunosuppressive potential*. In extension of these > 10-year-old experiments it might by very interesting to learn where to place the present wonder drug cyclosporin A in similar comparative studies. (4) The *tolerance mechanism* was meanwhile shown to be of the *suppression type*. Even killer lymphocytes were demonstrated, but only in the spleen and not in the transport compartment blood (where they could be found in control experiments without cytotoxic drugs). This *locking-in phenomenon*, as it was termed, has very recently also been found in our immunological evaluations of rat small-bowel and liver transplantation experiments, as reported by Ulrichs and by Engemann in this volume (p. 153).

The location of a graft other than an organ with surgically provided vascular anastomosis may be of great immunobiological relevance because of the dependence of both the initiation of sensitization and the destructive efficiency of sensitization on *vascularization of the graft bed*. This can best be studied in *cornea grafts*. To my knowledge the most systematic investigations with this normally avascular tissue were performed in the 1970s by the Kiel ophthalmologist Gronemeyer, who grafted interlaminar 2.6-mm cornea disks either close to the limbus (and its circular capillary) or centrally, using different inbred rat strain combinations. The outcome of the graft has been found to be clearly dependent on three parameters: *location* of the graft, grade of *histoincompatibility* between donor and recipient, and degree of *presensitization* of the recipient (e.g., permanent survival of even MHC-different grafts when applied centrally, as long as presensitization by blood transfusion etc. is avoided). Furthermore, we were able to demonstrate that acentrally placed grafts are accepted without sensitization, provided the surgical intervention is followed by 2 weeks of anti-inflammatory treatment with corticosteroids. Thus, the prototypical *privileged site* cornea has been shown to be governed by different variables and to be far from an all-or-none situation.

Recently rat *trachea transplantation* models have attracted the interest of otorhinolaryngologists as well as immunologists. A particular feature of this organ appears to consist in the fact that it is mainly composed of another avascular tissue, namely cartilage, and mucosa. From the first series of results after grafting four, six, or eight tracheal ring segments, which were obtained by Beigel (Department of Otorhinolaryngology of the University of Kiel) we may already conclude the following: Even in MHC-allogeneic donor-recipient combinations the cartilage rings are accepted, whereas the epithelium is easily rejected by the sensitized host. Nevertheless, if the rejection process is not too intense (as it is in presensitized recipients) the animals survive and manage to replace the host-derived epithelial layer by their own cells, as shown by Beigel, using immunoperoxidase-labeled antibodies against the reciprocal transplantation antigens. Findings of this sort demonstrate the importance of *biologically adequate evaluation* of transplantation experiments: Clearly developed immunological sensitization (as expressed in the above experiments, e.g., by strongly accelerated skin graft rejection) must not lead to a breakdown of the function of a graft. Basically similar findings were obtained several years ago by Thiede et al. when studying large-vessel grafts in rats.

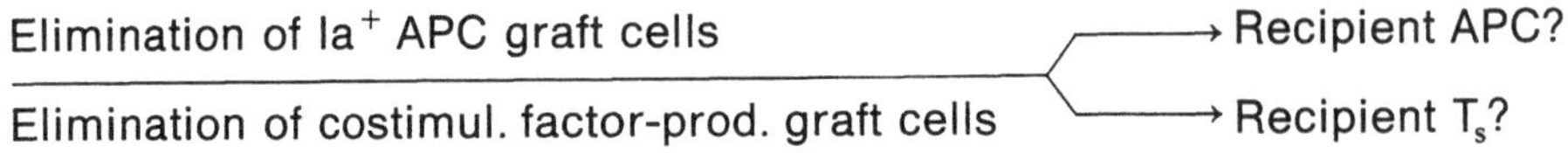

Fig. 1. Outline of the tentative role of antigen expression and metabolic activities of grafted cells

Table 2. Relevance of rat organ transplantation

1. Microsurgery in rats means that larger, but less relevant (noninbred) laboratory animal species are not necessary
2. Transplantation immunological parameters become analyzable (and comparable) for a large variety of tissues and organs, with regard to their individual role and their interrelationships:
Histoincompatibility (MHC and subregion, non-MHC, tissue restricted)
Antigen exposure (in the tissue, on the cell – means of manipulation)
Effector T cells vs antibodies in sensitization
Independently induced specific sensitization (pre- or postgrafting)
Susceptibility to unspecific or specific sensitization blockade mechanisms
Functional relevance of sensitization

Another tissue or organ which is outside the scope of surgeon's interest but, nevertheless, of high basic as well as clinical interest in transplantation immunobiology is *bone marrow*. In a large series of experiments in rats and mice we have been able to establish permanent *hemopoietic chimerism in MHC fully allogeneic* combinations. It is easily possible to avoid both early and delayed GVHR in such deliberately mismatched combinations and to restore normal immune functions but establish abolition-type *immunotolerance* of bone marrow donor transplantation antigens, provided the T-lymphocytes, including their committed prethymic precursors, have definitely been eliminated from the bone marrow inoculum by short-term in vitro incubation in an adequately specified antilymphocyte serum plus complement. Among the various areas of theoretical and practical interest in this sort of experimentation only one may be mentioned here: We may well envisage the future combination of any kind of tissue or organ transplantation with bone marrow transplantation (from the same donor) in order to establish the wanted specific immunotolerance without longer-lasting nonspecifically immunosuppressive protocols by establishing hemopoietic chimerism.

Coming back to the overview of Table 1, it may be stated that most of the parameters have at least been briefly characterized in terms of their immunobiological role in the foregoing paragraphs. This is, however, not true with regard to antigen expression on cell surfaces and metabolic activities of cells, i.e., with regard to two aspects which recently have gained particular interest. They will be covered specifically in this volume by two papers dealing with pancreatic islet transplantation, namely those of Hardy and of Ulrichs (pp. 205, 65). I may restrict myself to characterizing the present line of reasoning, as outlined in Fig. 1: The *immunogenicity* of a graft may be substantially reduced by *eliminating* from the grafted tissue its MHC class II (Ia-like) antigen-expressing cells and/or by eliminating cells (such as lymphocytes) which provide via their metabolism costimulative factors, i.e., factors which serve as second signals required for the initiation of sensitization. If such elimination can be achieved, the recipient's own antigen-presenting cells (APC) may be left as the only and much weaker antigen-handling cells, and/or T-suppressor lymphocytes may be preferentially activated.

The relevance of rat organ transplantation in transplantation research may be summarized in Table 2. It must be repeated that many of the aspects outlined cannot be studied in mice for obvious methodological reasons. And it may be added that none of these aspects can be satisfyingly studied without large-scale cooperation between so differently minded people as surgeons and immunologists.

Immunogenetic Aspects of Organ Transplantation in the Rat

E. Günther

Introduction

In this contribution some genetic aspects will be discussed which are relevant when studying organ graft rejection in the rat. The more general considerations will be illustrated mainly by results in the production of which the author has been involved, and no review (Günther and Štark 1977) of the literature is intended.

Graft rejection is due to a thymus-dependent immune response to antigenic determinants which are present on the graft and absent from the recipient. The antigens which elicit this response and are the target of the response are called histocompatibility (H) antigens. They are detectable in a species when at least one allelic variant exists, so that the donor carries a molecular configuration that is recognized as foreign by the host. Thus a more or less high degree of polymorphism is characteristic of H-antigens. Graft rejection, either of skin, which is the most commonly used type in experimental transplantation, or of organ grafts, which are more relevant with respect to clinical application, is only one manifestation of the alloimmune response to histocompatibility antigens. Other in vivo assays are the graft-versus-host and delayed-type hypersensitivity reactions to H-antigens. Some H-antigens induce antibody production or proliferation of T-lymphocytes in vitro and most appear to be detectable by cytotoxic T-lymphocytes in vitro.

A discussion of immunogenetic aspects of organ transplantation has to start with the identification of the genes which code for H-antigens and the description of their products. Of particular relevance are quantitative and qualitative aspects of their expression, i.e., on which cells they are expressed and to what amount and how expression is regulated. As is well known, we distinguish between major and minor H-antigens. This classification has first been based on empirical criteria concerning the strength of the graft rejection response, which in the case of skin grafts is expressed by the number of days it takes to reject the graft. Later on this distinction turned out to be a fundamental one. The major H-antigens are the gene products of a cluster of closely linked genes (major histocompatibility complex, MHC), the main function of which is the regulation of the T-lymphocyte response to antigens by creating a specific cell surface context in which antigens are recognized by T-lymphocytes. Whereas major H-antigens are recognized by

Microsurgical Models in Rats
for Transplantation Research
Eds. A. Thiede et al.

T-lymphocytes directly, it is characteristic of minor H-antigens that T-lymphocytes see them only in association with major H-antigens.

Genes and Gene Products of the Rat MHC

The exceptional role of the MHC-determined H-antigens – the RT1 antigens in the rat – is demonstrated by the rule that skin when grafted across the RT1 barrier is usually rejected in an acute fashion, and that kidney transplants are only rejected in case of RT1 incompatibility [for further details and discussion see Günther and Štark (1977)]. There are, however, many examples of successful organ grafting across the RT1 barrier, including kidney (Paris et al. 1978), liver (Zimmermann et al. 1983), and heart (Klempnauer, personal communication) grafts. In order to analyze why the RT1 system plays such a predominant role in transplantation and why rejection does not always occur, a more detailed knowledge of the genetic fine structure of the rat MHC is necessary. Since the MHC is built up of several closely linked genes, the first step is to identify these genes, their arrangement on the chromosome, and their products.

As is known from studies in mouse and man, three different kinds of products can be assigned so far to genes which map to the MHC chromosomal region: (1) cell surface molecules, which are the most important in this context and according to most people's view the MHC gene products proper, (2) genes coding for complement components of the activation pathway, and (3) some enzyme genes. The cell surface molecules which are determined by the MHC belong to two different classes, I and II. Class I type MHC antigens consist of a heavy chain of about 45000 molecular weight (MW) which is associated with β_2-microglobulin (MW about 12000). Class II antigens ("Ia antigens") are also built up of two chains, the α-chain (MW about 35000) and the β-chain (MW about 28000). Two types of class II molecules, I-A- and I-E-like, are distinguishable by their α- and β-chains serologically and biochemically by two-dimensional gel electrophoresis.

The RT1 genes encompass about 2–3 centimorgan on a rat chromosome which so far has not been identified but might be chromosome 14 according to banding pattern homology with chromosome 17 of the mouse (Nesbitt 1974), where the MHC of this species is located. Figure 1 tries to integrate our present knowledge about these genes and gene products in the rat. No data are available so far on complement genes mapping to the RT1 complex. With regard to enzyme genes, glyoxalase I (Stolc et al. 1980) and neuraminidase (VandeBerg et al. 1981) activity have been shown to be controlled by genes closely linked to RT1 genes. The genes determining cell surface antigens can be assigned to three different RT1 regions which are separable by recombination. The *A* and *C* regions determine class I antigens, RT1.*A* products being expressed ubiquitously, and RT1.*C* antigens being detectable so far only on all B- and T-lymphocytes, subpopulations of macrophages, and cortisone-resistant thymocytes, but not on red blood cells or liver and brain tissue (Stock and Günther 1982). Class I genes flank the region encoding class II antigens. The I-A-like class II molecules are RT1.*B* and the I-E-like molecules are RT1.*D* products (Blankenhorn et al. 1983). It should be mentioned

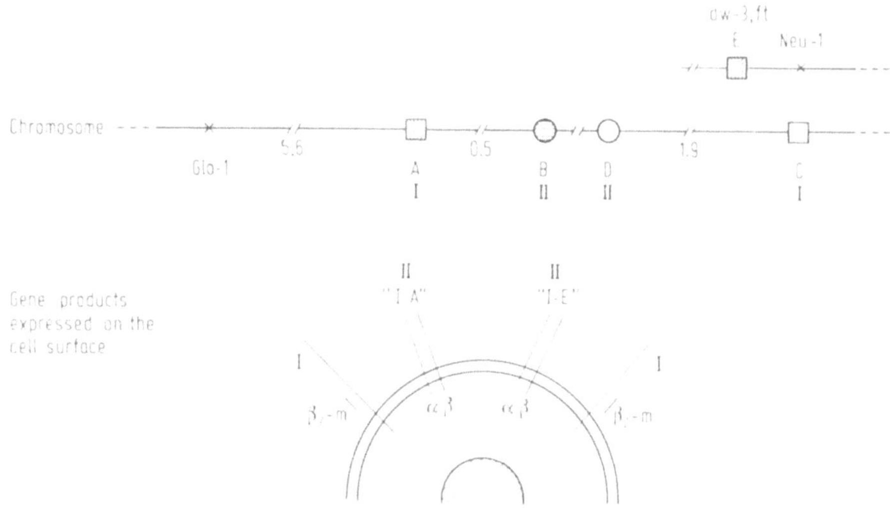

Fig. 1. The MHC of the rat. The *numbers in the upper part* refer to approximate recombination frequencies. References: *Glo-1* (Stolc et al. 1980), *A/B* (Butcher and Howard 1977; Günther et al. 1978; Štark et al. 1977), *B/D* (Günther et al. 1978; Lobel and Camer 1981), *C* (Kohoutová et al. 1980; Stock and Günther 1982), *E* (Kunz et al. 1982), *dw-3* and *ft* (Kunz et al. 1980), *Neu-1* (VandeBerg et al. 1981). The *lower part* is based on Blankenhorn et al. (1983), Haustein and Günther (1980), and Haustein et al. (1982)

that class I genes have been described which appear to resemble RT1.*A* products much more than RT1.*C* antigens with respect to tissue distribution and which map to the RT1.*D* side in a region called RT1.*E* (Kunz et al. 1982). The relation between RT1.*C* and *E* regions is so far unclear. In addition two mutant genes which affect body size (*dw-3*), and fertility (*ft*) have been assigned to this part of the rat MHC (Kunz et al. 1980).

The strains used in several of the experiments referred to belong to the LEW congenic series, which carry different RT1 haplotypes on the common LEW strain genetic background. From two of these strains, LEW.1A and LEW.1W, carrying $RT1^a$ and $RT1^u$ respectively, recombinants have been bred whose MHC is composed of permutations of the *A, B,* and *C* regions of $RT1^a$ and $RT1^u$ (Table 1). Three further recombinants are included here, which involve also the $RT1^u$ haplotype and in addition $RT1^l$ of the LEW strain and which have been identified by the use of the $RT1^a$/$RT1^u$ recombinants (Wonigeit, personal communication). The advantage of the set of recombinants presented in Table 1 resides in the fact that the *A, B,* and *C* regions they carry are derived from the same group of parental haplotypes and should be identical by descent and not only identical by nature, i.e., because they react similarly with typing reagents. Since all strains share the LEW strain genetic background, differential effects between them indicate association with MHC genes.

The tissue distribution of class I and II antigens has not yet been studied in great detail. RT1.*A* antigens appear to be present on nearly all tissues and their level of expression is controlled by MHC (Begovich et al. 1983; Howard et al.

Table 1. RT1 recombinants (Prague-Freiburg-Hannover "family")[a]

Strain	RT1	Region				
		A		*B*		*C*
LEW.1A	*a*	*a*		*a*		*a*
LEW.1AR2	*ar2=r3*	*a*		*a*	/	*u*
LEW.1AR1	*arl=r2*	*a*	/	*u*		*u*
LEW.1WR2	*wr2=r6*	*u*	/	*a*		*a*
LEW.1WR1	*wrl=r4*	*u*		*u*	/	*a*
LEW.1W	*u*	*u*		*u*		*u*
LEW.1R14	*r14*	*u*		*u*	/	*l*
LEW.1R15	*r15*	*l*		*l*	/	*u*
LEW	*l*	*l*		*l*		*l*
LEW.1LM1	*lml*	*l*		*l*		[b]

[a] The recombinants have been produced or characterized in the laboratories of Dr. O. Štark, Charles University Prague, Czechoslovakia, Dr. K. Wonigeit, Medizinische Hochschule Hannover, Federal Republic of Germany, and the author
[b] Presumably a mutant RT1. *C* region (K. Wonigeit, personal communication)

1979) as well as non-MHC genes (Howard et al. 1979). Of particular interest is the observation that MHC genes of the RT1.*B*/*D* region are able to modify the expression of RT1.*A* antigenic determinants qualitatively (Livingstone et al. 1983; Livingstone, personal communication). The expression of class II antigens, which are commonly found on B-lymphocytes and accessory cells like dendritic cells and some macrophage subpopulations, is inducible in certain epithetial cells by factors produced during the immune response (Mason et al. 1981). With regard to organ transplantation, detailed quantitative studies on the expression of class I and II antigens on the different cell types present in the various organs, especially on cells lining blood vessels, are desirable, but still incomplete (Barclay et al. 1983; Fabre 1982; Paul et al. 1982). The importance of the amount of MHC antigens present on the cells is demonstrated by the frequent observation that haploidentical, i.e., F_1-hybrid grafts, survive longer in parental recipients than grafts carrying the double dose of RT1 antigens.

Homology Between Rat, Mouse, and Human MHC

One of the major reasons for performing experimental organ grafting in the rat is to recognize and solve problems which are relevant for clinical transplantation. It is therefore appropriate to consider also the degree of homology between the MHC of rat and man. Figure 2 presents schematically the MHC-carrying chromosomes of man, rat, and, in addition, mouse, which in some sense serves as the paradigm of an MHC. It is obvious that the overall structure is similar and the same types of genes occur. There are, however, differences in gene order and probably in gene number. In mouse and rat the class II genes are flanked by class I genes, but not in man. More class II genes appear to exist in man than in the mouse and rat. Compared with mouse and man, the rat may have fewer expressed

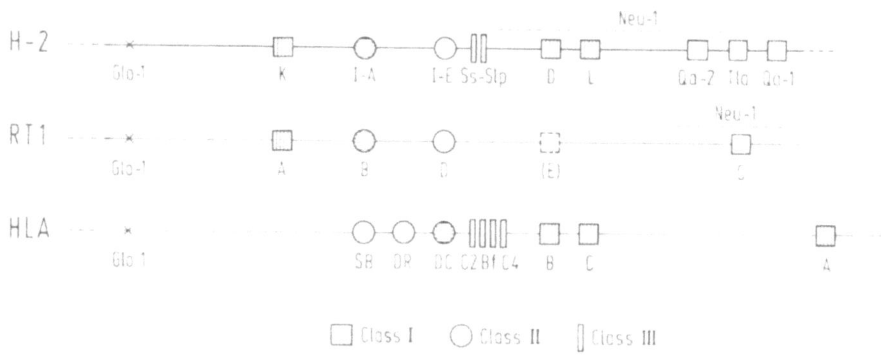

Fig. 2. Schematic comparison of the MHC chromosomal regions in the mouse (H-2 system), rat (RT1 system), and man (HLA system). The arrangement is matched with respect to *Glo-1* and the class II genes and is based on classic genetic data. The distances between the various genes or regions are not always proportional to the actual recombination frequencies. The order of some genes is not yet clearly determined (*D*/*L*, *Qa*/*Tla* in the mouse) and the relation between *E* and *C* in the rat is unclear. For the mouse data see (Klein et al. 1983); the rat data are taken from Fig. 1, and the HLA data have been compiled from various recent publications

class I genes. So far no evidence exists for more than one gene being present in the *A* region and it is open to question whether the RT1.*C* gene(s) are homologous to mouse *Qa* genes or represent what in the mouse is the H-2D,L gene group which is thought to be homologous to HLA-B,C. Only few data have been reported so far which indicate the existence of *Qa*/*T1a*-like antigens in man (Gazit et al. 1980). There is no evidence that variability in gene order or number implies gross functional differences.

The high degree of homology between the MHC of man, rat, and mouse can also be demonstrated at the phenotypic level when comparing the amino acid sequences of class I and class II molecules, respectively (Orr 1982), and by the many examples of serological cross-reactivity between class I or class II antigens of these species. The homology observed is clearly the consequence of the common evolutionary origin of the MHC in the different species (Klein 1982).

Restriction Function

The main function of the MHC cell surface molecules, as presently seen, is to form the context in which T-lymphocytes recognize foreign antigens. The polymorphism of the MHC molecules, which is the reason for the existence of the MHC-mediated histoincompatibility effect, is related to this so-called restriction in that not every antigen can be recognized by T-lymphocytes in the context of just any MHC molecule. From studies in mouse and man it is well-known that the cytotoxic T-lymphocyte (CTL) response to virus-infected or hapten-modified cells and to minor histocompatibility antigens is restricted by the ubiquitously expressed class I antigens, whereas helper T-lymphocytes are restricted by class II molecules. In the rat, immune responses to soluble antigens are controlled by the

RT1.*B*/*D* genes (Günther and Štark 1977), meaning that their products restrict the helper T-lymphocytes involved. In the course of studies on the restriction function of RT1 antigens it turned out that CTls against minor H antigens of the LEW strain genetic background, the H-Y antigen, and TNP-modified syngeneic lymphoid cells are restricted by the RT1.*A* and not the RT1.*C* region (Günther and Wurst 1984). Thus, RT1.*C* molecules do not appear to play a dominant role in restricting CTLs – although of course many more antigens will have to be tested – and resemble in this respect the mouse Qa antigens. It is interesting to speculate on the existence of class I MHC antigens which have no major restriction capacity. Is their cell surface density too low, or do they possess only a small repertoire of restricting elements, or is their action suppressed or overgrown by RT1.*A*-restricted T-lymphocytes?

Alloimmunue Response to RT1 Antigens

As is summarized in Table 2 alloantibodies can be induced against RT1.*A*,*B*, and *C* antigens; monoclonal antibodies have been produced against RT1.*A* and *B* antigens in many laboratories and recently also against RT1.*C* antigens (own un-

Table 2. Phenotypic characterization of the RT1 regions[a]

	Region		
	A	*B*/*D*	*C*
Class	I	II, II	I
Expression	Ubiquitous	B-lymphocytes Macrophages Dendritic cells Epithelial cells[b] Endothelial cells[c]	B-lymphocytes T-lymphocytes Macrophages Fibroblasts[d]
Density on lymphocytes	High	High	Low
Restriction function	Yes	Yes	No?
Control of immune responsiveness	Yes	Yes	?
H-2 homology	K,D	I-A, I-E	Qa?
Alloimmune response:			
MLR	+	++	–
CTL	+	+	+
Antibody	+	+	+
Graft rejection: Skin	+	+	+
Kidney	+	+	–
Pancreas[e]	+	+	–
Heart[e]	+/–	+	–

[a] Based on own results exept for data indicated
[b] See Barclay et al. (1983) and Mason et al. (1981)
[c] See Paul et al. (1982)
[d] K. Wonigeit (personal communication)
[e] See Klempnauer et al. (1936) and Klempnauer (personal communication)

published results). CTLs are also inducible against all three types of antigens, whereas mixed lymphocyte reactivity (MLR) appears to be stimulated only by RT1.*A* (weakly) and RT1.B/*D* antigens (Günther and Štark 1978). With respect to graft rejection, skin is destroyed by immune responses against RT1.*A*, as well as RT1.*B* or *C* incompatibilities (see also Table 3). It is interesting that kidney (Paris and Günther, unpublished), heart (Klempnauer, personal communication), and pancreas (Klempnauer et al. 1983b) grafts are not rejected, when grafted across an RT1.*C* barrier. These results again point to a difference between RT1.*A* and RT1.*C*-determined class I antigens. Organ graft rejection however, can occur across *A* or *B* region differences. In some strain combinations rejection is faster in the case of *A* incompatibility, in others of *B* incompatibility (see e.g., Klempnauer et al. 1983b; Lowry and Gurley 1983; Rozing et al. 1983), a variability which presumably depends upon the incompatible antigenic determinants involved and the responder status of the recipient (see below).

Grafting experiments with RT1-recombinant congenic strains not only enable the histoincompatibility effect mediated by each region to be studied separately, but will also be helpful in recognizing synergistic effects. An example of such a phenomenon is the observation that RT1.*A*,*B*-incompatible pancreas or heart grafts do not survive as long as RT1.*A*,*B*,*C*-incompatible organs (Klempnauer, personal communication).

Minor H Antigens and Organ Graft Rejection

Only little is known about organ graft rejection induced by minor H-antigen incompatibility. H-4(=RT4) (Kohoutovà and Štark 1981), H-5 (Křen et al. 1978), H-Y (Mullen and Hildemann 1972), and H-X (Mullen and Hildemann 1972) appear to be the only minor H-antigens defined in the rat; H-4, H-Y, and H-X incompatibility lead to skin graft rejection (Mullen and Hildemann 1972, see also below) but H-Y and H-X do not induce organ graft rejection (Mullen and Hildemann 1972). Most often minor H-antigens are studied in toto by using RT1-identical strains (see, e.g., Paul and Carpenter 1983). For the RT1.*A*-identical combination AVN/LEW.1A it can be shown that pancreas grafts are acutely rejected, whereas heart grafts survive more than 100 days (Klempnauer et al. 1983a; Klempnauer, personal communication). Thus minor H-antigens are of significance, for some types of graft at least.

High and Low Responsiveness to H-Antigens

As decisive for the fate of a graft as the H-antigens of the donor is the capacity of the host to react to the antigens it is confronted with. Differences between the reactivity of inbred strains to foreign H-antigens point to a genetic control of the response. Antibody production to RT1.A^{a} antigens and rejection of RT1.A^{a}-incompatible skin grafts have been studied in detail genetically and could be shown to be controlled quantitatively by RT1-linked genes (Butcher et al. 1982; Heslop and Jolly 1979). In many strain combinations, e.g., LEW.1A/LEW.1W, graft rejection is of unequal strength in both directions; LEW.1A is a low responder to

Table 3. Asymmetry of the LEW.1A ($RT1^{a}$)/LEW.1W ($RT1^{u}$) combination

RT1 Incom-patibility	RT1.*ABC*		Graft survival (days, mean ± sd)			
	Donor	Recipient	Skin[a]	Kidney[b]	Heart[c]	Pancreas[c]
ABC	aaa	uuu	11.3±1.5	7.0± 1.0	7.2±0.8	9.3±0.5
	uuu	aaa	12.1±0.7	59.3±48.2	14.0±0	17.0±4.8
A	*a*aa	uaa	13.3±1.2	nt	>100	14.0±2.8
	*a*uu	uuu	12.3±1.6	8.6±2.0	nt	nt
	*u*aa	aaa	19.5±9.8(20), >100(2)	nt	>28	>28
	*u*uu	auu	23.0±13.7(26), >100(5)	nt	nt	nt
C	aa*a*	aau	29.1±24.1(14), >100(3)	nt	>100	26, >100(5)
	uu*a*	uuu	24.4±8.6(39), >100(1)	nt	nt	nt
	aa*u*	aaa	17.1+6.4	nt	nt	nt
	uu*u*	uua	21.3±10.4	nt	nt	nt

nt, not tested
[a] See Kohoutovà et al. (1980)
[b] See Paris and Günther (1980)
[c] See Klempnauer et al. (1983b) and Klempnauer (personal communication)

$RT1^{u}$ antigens and LEW.1W a high responder to $RT1^{a}$ antigens (see also Butcher et al. 1982). This phenomenon is clearly evident when the survival times of kidney, heart, or pancreas grafts are compared, as is done in Table 3. $RT1^{u}$ grafts are rejected quicker than $RT1^{a}$ grafts. This difference is especially pronounced when skin graft survival across the RT1.*A* barrier is determined. Again $RT1^{u}$-carrying grafts are rejected significantly slower than $RT1^{a}$ grafts. Interestingly enough, this difference is apparently not influenced by the alleles at the *B* and *C* regions since, e.g., $RT1.B^{u}$ as well as $RT1.B^{a}$ is associated with slow rejection of $RT1.A^{u}$ skin. These data do not favor the idea that the graft response to RT1.*A* antigens is controlled ("restricted") by RT1.*B* genes, as has been postulated (Butcher and Howard 1982). It is worth mentioning that the response against RT1.*C*-incompatible skin grafts shows just the reverse asymmetry as that observed for RT1.*A*-incompatible grafts (Table 3). Many more such asymmetrical strain combinations have been described in organ grafting experiments. The analysis of this asymmetry might be helpful for the understanding of similar phenomena in man. As an example, HLA-DRw6-positive kidney grafts survive longer than HLA-DRw6-negative grafts, but HLA-DRw6-positive recipients are strong rejectors (Hendriks et al. 1983). So far the cellular and molecular basis of this type of genetic control is not yet understood.

Asymmetrically histoincompatible strain combinations like that of the LEW.1A and LEW.1W rats will also be of help in analyzing further the beneficial effect of pretransplantation blood transfusions in clinical transplantation. Only in the "weak" LEW.1A anti-LEW.1W combination is enhancement inducible by blood injected 7 days before heart transplantation, whereas the same regimen leads to accelerated rejection of LEW.1A hearts by the high-responder LEW.1W rats (Soulillou et al. 1984). This observation links the responder status of the recipient with the effect of pretransplantation blood transfusion.

Table 4. Genetic control of anti-H-Y-responsiveness

RT1	*ABC*	Skin graft survival (days)		HVGR[a]	CML[b]
		1°	2°	(mg)	(%)
a	aaa	>100	28.3±4.7	33	33
ar2	aau	nt	15.2±5.5	nt	35
wr2	uaa	nt	nt	nt	1
u	uuu	>100	>100	4	3
n	nnn	> 60	nt	42	nt

1°, primary graft; 2°, secondary graft; nt, not tested

[a] Host-versus-graft reaction; 15×10^6 lymphoid cells are injected into the footpads of female rats of the same strain, and 10 days later the enlargement due to injection of male compared with female cells is determined

[b] Cell-mediated lympholysis (Günther and Wurst, 1984). Male-specific CTLs are generated after in vivo priming of female rats and in vitro restimulation with male lymphocytes of the same strain and assayed against ^{51}Cr-labeled male lymphoblasts. Percentage specific ^{51}Cr release is given

The immune response to minor antigens has been mentioned above to be MHC restricted and the restricting MHC molecules have been assigned some specificity for the antigen the response to which they restrict. Table 4 illustrates both facts. CTLs against the H-Y antigen can only be induced in LEW.1A, but not in LEW.1W rats after in vivo priming and in vitro restimulation (Günther and Wurst 1984). Since both strains are RT1 congenic, the low response must be controlled by MHC genes; since LEW.1WR2 rats are low responders in contrast to LEW.1AR2, low responsiveness must be associated with RT1.A^u. The induction of a significant host-versus-graft response in female LEW1.A, but not LEW.1W, rats and the fact that LEW.1A and LEW.1AR2, but not LEW.1W, rats reject male skin after priming are further signs of the low-anti-H-Y responder status of LEW.1W rats. Such models should be helpful in studies aiming at the identification of the cells responsible for graft rejection (Dallmann and Mason 1983; Loveland and McKenzie 1982), since various types of immune responses are affected in which different cell types are involved.

Concluding Remarks

Experimental organ transplantation in the rat is relevant in its own right, but gets special significance from being a model for clinical transplantation. The rat is especially useful in this respect, since no other species is suitable for microsurgery and at the same time has been studied immunologically and genetically in such detail. It might turn out to be advantageous to stick to a few strain combinations and choose strains where the effect of defined genes can be tested separately and in combination. Genetics and immunology of graft rejection are complex, but this complexity has to be revealed by the aid of simplified systems. The availability of such a powerful and selectively acting immunosuppressive drug as cyclosporin A is not an argument against this postulate. On the contrary, the recognition of the desired selectivity of action is based on our insight into the various cells and factors which interact during the immune response. We have known for a long time that acceptance or rejection of grafts occurs; our aim is to find out why rejection occurs and how we can predict it.

A completely new aspect of the immunogenetics of transplantation has been introduced by the availability of methods which enable histocompatibility genes to be studied directly at the DNA level. Studies have been already published on the molecular genetics of the human and mouse MHC (Hood et al. 1983; Orr 1982) and are currently under way on the RT1 system in several laboratories.

References

Barclay AN, Mason DW, Mayrhofer G (1983) The distribution of Ia antigens in the rat. Transplant Proc 15:1587–1588

Begovich AB, Blankenhorn EP, Winters BA (1983) Genetic control of the expression of class I molecules on rat erythrocytes. Immunogenetics 18:45–56

Blankenhorn EP, Symington FW, Cramer DV (1983) Biochemical characterization of Ia antigens encoded by the RT1.B and RT1.D loci in the rat MHC. Immunogenetics 17:475–484

Butcher GW, Howard JC (1977) A recombinant in the major histocompatibility complex of the rat. Nature 266:362–364

Butcher GW, Howard JC (1982) Genetic control of transplant rejection. Transplantation 34:161–166

Butcher GW, Corvalan JR, Licence DR, Howard JC (1982) Immune response genes controlling responsiveness to major transplantation antigens. Specific major histocompatibility complex-linked defect for antibody responses to class I alloantigens. J Exp Med 155:303–320

Dallmann MJ, Mason DW (1983) Cellular mechanism of skin allograft rejection in the rat. Transplant Proc 15:335–338

Fabre JW (1982) Rat kidney allograft model: Was it all too good to be true? Transplantation 34:223–225

Gazit E, Terhorst C, Yunis EJ (1980) The human "T" genetic region of the HLA linkage group is a polymorphism detected on lectin-activated lymphocytes. Nature 284:275–277

Günther E, Štark O (1977) The major histocompatibility system of the rat (Ag-B or H-1 system). In: Götze D (ed) The major histocompatibility system in man and animals. Springer, Berlin Heidelberg New York, pp 207–253

Günther E, Štark O (1978) At least two loci of the major histocompatibility complex can determine mixed lymphocyte stimulation in the rat. Tissue Antigens 11:465–470

Günther E, Wurst W (1984) Cytotoxic T lymphocytes of the rat are predominantly restricted by RT1.A and not RT1.C – determined major histocompatibility class T antigens. Immunogenetics 20:1–12

Günther E, Štark U, Koch C (1978) Genetic definition of I region-determined antigens of the rat major histocompatibility complex. Eur J Immunol 8:206–212

Haustein D, Günther E (1980) Biochemical analysis of gene products of major histocompatibility recombinant haplotypes in the rat. Eur J Immunol 10:615–620

Haustein D, Stock W, Günther E (1982) Rat major histocompatibility RT1.C antigens of restricted tissue distribution consist of two polypeptide chains with molecular weights of about 42000 and 12500. Immunogenetics 15:271–277

Hendriks GFJ, Persijn GG, Lansbergen Q, d'Amaro JD, Schreuder GMT, Cohen B, Rood JJ van (1983) Excellent outcome after transplantation of renal allografts from HLA-DRw6-positive donors even in HLA-DR mismatches. Lancet 2:187–189

Heslop BF, Jolly KD (1979) Genetic control of haemagglutinin production against an MHC determined alloantigen in rats. Immunogenetics 8:567–570

Hood L, Steinmetz M, Malissen B (1983) Genes of the major histocompatibility complex of the mouse. Annu Rev Immunol 1:529–658

Howard JC, Butcher GW, Galfré G, Milstein C, Milstein CP (1979) Monoclonal antibodies as tools to analyse the serological and genetic complexities of major transplantation antigens. Immunol Rev 47:139–174

Klein J (1982) Evolution and function of the major histocompatibility complex. In: Parham P, Strominger J (eds) Histocompatibility antigens: structure and function. Chapman and Hall, London, pp 223–239 (Receptors and recognition. Series B, vol 14)

Klein J, Figueroa F, Nagy ZA (1983) Genetics of the major histocompatibility complex: the final act. Annu Rev Immunol 1:119–142

Klempnauer J, Wonigeit K, Günther E, Pichlmayr R (1983a) Pancreas whole organ transplantation in the rat. Evidence for a strong effect of non-MHC incompatibilities. Transplant Proc 15:1649–1651

Klempnauer J, Wonigeit K, Steiniger B, Günther E, Pichlmayr R (1983b) Pancreas whole organ transplantation in the rat. Differential effect of individual MHC regions. Transplant Proc 15:1308–1310

Kohoutová M, Štark O (1981) A new congenic strain LEW.C4H with a weak H-locus. Transplant Proc 13:1319–1320

Kohoutová M, Günther E, Štark O (1980) Genetic definition of a further gene region and identification of at least three different histocompatibility genes in the rat major histocompatibility system. Immunogenetics 11:483–490

Křen V, Bílá V, Kršiaková M (1978) Alleles of the H-5 locus linked to *lx* locus in different rat strains. Folia Biol 24:376–377

Kunz HW, Gill TJ III, Dixon BD, Taylor FH, Greiner DC (1980) The growth and reproduction complex in the rat: genes linked to the major histocompatibility complex which affect development. J Exp Med 152:1506–1518

Kunz HW, Gill TJ III, Misra DN (1982) The identification and mapping of a second class I locus in the major histocompatibility complex of the rat. J Immunol 128:402–408

Livingstone AM, Butcher GW, Howard JC (1983) Recombination within the MHC can change the specificity of class I major transplantation antigens. Transplant Proc 15:1557–1559

Lobel SA, Camer DV (1981) Demonstration of a new genetic locus in the major histocompatibility system of the rat. Immunogenetics 13:465–473

Loveland BE, McKenzie IFC (1982) Which T cells cause graft rejection? Transplantation 33:217–221

Lowry RP, Gurley KE (1983) Rejection of organ allografts transplanted across MHC subregion disparity (RT1.B). Transplant Proc 15:1643–1646

Mason DW, Dallmann M, Barclay AN (1981) Graft-versus-host disease induces expression of la antigens in rat epidermal cells and gut epithelium. Nature 293:150–151

Mullen Y, Hildemann WH (1972) X- and Y-linked transplantation antigens in rats. Transplantation 13:521 529

Nesbitt MN (1974) Evolutionary relationship between rat and mouse chromosomes. Chromosoma 46:217–224

Orr H (1982) The major histocompatibility complex: analysis at the protcin and DNA levels. In: Parham P, Strominger J (eds) Histocompatibility antigens: structure and function. Chapman and Hall, London, pp 3–51 (Receptors and recognition. Series B, vol 14)

Paris A, Günther E (1980) Kidney grafting between rats which carry recombinant major histocompatibility haplotypes. Immunogenetics 10:205–209

Paris A, Bishop C, Festenstein H, Günther E (1978) Non-major histocompatibility system immunogenetic influences in rat renal allograft survival. Transplantation 25:252–254

Paul LC, Carpenter CB (1983) Non-MHC transplantation antigens in the rat. Transplant Proc 15:217–219

Paul LC, Paradysz JM, Milford EL, Kunz HW, Carpenter CB (1982) Expression of RT1.A and RT1.B/D antigens on endothelium of rat kidneys. Transplantation 34:121–128

Rozing J, Bonthuis P, Joling P, Vaessen LMB, Lameijer LDF (1983) The influence of RT1 subregion differences on cardiac allograft survival. Transplant Proc 15:1647–1648

Soulillou JP, Blandin F, Günther E, Lemoine V (1984) Genetics of the blood transfusion effect on heart allografts in rats. Transplantation 38:63–67
Štark O, Günther E, Kohoutová M, Vojčík L (1977) Genetic recombination in the major histocompatibility complex (H-1,Ag-B) of the rat. Immunogenetics 5:183–187
Stock W, Günther E (1982) Serologic and cellular characterization of products of a new major histocompatibility gene region, RT1.C, of the rat; possible homology to mouse H-2Qa. J Immunol 128:1923–1928
Stolc V, Kunz HW, Gill TJ III (1980) The linkage of glyoxalase I to the major histocompatibility complex in the rat. J Immunol 125:1167–1170
VandeBerg JC, Bittner GN, Meyer GS, Kunz HW, Gill TJ III (1981) Linkage of neuraminidase and α-mannosidase to the major histocompatibility complex in the rat. J Immunogenet 8:239–242
Zimmermann FA, Knoll PP, Davies H ffS, Gokel JM, Schmid T (1983) The fate of orthotopic liver allografts in different rat strain combinations. Transplant Proc 15:1272–1275

Venovenous Allotransplantation in the Rat

T. Schang, B. von Gaudecker, and *A. Thiede*

In previous studies Engemann et al. (1977) and Körner (1979) were able to demonstrate that vein grafts transplanted into the venous system have the capacity to sensitize allogeneic recipients. Vein grafting led to accelerated rejection of subsequent donor-type skin grafts and production of cytotoxic antibodies to donor-type tissue. But all allogeneic vein grafts showed long-term patency, and histologically they appeared to be intact when investigated between day 20 and day 100 postoperatively. Especially no mononuclear cellular infiltrations could be found within this period. Allogeneic vein grafts in nonimmunosuppressed recipients were not distinguishable from syngeneic vein grafts 100 days postoperatively. These findings were unexpected, because arterial allografts (Thiede et al. 1979; Timmermann and Thiede 1981) as well as venous allografts (Deltz 1977) used for arterial replacement in the rat underwent a marked cellular infiltration even if long-term patency was achieved. Therefore the aim of our study was to compare syngeneic and allogeneic venovenous grafts by means of morphometric methods. Especially morphological alterations of the vein grafts occurring in the early postoperative phase needed investigation.

Material and Methods

Male inbred F344 rats ($RT1^l$) served as recipients and syngeneic vein graft donors. RT1-incompatible vein grafts were obtained from inbred male CAP rats ($RT1^c$). All animals were used at the age of 3 months, with a body weight of about 200 g.

The thoracic portion of the inferior caval vein (IVC) served as the vein graft. The graft was stored in 0.9% saline at 4 °C during the preparation of the recipient for about 20 min. A segment of the recipient's abdominal IVC below the renal veins about 1 cm long was excised and replaced by the vein graft. Both end-to-end vessel anastomoses were performed by the continuous suture technique using 8-0 Ethilon. Donor and recipient operations were carried out with the animal under ether anesthesia and under sterile conditions. No drugs were administered. Experimental groups were formed as shown in Table 1. Five, 10, 20, and 50 days postoperatively the vein grafts were fixed by perfusion using Bouins fluid. The tis-

Microsurgical Models in Rats
for Transplantation Research
Eds. A. Thiede et al.

Table 1. Experimental groups

Group	Strain combination	Immunogenetic difference	Days postoperatively	*n*
I	F 344	– (normal vein)	–	3
II	F 344–F 344	Syngeneic	5	5
		$RT1^l$–$RT1^l$	10	10
			20	10
			50	4
III	CAP–F 344	Allogeneic	5	10
		$RT1^l$–$RT1^l$	10	10
			20	10
			50	10

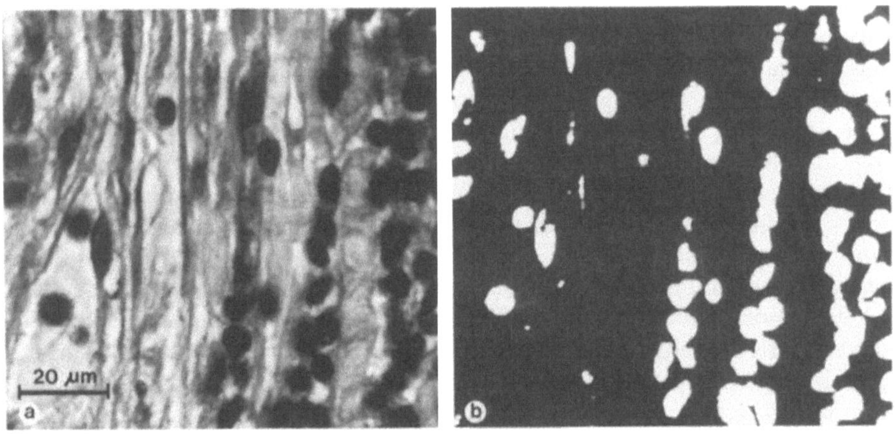

Fig. 1. **a** Measurement area seen on a monitor. **b** The same area at the discriminating threshold. The area of cell nuclei per measurement area is counted by the microvideomat

sue was embedded in Araldite, and 5-µm sections were stained by H & E and Goldner's trichrome stain combined with an elastic fiber stain.

The grade of cellular infiltration of the vein grafts was investigated by means of a computerized microvideomat (Zeiss, Oberkochen, W.-Germany), which measured gray level indexes (GLIs) of histological sections with only cell nuclei stained by Weigert's iron-hematoxylin. GLIs can serve as a parameter for projections of cell nuclei within a distinct measurement area, measured at a certain discriminating threshold (Zilles et al. 1978). All areas within a measurement area darker than a certain discriminating threshold are counted (Fig. 1). So GLIs cor-

Figs. 2–5.
Fig. 2. Isograft 5 days postoperatively. **Fig. 3.** Isograft 50 days postoperatively. **Fig. 4.** Allograft 5 days postoperatively. **Fig. 5.** Allograft 50 days postoperatively.
En, endothelium; *M*, smooth muscle cell; *E*, elastic fibers; *K*, blood capillary; *F*, fibroblast; *arrow*, mononuclear hump at the endothelium

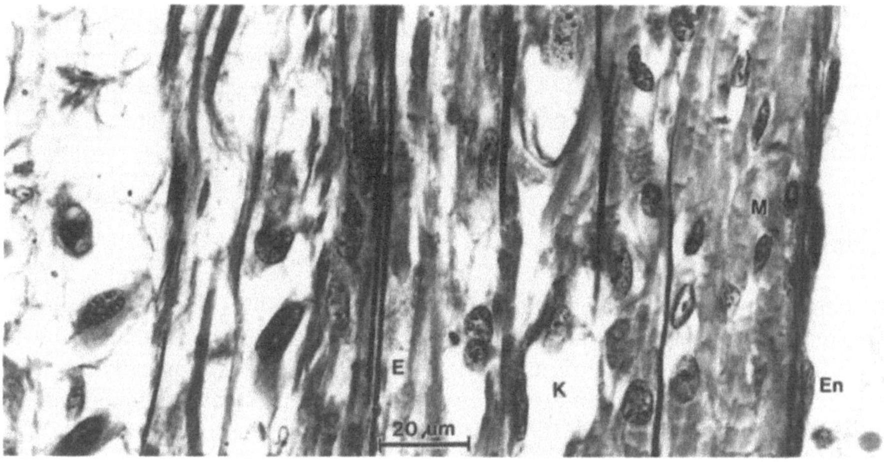

▲ **Fig. 2**

▼ **Fig. 3**

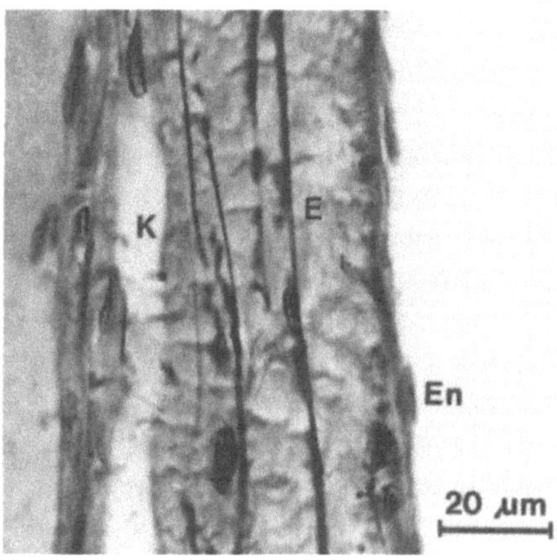

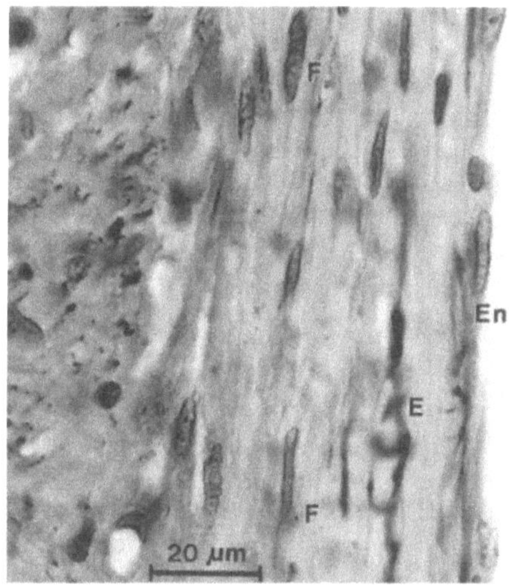

▼ **Fig. 4**

Fig. 5 ►

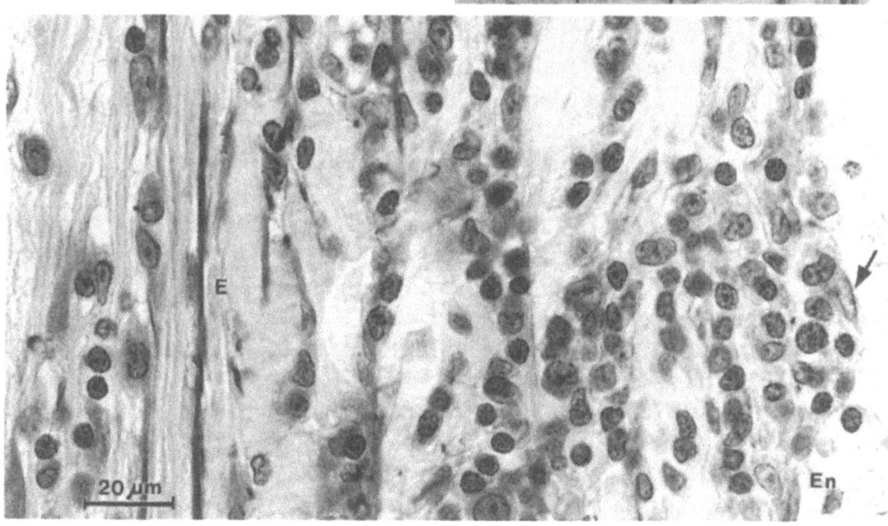

relate positively with volumes of cell nuclei within a certain volume of the histological section. Twelve to 87 ($\bar{x}$, 46) measurement areas were chosen in each of the histological sections. GLIs of normal thoracic IVC, syngeneic vein grafts, and allogeneic vein grafts were compared statistically using the Wilcoxon-Mann-Whitney test.

Results

All syngeneic vein grafts were patent within the observation period. Only 1 out of 40 allogeneic vein grafts had been closed by a thrombus when the animal was killed for histological evaluation 5 days postoperatively. Histologically the wall of a normal thoracic IVC consists mainly of collagenous fibrous tissue braced by four to six layers of strong elastic fibers. Single smooth muscle cells lie in a subendothelial position without forming a continuous layer. Vasa vasorum can be found in the vascular wall.

Five days postoperatively the vascular wall of syngeneic vein grafts is thickened about threefold and shows a loose structure. Many blood capillaries pass through the graft. Fibroblasts show a blastic transformation of their cell nuclei. Only a few macrophages can be found and no loss of endothelium is detectable (Fig. 2). Ten days postoperatively the histological appearance of syngeneic vein grafts does not change, but 20 days postoperatively the vein grafts again look like a normal vein. Fifty days postoperatively the grafts are not distinguishable from normal veins (Fig. 3).

Five days postoperatively allogeneic vein grafts are heavily infiltrated by lymphocytes and macrophages. The vascular wall is thickened about threefold. Destruction of elastic fibers is indicated by shrinking and fragmentation. In several places the endothelium is covered by mononuclear cells (Fig. 4). Ten days postoperatively the histological appearance of allogeneic vein grafts shows no further changes; the grafts are still markedly infiltrated by mononuclear cells. But 20 days postoperatively cellular infiltrations have vanished completely. Nevertheless the vascular wall is about two times thicker than normal. A smaller number of capillaries is found in the vascular wall. Fragmentated remnants of elastic fibers lie between collagenous tissue, which is populated by fibroblasts. The vascular lumen is covered by a continuous endothelial layer. Fifty days postoperatively no further changes can be found (Fig. 5).

Morphometric Findings. Five days postoperatively syngeneic vein grafts present significantly higher GLIs than normal veins (*P*, 0.05), due to blastic transformation of cell nuclei and infiltration of a few macrophages (Fig. 6). Ten days postoperatively the GLIs return to normal values.

Five days postoperatively cellular infiltration of allografts results in significantly higher GLIs than in corresponding isografts (*P*, 0.05). Ten days postoperatively the GLIs still reach the same levels, being significantly higher than corresponding isografts (*P*, 0.01). Twenty days postoperatively the GLIs are significantly reduced (*P*, 0.05), due to the total disappearance of cellular infiltration. Fifty days postoperatively the GLIs of allografts show no further changes. They

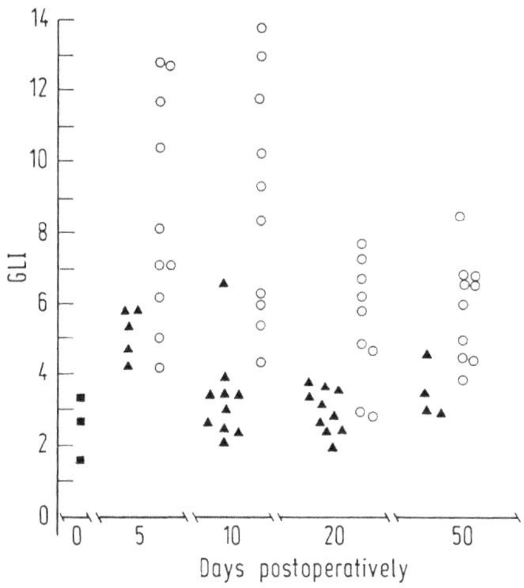

Fig. 6. Gray level indexes (GLIs) as parameters for areas of cell nuclei per measurement area in histological sections. ■, group I: normal veins; ▲, group II: isografts; ●, group III: allografts

do not return to normal levels (*P*, 0.02) due to a greater number of fibroblasts in the vascular wall.

Discussion

Our results demonstrate a marked cellular immunoreaction in venovenous RT1-incompatible allografts. But strikingly these cellular infiltrations are confined to the early postoperative phase. Twenty days postoperatively cellular infiltrations have totally disappeared from venovenous allografts. Histological findings could be objectified by amorphometric study of the time course of cellular infiltrations.

The most simple explanation for the disappearance of cellular infiltrations would be complete elimination of donor-type antigens in the graft. In this case cells of host origin would have repopulated the graft simultaneously. As demonstrated by Williams et al. (1975) arterial allografts possess at least endothelial cells of host origin. Gronemeyer (1979) described repopulation of cornea grafts by host cells in the rat. Caval vein grafts mainly consist of collagenous fibrous tissue and only very few other cells. So elimination and repopulation could be performed very quickly.

On the other hand, Körner (1979) was able to induce mononuclear infiltration in allogeneic venovenous grafts lasting more than 20 days postoperatively by repeated presensitization by donor-type skin grafts. Allogeneic vein grafts used for arterial replacement (Deltz et al. 1977) were heavily infiltrated by mononuclear cells for more than 100 days postoperatively so the vascular wall of vein grafts seems to contain targets for a cellular immunoreaction for a much longer period than only 10 days postoperatively.

For this reason some sort of immunological enhancement cannot be excluded for venovenous allografts. In any case it has been demonstrated that allogeneic vein grafts have a different fate, depending on the site of transplantation. If placed into the venous system the graft undergoes a milder reaction than when placed into the arterial system. MHC-incompatible vein grafts could be used for venous replacement without any subsequent immunosuppression, if autologous vein grafts are not available.

References

Deltz E, Sonntag H, Thiede A (1977) Funktionelle und morphologische Untersuchungen bei allogenen Venentransplantaten. Vasa 6:211–214

Engemann R, Körner H, Thiede A (1977) Microsurgical venous replacement in allogeneic rat strain combinations. Twelfth Congress of the European Society of Surgical Research, Warsaw

Gronemeyer U (1979) Zur Immunbiologie der Hornhauttransplantation. Modelluntersuchungen an Inzuchtratten. *Habilitationsschrift*, University of Kiel

Körner H (1979) Veno-venöse Transplantationen in standardisierten Ratteninzuchtstammkombinationen. Die Bedeutung der Sensibilisierung für das Transplantatverhalten. Thesis University of Kiel

Thiede A, Engemann R, Körner H, Müller-Ruchholtz W (1979) Comparison of the immunological reactions of arterial transplants in the arterial system and of venous transplants in the venous system using inbred strains of rats. Transplant Proc 11:603–609

Timmermann W, Thiede A (1981) Histologische und funktionelle Untersuchungen an Femoralarterientransplantaten bei Ratten. Vasa 10:141–147

Williams GM, terHaar A, Krajewski C, Parks LC, Roth J (1975) Rejection and repair of endothelium in major vessel transplants. Surgery 78:694–706

Zilles K, Schleicher A, Kretschmann HJ (1978) A quantitative approach to cytoarchitectonics. The areal pattern of the cortex of *Tupaia belangeri*. Anat Embryol (Berl) 153:195–212

The Rejection Response to Segmental Grafts of the Femoral Artery in Inbred Rat Strains

W. Timmermann and *A. Thiede*

Introduction

The femoral artery represents a peripheric vessel with a relatively high percentage of muscle cells in its wall compared with the aorta. The rejection of aortic segmental grafts has been described extensively by Thiede in 1977. He found the strength of the rejection process and a consecutive formation of aneurysms to be correlated with the histocompatibility between donor and recipient. In this study, we wanted to describe the microscopic pattern of the rejection process of segments of the femoral artery and the functional alterations such as thrombosis or aneurysms of these grafts in dependence on the immunogenetic difference between donor and recipient.

Segments of femoral arteries were grafted within the same strain of inbred rats to evaluate technically caused alterations of these grafts and their long-term function. Additional influences of the histoincompatibility were studied using one donor-recipient combination differing in all major histocompatibility complex (MHC) antigens and non-MHC antigens and one combination differing only in non-MHC antigens and the *C* region of the MHC.

Material and Methods

An 8- to 10-mm segment of the donor's femoral artery (internal diameter, 0.4 mm) was implanted orthotopically in the recipient. The anastomoses were performed by a running suture with two corner stitches. A microscope with × 12 magnification was used and 11-0 Ethilon served as suture material. No anticoagulants or vasoactive substances were administered at any time. Animals of the inbred strains F344 ($RT1^{lv1}$) were used for all recipients (Table 1). A syngeneic combination F344→F344 was formed as technical control (group A, $n=38$). In the allogeneic groups either LEW ($RT1^{l}$) rats were used, resulting in a combination differing in non-MHC antigens and *C* region antigens of the MHC (group B, $n=35$), or CAP ($RT1^{c}$) rats were used as donors, resulting in a combination differing in non-MHC and all MHC antigens (group C, $n=35$). Twenty, 50, and 100 days after the operation, one-third of the animals of each group were killed,

Microsurgical Models in Rats
for Transplantation Research
Eds. A. Thiede et al.

Table 1. Numbers, strain combination, and functional results of segmental grafts of the femoral artery

Group	Strain combination	Days after operation	Total number of grafts	Thrombosed grafts
A	F344 ($RT1^{lv1}$)	20	17	7
$n=38$	→ F344 ($RT1^{lv1}$)	50	11	1
		100	10	0
			38 = 100%	8 = 21%
B	LEW ($RT1^{l}$)	20	11	1
$n=35$	→ F344 ($RT1^{lv1}$)	50	12	4
		100	12	3
			35 = 100%	8 = 29%
C	CAP ($RT1^{c}$)	20	14	5
$n=40$	→ F344 ($RT1^{lv1}$)	50	13	2
		100	13	3
			40 = 100%	10 = 25%
		Total	113 = 100%	26 = 23.1%

and the grafts were inspected macroscopically and finally taken out for histological examination.

Results

Thrombosis was found in 26 of the 113 grafts of all three groups (Table 1). In the syngeneic group (A) the rate of thrombosis was 21%, in the partially histoincompatible combination (group B) it was 29%, and in the totally histoincompatible combination it was 25%. In none of the groups was the rate of thrombosis dependent on the time of examination after the operation.

A significant influence of the immunogenicity of the grafts on the rate of thrombosis cannot be observed; time-dependent alterations in the grafts resulting in thrombosis can be excluded as well. Aneurysms resulting from destruction of the vascular wall were not found in the grafts at any time.

In histological examination, the isografts (group A) showed no cellular infiltrations of the vascular wall, a well-preserved medial layer and the formation of a neointima. The neointima is found as soon as 20 days after the operation and appears to be unchanged over the period of observation (Fig. 1 a–c).

The alterations of the allografts were cellular infiltrations of the vascular wall and progressive destruction of the medial layer. A neointima was found at all examinations as in the isografts. In the partially histoincompatible combination (group A) 20 days after the operation the medial layer was only slightly affected in 90% of all grafts; 50 days postoperatively it was severely affected in over 60% of the grafts and after 100 days it was totally destroyed in 78% of the cases (Fig. 2 a–c).

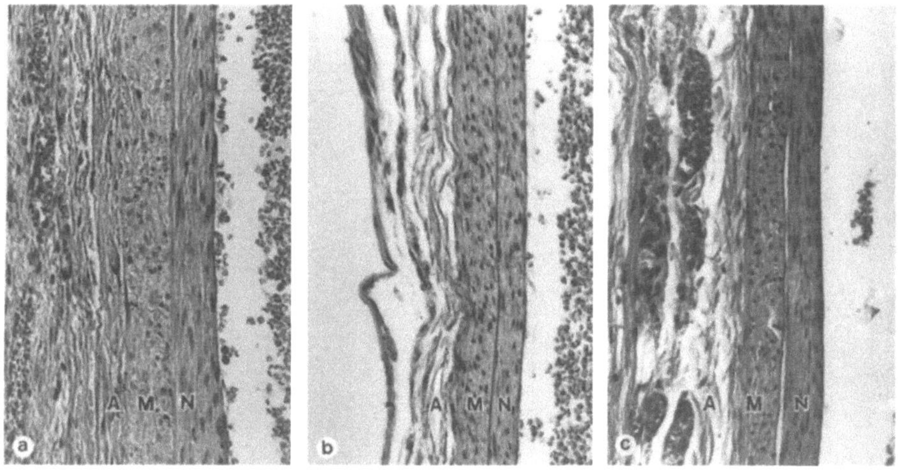

Fig. 1 a–c. Isografts, group A. *A*, adventitia; *M*, medial layer; *N*, neointima. H & E, ×200. **a** 20 days, postoperatively; **b** 50 days postoperatively; **c** 100 days postoperatively

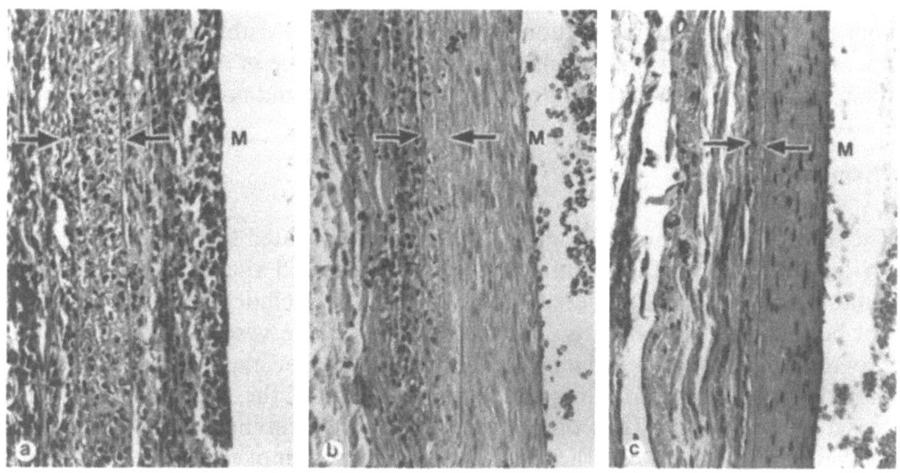

Fig. 2 a–c. Allografts, partially histoincompatible, group B. > < *M*, medial layer. H & E, ×200. **a** 20 days postoperatively; **b** 50 days postoperatively; **c** 100 days postoperatively

In the totally histoincompatible group B the medial layer of the vascular wall was significantly more strongly affected 20 days after the operation, resulting in an at least partial destruction of the medial layer in all grafts. Fifty and 100 days postoperatively no differences to group A could be noticed (Fig. 3 a–c).

All allografts were infiltrated by mononuclear cells at any time after the operation. Twenty days postoperatively in 80% of all grafts all layers of the vascular wall were infiltrated. After 50 days, no cells could be found in the neointima but most grafts had broad infiltrates between the medial layer and the surrounding

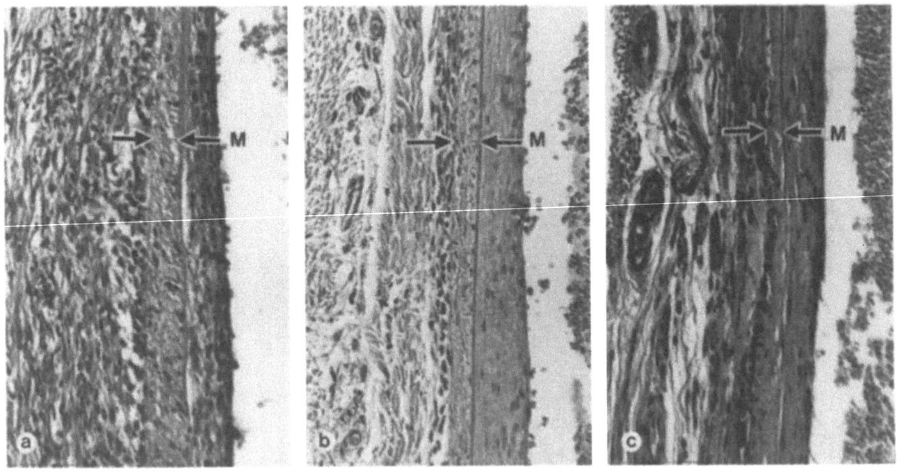

Fig. 3 a–c. Allografts, totally histoincompatible, group C. > < *M*, medial layer. H & E, ×200. **a** 20 days postoperatively; **b** 50 days, postoperatively; **c** 100 days postoperatively

connective tissue. The infiltrations are smaller but still visible 100 days after the operation (Fig. 2 a–c, Fig. 3 a–c). A significant difference in the degree of the cellular infiltration between the two allogeneic groups could not be found.

Discussion

The transplanted femoral allografts are heavily infiltrated by mononuclear cells on day 20; residues of these infiltrates can still be found after 100 days. Non-immunological reasons for these infiltrations can be excluded, because they cannot be found in the isografts. The medial layer of the vascular wall undergoes progressive destruction which can be observed over the whole observation period. An interesting finding is the homogeneity of the rejection process between groups B and C. The only difference is more prominent destruction of the medial layer in group B, indicating a faster destruction of these cells. The grafts in group B are identical with the recipient in the *A* and *B* regions of the major histocompatibility complex (MHC) of the rat whereas the grafts in group C differ in MHC and non MHC-genes. The results of this study indicate that differences in only non-MHC antigens and one subregion of the MHC (group B) are able to induce a rejection response almost as strong as a difference in all MHC and non-MHC antigens.

In transplanted aortic segments (Thiede 1977) the rejection process in partially histoincompatible grafts is clearly weaker than in totally histoincompatible ones. The macroscopic equivalent of the rejection of aortic segments is the formation of aneurysms; their incidence is also dependent on the histoincompatibility between donor and recipient. The formation of aneurysms or any other influence on the function could not be observed in femoral allografts independent of their immunogenicity.

In peripheral vessels the rejection response seems to be comparatively stronger than in the aorta for partially histoincompatible grafts, if one regards the cellular infiltration and the destruction of the medial layer of the graft. This observation could be explained by a high immunogenicity of muscle cells, which represent a major part of the wall of peripheral vessels. Two observations favor this theory. Aortic valves grafted together with heart muscle tissue proved to be more immunogenic than the valves themselves (Heslop et al. 1973) and veins produce a weaker rejection response than arterial grafts (Deltz et al. 1977; Thiede et al. 1979).

By contrast with aortic segments, the functional results of peripheral vascular allografts are not affected by their histoincompatibility and do not differ from those of isografts. To explain this result, the influence of the neointima should be considered. It could stabilize the vascular wall and prevent the formation of aneurysms although the medial layer is almost totally destroyed in allografts after 100 days.

References

Deltz E, Sonntag H-G, Thiede A (1977) Funktionelle und morphologische Untersuchungen bei allogenen Venentransplantaten. Vasa 6:211–214

Heslop BF, Wilson SE, Hardy BE (1973) Antigenicity of aortic valve allografts. Ann Surg 177:301–306

Thiede A (1977) Gefäßtransplantation. Die Bedeutung immunologischer Reaktionsmechanismen. Untersuchungen an standardisierten Ratteninzuchtkombinationen. Schattauer, Stuttgart (Ergebnisse der Angiologie, Vol 15)

Thiede A, Engemann E, Körner H, Müller-Ruchholtz W (1979) Comparison of the immunologic reactions of arterial transplants in the arterial system and of venous transplants in the venous system using inbred strains of rats. Transplant Proc 11:603–606

Factors Influencing Endothelial-Cell Viability During Procurement and Preservation of Valve Allografts

A. C. Yankah, G. Randzio, H. U. Wottge, and *A. Bernhard*

The viability of endothelial cells in relation to technique of procurement and preservation or aortic valve allografts is not well established. This is partly because endothelial cells have not been considered in the extensive viability studies of valve allografts (Mohri et al. 1968; Reichenbach et al. 1971; Lockey et al. 1972; Angell et al. 1972; Al-Janabi and Ross 1973; Parker et al. 1978; Henney et al. 1980; van der Kamp et al. 1981).

Living fibroblasts have been used as an index of viability of valve allografts, but the role of endothelial-cell function on valve integrity and valve function has been neglected (Ines et al. 1969). Viability of valve allografts therefore needs to be redefined with endothelial cells in order to reconsider the immunological consequences in clinical practice.

Our research program in orthotopic transplantation of aortic valve allografts therefore deals with endothelial cells and their viability and fate. The search for unfixed cells for viability studies on the rat model led us to the modification of the tissue-stripping technique by flash freezing first described by Warren 1965. This method needs prefixation of the tissue with 4% formaldehyde vapor. Preparation of monolayer endothelial cells (*Häutchen*) was first described by Zahn (1884) and later by others (Poole 1964; Obaze et al. 1968; Pugatch and Saunders 1968; Sinapius 1952; Freudenberg et al. 1983). The endothelial cells obtained were frequently small and fragmentary of already fixed and hence unsuitable for dye tests or immunological investigation.

The initial stage of our study is directed toward answering a basic question regarding viability of endothelial cells: Are the endothelial cells of valve allografts obtained within 40 h postmortem and preserved in nutrient medium viable before transplantation? This report also presents a modified technique for preparing monolayer endothelial cells employed for viability studies of the valve allograft.

Methods and Materials

Aortic valve allograft were dissected from 40 experimental rats (Wistar, BN, and LEW, weighing 300–400 g)

Microsurgical Models in Rats
for Transplantation Research
Eds. A. Thiede et al.

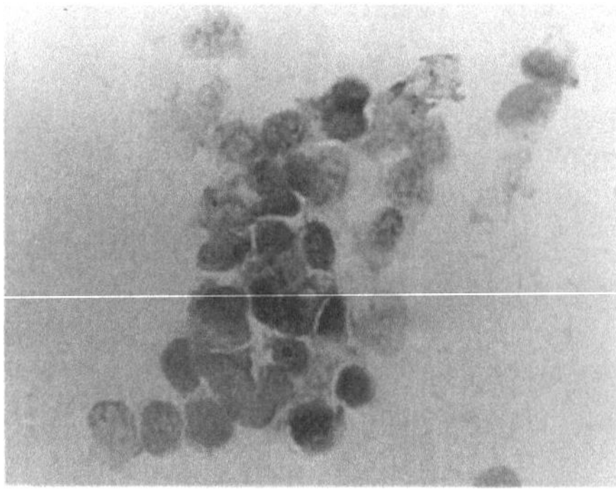

Fig. 1. Monolayer of endothelial cells from aortic valve of a rat obtained 40 h after death. Alcian blue stain, ×1120

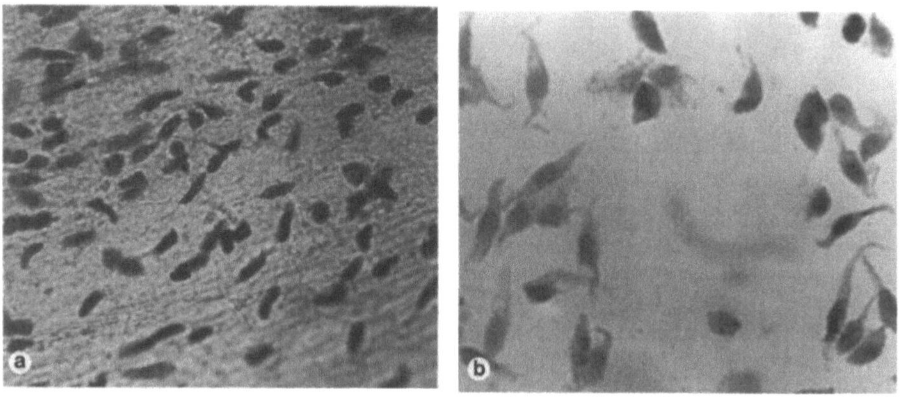

Fig. 2 a, b. Monolayer of fibroblasts from aortic valve of a rat. **a** Hematoxylin and eosin stain, ×700. **b** Alcian blue stain, ×1120

Modified Häutchen *Technique*. The dissected aortic valve was placed between two grease-free glass slides and dipped in liquid nitrogen for 10–12 s. The endothelial peel *Häutchen* remained adherent to the lower glass slide after quick separation of the two glass slides (Fig. 1). The technique can also be applied to an isolated monolayer of fibroblasts from the aortic valve (Fig. 2).

Histology. The *Häutchen* was fixed by drops of absolute alcohol or 4% formalin stained with hematoxylin and eosin.

Viability Dye Test. The *Häutchen* was fixed with 4% glutaraldehyde in 0.05 *M* cacodylate buffer (ph 7.4). Alcian dye was used for the viability test. The staining technique is described elsewhere (Thornthwaite and Leif 1978).

The modified *Häutchen* technique has been employed to assess and compare viability of endothelial cells of aortic valve allografts under different procurement techniques at variable temperatures and storage times. The experimental rats

Table 1. Procurement and preservation technique of rat aortic valve allografts

Group	Immediate postmortem storage temperature	Time interval between death and cold storage at 4 °C	Cold storage time before autopsy
I	4 °C	0	36 h
II	4 °C	0	40 h
III	Room temperature	2 h	36 h
IV	Room temperature	2 h	40 h

were divided into four groups (I–IV), with an additional group serving as control (Table 1).

In groups I and II the rats were stored at 4 °C immediately after death for 36 and 40 h respectively and in groups III and IV the rats were kept at room temperature for 2 h after death and then stored in a refrigerator at 4 °C for 36 and 40 h respectively before autopsy. The valves were dissected and sterilized in antibiotic-nutrient medium and fetal calf serum and stored for 5 and 10 days.

Results

The percentage viability of the ratio of the number of unstained cells to the total cells times 100. After 10 min more than 90% of the viable cells took up the stains. Figure 1 is a light micrograph of endothelial cells before preservation and Fig. 3 shows the same cells after preservation. The viable cells are unstained or slightly stained, while the dead cells are recognized by deep staining. Since alcian blue is an effective stain for glycoproteins (Wardi and Michos 1972), slight staining of the outer membrane does not mean the cell is classified as nonviable. However,

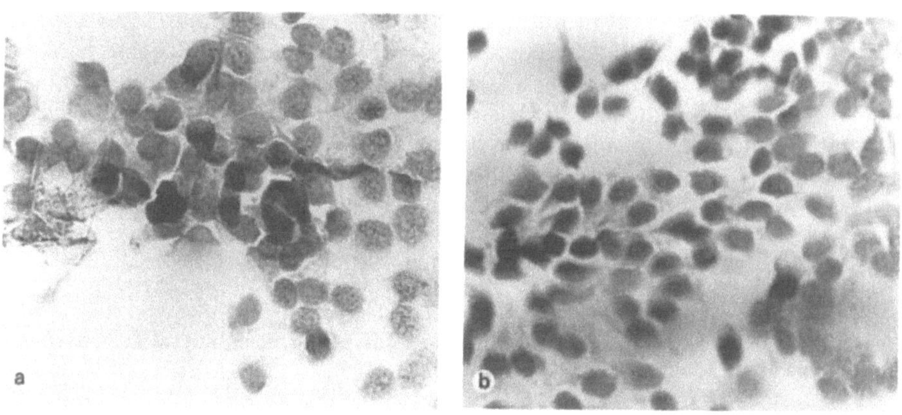

Fig. 3 a, b. Glutaraldehyde-fixed and alcian-blue-stained monolayer of endothelial cells obtained from allograft stored for 10 days in nutrient medium. **a** Slightly tinted and outer-membrane-stained viable cells. **b** Completely stained and nuclear-stained nonviable cells, × 1120

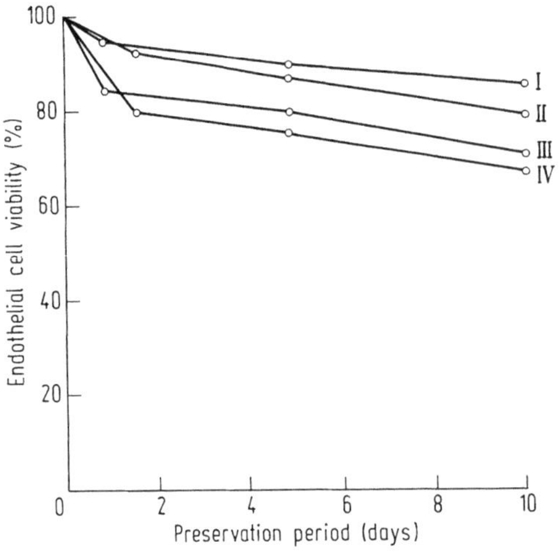

Fig. 4.
Viability of heart valve allografts

if the stain penetrates the cell and stains it completely or penetrates and stains the nucleus, the cell is then classified as nonviable.

The loss of viability in the four groups is shown in Fig. 4. In groups I and II the viability was 93%–95% within 36–40 h and 88%–90% and 80%–86% on the 5th and 10th days of storage at 4 °C respectively; while in groups III and IV the viability was 75%–80% within 36–40 h and 76%–80% and 68%–72% on the 5th and 10th days of storage at 4 °C respectively. The most significant factors causing cell death in allograft valves are technique of procurement, duration, and conditions of storage (Angell et al. 1972). As a result of our observations it would seem that, depending upon the procurement technique, valve allografts lose their viable endothelial cells during the procurement and preservation period by 28%–32% on the 10th day of storage.

On the basis of our findings, our aortic valve allografts are currently stored at 4 °C in antibiotic-nutrient medium for 7–10 days or deep frozen if they are not used within this period. Our method of storage provides an adequate bank of clinically useful fresh valve allografts with more viable endothelial cells. Maintenance of viable endothelial cells has consequently raised the question of tissue typing in orthotopic transplantation of valve allografts. The *Häutchen* preparation, which is a simple technique, is therefore useful and suitable for immunological studies.

References

Al-Janabi N, Ross DN (1973) Entranced viability of fresh aortic homografts stored in nutrient medium. Cardiovasc Res 7:817–822

Angell W, Shumway NE, Kosek JC (1972) A 5-year study of viable aortic valve homografts. J Thorac Cardiovasc Surg 64:329–339

Freudenberg N, Riese KH, Freudenberg MA (1983) The vascular endothelial system. Fischer, Stuttgart

Henney AM, Parker DJ, Davies MJ (1980) Estimation of protein and DNA synthesis in allograft organ cultures as a measure of cell viability. Cardiovasc Res 14:154–160

Innes BJ, Thomson NB, Ayers W (1969) Postmortem changes in endothelial cells of aortic valve homografts. J Thorac Cardiovasc Surg 58:416–423
Lockey E, Al-Janabi N, Gonzalez-Lavin L, Ross DN (1972) A method of sterilizing and preserving fresh allograft heart valves. Thorax 27:398–400
Mohri H, Reichenbach DD, Barnes RW, Merendino KA (1968) Homologous aortic valve transplantation. J Thorac Cardiovasc Surg 56:767–774
Obaze DE, Wright HP (1968) A modified technique for producing "en face" (*Häutchen*) preparations of endothelium for autoradiography. J Atheroscler Res 8:861–863
Parker R, Randev W, Wain HW, Ross DN (1978) Storage of heart valve allografts in glycerol with subsequent antibiotic sterilization. Thorax 33:638–645
Poole JCF (1964) Regeneration of aortic tissues in fabric grafts of the aorta. Symp Zool Soc Lond 11:131
Pugatch EMJ, Saunders AM (1968) A new technique for making *Häutchen* preparations of unfixed aortic endothelium. J Atheroscler Res 8:735–738
Reichenbach DD, Mohri H, Merendino KA (1971) Viability of connective tissue cells following storage of aortic valve leaflets. J Thorac Cardiovasc Surg 62:690–695
Sinapius D (1952) Über das Aortenendothel. Virchows Arch [Pathol Anat] 322:662
Thornthwaite JT, Leif RC (1978) A permanent cell viability assay using alcian blue. Stain Technol 53:199–204
Van der Kamp AWM, Visser WJ, Dongen JM van, Nauta J, Galjaard H (1981) Preservation of aortic heart valves with maintenance of cell viability. J Surg Res 30:47–56
Wardi AH, Michos GA (1972) Alcian blue staining of glycoproteins in acrylamide dis electrophoresis. Anal Biochem 49:607–609
Warren BA (1965) A method for the production of "en face" preparations on cells in thickness. J R Microsc Soc 84:407
Zahn FW (1884) Untersuchung über die Vernarbung von Querrissen der Arterienintima und -media nach vorheriger Umschnürung. Arch Pathol Anat 96:1

The Rat Model as Precursor of Clinical Lymph Vessel Transplantation

R. G. H. Baumeister, J. Seifert, H. Liebich, B. Wiebecke, Ch. Gabka, and *U. Goldmann*

Transplantation of lymph vessels is a new method for causal therapy of secondary and of special forms of primary lymphedema (Baumeister et al. 1981 a). The aim of this new method is the reconstruction of the lymphatic pathway in the case of a local blockage: for example, after lymph node dissection, radiation, and trauma, and also in the case of unilateral atresia of the lymphatic system in the iliac region (Baumeister et al. 1981 b). In this way the diminished transporting capacity in lymphedema (Földi 1977) should be restored or at least improved.

Since lymph vessel transplantation had never been used, all the basic research had to be done before it was used in therapy. Ways of handling the lymph vessels

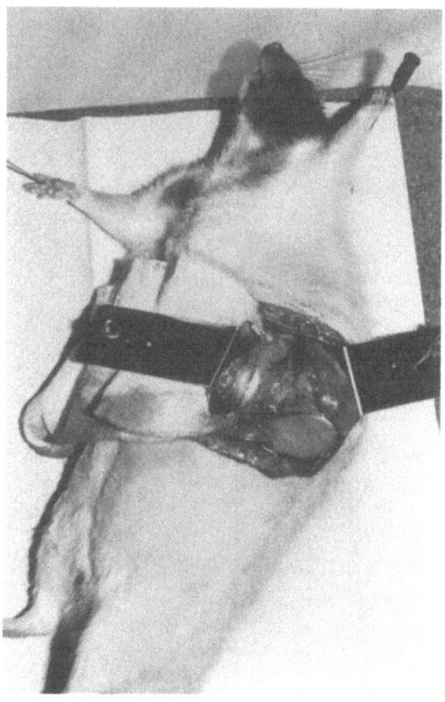

Fig. 1. Exposure of the abdominal thoracic duct in the rat

Microsurgical Models in Rats
for Transplantation Research
Eds. A. Thiede et al.

and of creating safe anastomoses had to be developed. Function after transplantation had to be ensured, behavior after storage had to be checked, and the feasibility of allogeneic transplantation had to be confirmed. These basic investigations have been done in the rat model. A short review of these investigations is given in the rest of this paper.

Material and Methods

Long Evans, Lewis, and BD IX rats were used for the several studies we carried out. The animals were anesthesized initially by ether, followed by chloral hydrate (3.6% solution, 32 mg/100 g bodyweight): As lymph vessels the abdominal thoracic duct between the Cisterna chyli and the diaphragm was studied. For the operation the animals were placed on their right side and the abdomen was opened by a subcostal incision on the left side. The left kidney was shifted to the right side and fixed together with the intestine by retractors. The aorta was freed from surrounding connective tissue, thus exposing the abdominal thoracic duct on the left side (Fig. 1). Polyglactin 910 and Polyamide 6.6, size 10/0–12/0 with a BV 6–8 needle, were used as suture material.

The patency of the lymphatics was proved by intraoperative inspection, dye injection (Patent blue), and histological examination. Function was investigated by injection of the isotope ^{131}I-gammaglobulin. Using scanning and transmission electron microscopy, the ultrastructure of the lymph vessels was studied.

Suture Technique and Suture Material

A so-called tension-free anastomosing technique was developed to handle the fragile lymph vessel wall. The vessels remain in situ during the whole anastomosing procedure. First the corner suture opposite to the surgeon is made. In order

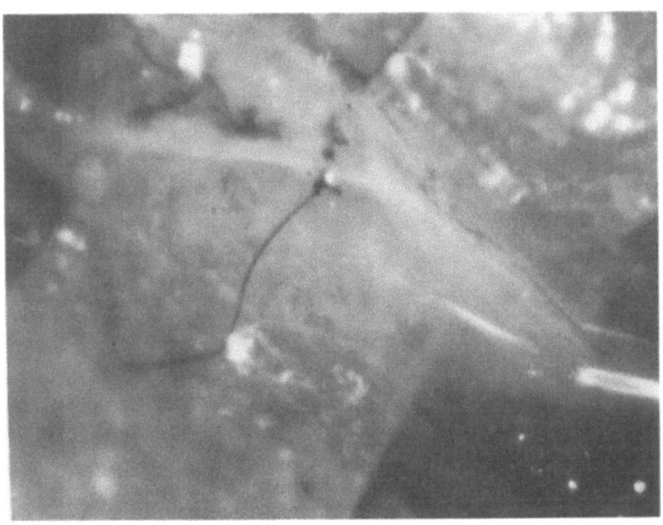

Fig. 2. Tension-free anastomosing technique, backwall finished

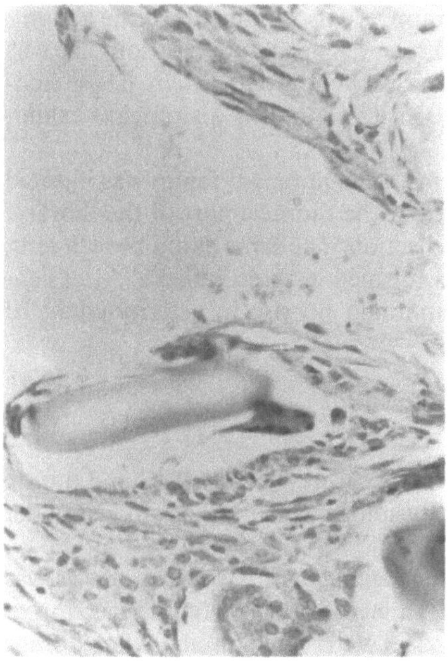

Fig. 3

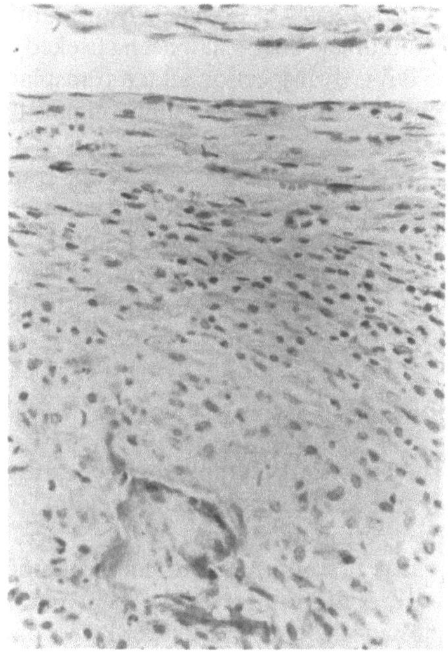

Fig. 4

Fig. 3. Thoracic duct anastomosis, sutured with nonabsorbable suture material after 4 weeks

Fig. 4. Thoracic duct anastomosis sutured with absorbable suture material after 6 weeks

to sew the backwall with single stitches, the wall is lifted only as far as the needle needs space (Fig. 2). Then the front wall can be sewed with single stitches. The patency of lympholymphatic anastomoses of the thoracic duct using this suture technique and the influence of the suture material was proved in 26 anastomoses.

The mean observation time was 65 days. Thirteen anastomoses were created using reabsorbable Polyglactin 910 (group 1) and 13 using nonabsorbable Polyamide 6.6 (group 2) (Figs. 3 and 4).

During reoperation all anastomoses were classified as patent by clinical inspection. Eleven out of 13 in group 1 and 10 out of 13 in group 2 were proved as patent by dye injection. In the histological examinations all anastomoses sewed with reabsorbable material have been patent in contrast to only 10 of 13 anastomoses sewed with nonabsorbable material.

In the histological examinations a heavy foreign body reaction with giant cells surrounded the nonabsorbable suture material. This could be seen during the whole observation time (Fig. 3). With reabsorbable material the foreign body reaction diminished during the time of observation (Fig. 4).

Patency and Function of Lymphatic Grafts

After clarifying the anastomosing technique and the use of the suture material, the problem of the transplantation of lymphatic vessels had to be solved.

The patency of isogeneic transplanted lymph vessels – segments of the abdominal thoracic duct – had been checked in ten rats. The mean observation time was 65 days. By inspection all ten transplants have been judged as patent. These findings were confirmed in nine animals by dye injection and by histological examination.

To estimate the function of the transplants, ^{131}I-gammaglobulin was injected into the cisterna chyli. Activity was counted at the thoracic part of the thoracic duct. Normally activity rises quickly after injection. The same could be seen after thoracic duct transplantation, which was confirmed in three animals. After ligation of the thoracic duct beneath the diaphragm no rise in activity could be seen.

Storage of Lymph Vessels

The behavior of lymph vessels during short storage at 4 ° and 20 °C and during middle- and long-term storage at 4 °C in physiological saline was studied under the electron microscope.

In a first series the influence of storage on segments of lymphatic vessels was studied. Storage for between 30 min and 10 h leads to increasing destruction of the endothelium. The deeper layers of the vessel wall remain intact. After storage for 8 days subendothelial parts of the vessel wall are also partly destroyed. After 30 days storage only the deeper parts of the vessel wall remain free from damage.

In a second series stored lymphatics were transplanted. After transplantation the endothelium and subendothelial tissue were rebuilt. In the first days the inner surface of the vessel wall was built by a network of fibrin. Between days 8 and 40 after transplantation the restoration of the endothelium could be seen.

These investigations showed an increasing loss of endothelium and subendothelial tissue with the duration of storage. Nevertheless the damage can be reversed by restoration processes if the lymphatic segments are transplanted. This does not affect the high patency rate of transplants. Similar patency rates were detected after isogeneic lymph vessel transplantation in rats. Therefore first experimental hints for possible allogeneic transplantation with storage of the transplants for some time were able to be obtained.

In the rat model so far not only have the first experimental steps been taken to develop clinical lymph vessel transplantation, which has already been applied to more than 25 patients with lymph edema, but at the same time new developments in lymphatic surgery have also been elaborated in this model.

References

Baumeister RGH, Seifert J, Hahn D (1981 a) Autotransplantation of lymphatic vessels. Lancet 1:147

Baumeister RGH, Seifert J, Wiebecke B, Hahn D (1981 b) Experimental basis and first application of clinical lymph vessel transplantation of secondary lymphedema. World J Surg 5:401

Földi M (1977) Physiology and pathophysiology of lymph flow. In: Clodius L (ed) Lymphedema. Thieme, Stuttgart, p 1

Microstructure of Allogeneic Lymph Vessel Transplants in the Rat

Ch. Gabka, R. G. Baumeister, H. G. Liebich, J. Seifert, and *W. Brendel*

Lymph vessel transplants are now used in the therapy of secondary lymphedema (Baumeister et al. 1981). Since the conception of bridging local lymph vessel defects by autogeneic grafts, Baumeister (1983) has performed 21 lymph vessel transplantations to date, mostly in women suffering from secondary lymphedema after mastectomy and postoperative irradiation. During on observation period of 5 months, an average reduction of 60% of the lymphedema was seen.

The application of allogeneic transplants could expand the clinical use of lymph vessel transplantations: for example, avoiding the need to prepare the donor site in the patient; shortening the operation time; applying the method to patients suffering from primary lymphedema of the lower extremities. For this reason and on account of rejection against allogeneic tissue, allogeneic lymph vessel transplantations have been performed in rats, and the function and microstructure of the grafts have been observed.

Under the operating microscope, segments of rat abdominal thoracic ducts were prepared between the cisterna chyli and the diaphragma. These grafts were allogeneically transplanted from the inbred rat strain DB 9 into the inbred rat strain Long Evans. (The latter reject skin allografts within 14 days on average.) In addition, syngeneic transplantations were carried out (Long Evans) as a control group. A total of 28 allogeneic transplantations were performed. The observation period extended from 2 to 80 days after operation. Examination and sampling of the grafts for further investigations took place as in Table 1.

Following the operation period the function of the transplants was evaluated by injection of blue dye (patent blue). Of the 28 transplants 25 were patent (Table 2). These results were very similar to those for the syngeneic transplantations. However, under the operating microscope no signs of rejections were observed at the vessel transplants during the entire observation period.

Table 1. Postoperative observation times

$n=$	2	2	2	2	2	3	5	5	5	
	2	4	6	8	10	15	40	60	80	days

Microsurgical Models in Rats
for Transplantation Research
Eds. A. Thiede et al.

Table 2. Patency of allogeneic lymph vessel transplants in the rat

Days after operation	*n*	Patent
2	2	2
4	2	2
6	2	2
8	2	2
10	2	2
15	3	2
40	5	5
60	5	4
80	5	4
Total	28	25

In order to judge the microstructure of the grafts, electron microscopic investigations were performed: 2 days after the operation the grafts still showed a normal formation of the wall: Under the entire endothelium a frame of collagen fibers (type I) forms the vessel wall, exposed in lateral and longitudinal axes. In between some fibrocytes may be observed. In the period from 4 to 6 days after transplantation alterations of the grafts took place that led to a partial loss of the

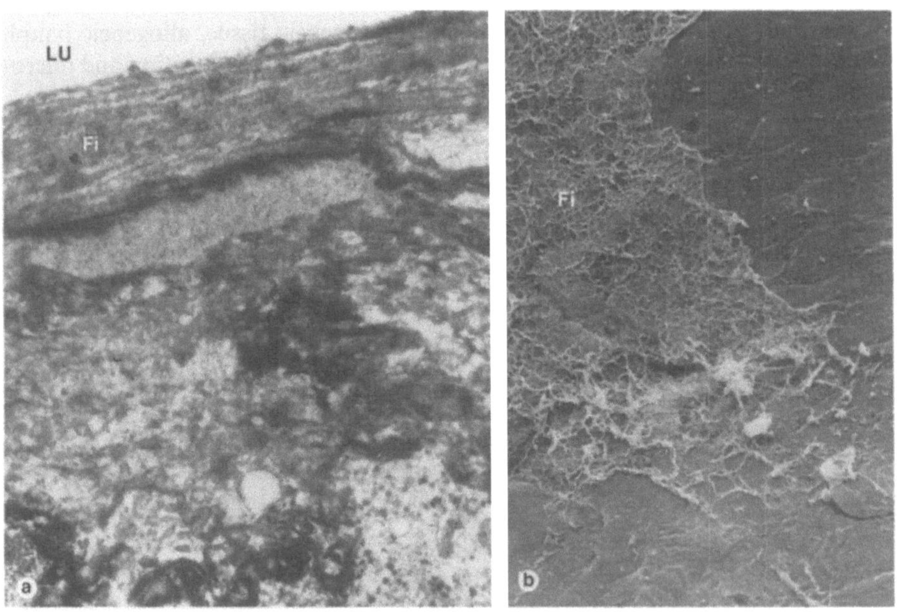

Fig. 1 a, b. Allogeneic lymph vessel transplant, 8 days after transplantation. **a** Transmission electron micrograph: After the loss of endothelium the surface of the graft is now covered by a layer of fibrin (*Fi*). *LU*, lumen. **b** Scanning electron micrograph: same situation as in **a**

endothelium 8 days after the operation (Fig. 1). The surface of the graft is now covered by a thick layer of fibrin. In addition the frame of subendothelial collagen fibers only shows light disintegration. Nevertheless, a complete replacement of the endothelium could be observed as soon as within 15 days after the operation. Not until 40–60 days after the operation did the subendothelial layer of collagen fibrils show a distinct alteration, i.e., loosening of their former shape and forming of small intramural cavities. Moreover, local infiltrations of the lymphocytes and plasma cells could be seen in the graft wall. Nevertheless, the function of the transplants was maintained (Table 1). This was the result of vessel reconstructions, probably directed by the host and taking place parallel to the graft alterations. Thus, 80 days after operation the grafts showed a more or less normal structure of the wall, expressing an entire endothelium and ordered subendothelial collagen fibrils (Fig. 2). An additional sign of the regeneration process was the observation of fibroblasts in the graft wall.

In conclusion, allogeneic lymph vessel transplants are subject to rejection, first leading to a loss of endothelium and later to alterations of the graft wall. Simultaneously, however, attempts at reconstruction of the affected structures can be observed. There is replacement of the endothelium within 15 days and, moreover, the appearance of fibroblasts points to stabilization and regeneration of the collagen fibers. Obviously the host itself uses the transplant as a structural guide for regeneration of the vessel wall. For this reason and because of the special condition of the lymph system, i.e., a low intravasal pressure, a low potential of clot-

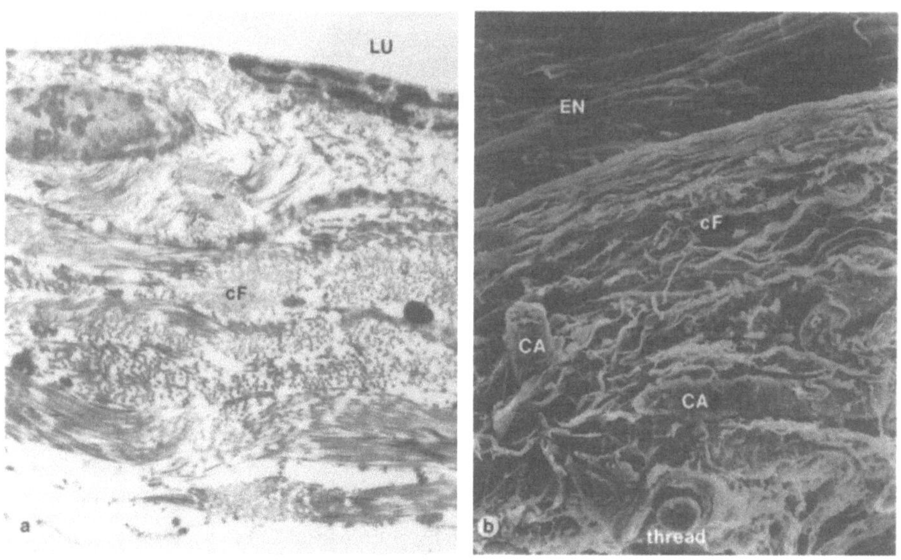

Fig. 2 a, b. Allogeneic lymph vessel transplant, 80 days after transplantation. **a** Transmission electron micrograph: Under an entire endothelium (*EN*) a frame of collagen fibers (*cF*) guarantees the stability of the graft wall. *Fi*, fibroblast; *CA*, capillary. **b** Scanning electron micrograph: same situation as in **a**

ting, and its quality as a bradytrophic tissue, the observed reactions will not lead to a loss of function of the grafts. Thus, these investigations in rats could represent the first step for the clinical use of allogeneic lymph vessel transplants.

References

Baumeister RG (1983) Microchirurgie des Lymphgefäßsystems. Chirurg 54:374

Baumeister RG, Seifert J, Wiebecke B, Hahn D (1981) Experimental basis and first application of clinical lymph vessel transplantation of secondary lymphedema. World J Surg 5:401

Immunogenetic Basis of Heart Transplantation *

J. Klempnauer, B. Steiniger, K. Wonigeit, and *E. Günther*

Introduction

There is a considerable need for heart transplantation. The widespread application of heart grafting, however, is limited not only by the shortage of appropriate donor organs but mainly by the destruction of the transplanted heart by alloimmune reactions. There are two principal clinical strategies to prevent the immune destruction of an organ allograft: general immunosuppressive of the recipient and reduction of graft immunogenicity prior to transplantation by prospective tissue typing. A beneficial effect of prospective HLA typing especially for the D/DR locus has been demonstrated only in clinical renal transplantation. Little is known about the effect of various HLA antigens in rejection of heart allografts. In order to determine the effect of various classes of histocompatibility antigens in the immune destruction of heart allografts heterotopic accessory heart transplantation was performed in congenic rat strains representing different classes of histocompatibility barriers. By using the appropriate donor recipient strain combinations it was possible to analyze the relative contribution of class I and class II major histocompatibility complex (MHC) alloantigens as well as of non-MHC histoincompatibilities. The data presented in this study may be helpful for a better understanding of the immunogenetic basis of heart transplantation.

Materials and Methods

Organization of the Rat MHC. The MHC of the rat is designated as the RT1 system and consists of several closely linked gene regions which have been identified on the basis of genetic recombination within the MHC. Class I histocompatibility antigens are encoded by the RT1.*A* region; class II antigens are encoded by the RT1.*B* region. The RT1.*C* region represents a further class I locus which resembles the *Qa* system of the mouse. All histocompatibility antigens not encoded within the MHC are summarized as non-MHC alloantigens (Günther, this volume, p. 83).

* This work was supported by Stiftung Volkswagenwerk AZ I/38 510

Microsurgical Models in Rats
for Transplantation Research
Eds. A. Thiede et al.

Table 1. Effect of MHC and non-MHC alloantigens in rejection of allogenic accessory heart grafts

Donor	RT1	Recipient	RT1	Histocompatibility barrier	Rejection times (days)
LEW.1A	*a*	LEW	*l*	MHC	8, 8, 8, 8, 9, 9
AVN	*a*	LEW.1A	*a*	non-MHC	100 × 6
AVN	*a*	LEW	*l*	MHC plus non-MHC	7, 7, 7, 10, 10, 10

Animals. The relative contribution of MHC and non-MHC alloantigens to rejection of heart grafts was investigated using the rat strains AVN, LEW, and LEW.1A. LEW.1A rats express the MHC of AVN on the genetic background of LEW rats. In appropriate strain combinations donor and recipient differed either at MHC or non-MHC histocompatibility antigens or both (Table 1). The different effect of class I and class II MHC histocompatibility antigens in heart allograft rejection was assessed in congenic and intra-MHC recombinant rat strains which differed only at the RT1.*A*, RT1.*B*, or RT1.*C* regions. For this purpose the congenic strains LEW.1A and LEW.1U and their recombinant strains LEW.1AR2, LEW.1WR1, and LEW.1WR2 were used in this study.

Technique of Heart Transplantation and Definition of Rejection. Accessory abdominal heart transplantation was performed according to Ono and Lindsay (1969). Graft function was determined by daily external palpation. Rejection was defined by cessation of palpable heart beats. The functional demonstration of rejection was always confirmed by histological examination.

Results

Effect of MHC and Non-MHC Alloantigens. Table 1 shows the relative contribution of MHC and non-MHC alloantigens to rejection of heart allografts. MHC-incompatible heart grafts from LEW.1A donors were rapidly rejected by LEW recipients within 9 days. Additional incompatibility at non-MHC alloantigens did not significantly alter the rejection times of AVN heart grafts in LEW recipients. Incompatibility at non-MHC antigens in the strain combination AVN→LEW.1A failed to induce rejection within the observation period of 100 days. The grafts retained strong pulsations and were macroscopically and histologically in distinguishable from isografts performed in LEW rats (*n*, 10).

Effect of Class I and Class II MHC Alloantigens. Fully MHC-incompatible heart grafts in the strain combination LEW1A→LEW.1U (*n*, 6) were rejected within 6–8 days after transplantation. The individual MHC regions differed markedly in their capacity to induce rejection of heart allografts. RT1.*A* disparate heart allografts from LEW.1A donors were not rejected by LEW.1WR2 recipients (*n*, 6) within 100 days. They retained strong pulsations and were histologically no different from LEW isografts. Incompatibility at the RT1.*B* region in the strain combination LEW.1WR2→LEW.1WR1 (*n*, 6) induced rapid rejection after a mean of 10 ± 0.5 days. A cumulative effect of the RT1.*A* and RT1.*B* regions was noted in the strain combination LEW.1A→LEW.1WR1 (*n*, 6), where all heart grafts were rejected after 7 days. Incompatibility at the RT1.*C* region in the strain combination LEW.1A→LEW.1AR2 (*n*, 6) failed to induce rejection of heart allografts which retained strong pulsations throughout the observation period of 100 days. Histologically focal subendocardial cellular infiltrates were demonstrated in all RT1.*C* region disparate grafts.

Discussion

Histoincompatibility between donor and recipient tissue forms the basis of allograft rejection. The effect of various classes of histocompatibility antigens in the rejection of heart allografts was studied in a model of accessory abdominal heart transplantation in genetically well-defined rat strains.

Accessory abdominal heart transplantation is an artificial but nevertheless useful and reliable method for immunological investigations on heart allograft rejection. In contrast to orthotopic transplantation heterotopically transplanted heart grafts have no life-supporting mechanical function. Rejection times of abdominal heart allografts were defined by complete cessation of palpable heart beats. It is therefore conceivable that the survival times after orthotopic heart transplantation might be generally shorter, since graft recipients will die earlier in the course of an alloimmune reaction from progressive heart failure. In all instances the functional demonstration of rejection was confirmed by histological examination of the rejected organ. No technical failures were encountered in this series of experiments.

The genetic isolation of specific histocompatibility antigens proved possible by the production of congenic and recombinant rat strains. The availability of such strains facilitated an assessment of the effect of different classes of histocompatibility antigens in heart allograft rejection by transplantation between rat strains differing exclusively at the respective antigen. Furthermore, genetic interactions could be investigated by comparison of the effect of combined histocompatibility differences to the effect of the respective individual gene regions.

The relative role of MHC and non-MHC histocompatibility genes was studied in strain combinations differing either at MHC or non-MHC antigens or both. Our results confirm the predominant effect of MHC antigens in rejection of heart allografts (Bildsoe 1972). All MHC-incompatible heart grafts were acutely rejected. In sharp contrast, heart allografts differing at non-MHC alloantigens only were not rejected within an observation period of 100 days. In combination with MHC alloantigens they did not produce a significant acceleration of rejection. The failure of non-MHC alloantigens to provoke rejection of heart allografts could be demonstrated in a total of six different non-MHC disparate donor recipient strain combinations (Klempnauer et al., in preparation). It is therefore both interesting and surprising that Katz et al. (1983) also reported regular rejection of heart allografts within 18 days in six different strain combinations when transplanted across non-MHC histocompatibility barriers. Furthermore, we were able to demonstrate rapid rejection of vascularized pancreas allografts (Klempnauer et al. 1983 a) when transplanted across the same non-MHC genetic barriers where heart grafts are not rejected.

These results demonstrate that non-MHC histocompatibility antigens undoubtedly have a great potential for inducing rejection of organ allografts. Unlike MHC antigens non-MHC histocompatibility antigens are still poorly defined. The remarkable differences in the capacity of non-MHC alloantigens to induce rejection of heart allografts may reflect the mere number and the cumulative effect of disparate individual non-MHC gene loci. It is conceivable that the non-MHC disparate strain combinations used by ourselves which are based on the *a*, *l*, and *u* haplotypes of the RT1 system differ in a smaller number of non-MHC genes than the strain combinations used by Katz et al. (1983) which are based on the *c* and *n* haplotypes. The remarkable organ-specific differences of the effect of non-MHC alloantigens in that using identical strain combinations pancreas but not heart allografts are rejected (Klempnauer et al. 1984) could be explained in terms of a different tissue distribution of the relevant non-MHC antigens on these organs. There may be functionally significant non-MHC alloantigens present on the pancreas but not on the heart which induce strong rejection responses. Alternatively the effect may be explained by strong suppressor-cell activity induced by certain non-MHC alloantigens present on the heart but not on the pancreas. Finally it cannot be excluded that the effect of non-MHC alloantigens may be influenced by the responder status of the graft recipient to the presented relevant non-MHC alloantigens.

Incompatibility for the MHC led to acute rejection of all heart allografts examined. The individual MHC regions differed markedly in their capacity to induce rejection. Class II histocompatibility antigens encoded by the RT1.*B* region were the single most important stimulus for heart graft rejection. RT1.*B*-region-incompatible grafts were rapidly rejected. On the contrary, class I alloantigens encoded by the RT1.*A* or RT1.*C* regions were not capable of inducing rejection on their own. Combined incompatibility at the RT1.*A* and RT1.*B* regions had a cumulative effect on heart graft rejection. No effect of additional incompatibility at the RT1.*C* region could be demonstrated.

The failure of RT1.*A*-region incompatibility to induce rejection of heart allografts cannot be explained by a lack of expression of RT1.*A* antigens on rat heart, since a cumulative effect of the RT1.*A* and RT1.*B* regions on heart allograft rejection is only possible when both antigens are in fact expressed on the transplanted heart. In RT1.A^c homozygotes an *Ir*-gene-mediated low responsiveness has been demonstrated for RT1.A^a grafts (Butcher and Howard 1982). A similar phenomenon may also explain the permanent acceptance of RT1.A^a-disparate heart grafts by RT1.A^u graft recipients observed in this study. *Ir*-gene-mediated differences in responsiveness against allografts would also be compatible with the findings of Rozing et al. (1983), who observed rejection of heart allografts across RT1.*A* gene region differences using both natural recombinants and congenic PVG rat strains. However, it is difficult to explain by an *Ir* gene effect that in identical strain combinations RT1.A^a-disparate pancreas (Klempnauer et al. 1983 b, this volume, p. 165) and kidney (Paris and Günther 1980) grafts were in fact rapidly rejected by RT1.A^u recipients.

No in vivo effect of RT1.*C* region gene products could be demonstrated in heart allograft rejection. However, the histological findings of subendocardial cellular infiltrates and the demonstration of alloantibodies directed against RT1.*C* antigens (unpublished observations) imply that the relevant antigens are in fact expressed on the transplanted heart. The significance of the RT1.*C* region in heart allotransplantation remains to be established. Class II histocompatibility antigens encoded by the RT1.*B* region were the only antigens to induce heart graft rejection on their own. Disparity at the RT1.*B* region caused rapid rejection of all heart allografts transplanted. This is in contrast to the permanent graft acceptance following transplantation across class I incompatibilities at the RT1.*A* or RT1.*C* region. A stimulus from class II alloantigens proved necessary for the induction of heart graft rejection in the strain combinations examined. This finding may have important implications in view of the passenger-cell concept of allograft rejection. Removal or inactivation of bone-marrow-derived interstitial cells carrying *Ia* or class II alloantigens has been shown to reduce graft immunogenicity and prolong allograft survival (Batchelor et al. 1979; Faustman et al. 1981). The results of experiments concerned with depletion of passenger cells, however, largely depend on the strain combination used and in most instances the degree of histoincompatibility was poorly defined. Organ transplantation in the congenic and recombinant rat strains used in this study appears to be a particularly suitable model for an in vivo study of the significance of passenger cells in allograft rejection, since class II disparities are mandatory for rejection and the effect of alterations of the passenger-cell population can be tested without interference from class I antigens.

In summary, the relative contribution of class I and class II MHC and non-MHC histocompatibility antigens in heart transplantation was studied using congenic and recombinant rat strains. A predominant effect of class II antigens could be demonstrated in contrast to the permanent survival of grafts disparate at class I antigens. Since interstitial dendritic cells are the major source of *Ia* or class II determinants in transplanted heart grafts (Steiniger et al. 1984), removal or inactivation of these cells may be especially effective in reducing heart allograft immunogenicity.

Acknowledgment. We gratefully appreciate the excellent technical assistance of Margot Giesecke and Petra Peschel.

References

Batchelor JR, Welsh KE, Maynard A, Burgos H (1979) Failure of long surviving, passively enhanced kidney allografts to provoke T-dependent alloimmunity. J Exp Med 150:455–464

Bildsoe P (1972) Organ transplantation in the rat. Acta Pathol Microbiol Scand [B] 80:221

Butcher GW, Howard JC (1982) Genetic control of transplant rejection. Transplantation 34:161–166

Faustman D, Hauptfeld V, Lacy P, Davie J (1981) Prolongation of murine islet allograft survival by pretreatment of islets with antibody directed to Ia determinants. Proc Natl Acad Sci USA 78:5156–5159

Katz SM, Liebert M, Gill TJ, Kunz HW, Cramer DV, Guttmann RD (1983) The relative roles of MHC and non-MHC genes in heart and skin allograft survival. Transplantation 36:96–101

Klempnauer J, Wonigeit K, Günther E, Pichlmayr R (1983 a) Pancreas whole organ transplantation in the rat. Evidence for a strong effect of non-MHC incompatibilities. Transplant Proc 15:1649–1651

Klempnauer J, Wonigeit K, Steiniger B, Günther E, Pichlmayr R (1983 b) Pancreas whole organ transplantation in the rat. Differential effect of MHC regions. Transplant Proc 15:1308–1310

Klempnauer J, Hoins L, Steiniger B, Günther E, Wonigeit K, Pichlmayr R (1984) Evidence for a differential importance of MHC and non-MHC alloantigens in pancreas and heart transplantation in the rat. Transplant Proc 16:778–780

Ono K, Lindsay ES (1969) Improved technique of heart transplantation in rats. J Thorac Cardiovasc Surg 57:225

Paris A, Günther E (1980) Kidney grafting between rats which carry recombinant major histocompatibility haplotypes. Immunogenetics 10:205–209

Rozing J, Bonthuis F, Joling P, Vaessen LMB, Memeijer LDF (1983) The influence of RT1 subregion differences on cardiac allograft survival. Transplant Proc 15:1647–1648

Steiniger B, Klempnauer J, Wonigeit K (1984) Phenotype and histological distribution of interstitial dendritic cells in the rat pancreas, liver, heart and kidney. Transplantation 38(2):169–175

Cyclosporin A in Heart Allografts in Rats

W. Konertz and *A. Bernhard*

Temporary treatment of heart-grafted rats with cyclosporin A (CyA) has resulted in indefinite allograft survival in certain combinations (Kavahara et al. 1980; Nagao et al. 1982). However, in the BN-LEW and DA-LEW combination we were unable to obtain graft acceptance after a course of 15 mg CyA p.o. given 14 days postoperatively. Liver grafts in these combinations and with identical treatment showed tolerance (Engemann et al. 1982; Konertz et al. 1984). This led us to change the treatment protocol stepwise and alter the dose of CyA and the route of administration, to apply donor and recipient pretreatment, and to use combinations with acetylsalicylate (ASA), which has shown its immunosuppressive effect in previous studies (Jamieson et al. 1979; Konertz et al. 1980).

Methods

Rats were obtained from commercial sources. Male BN ($RT1^{n}$) and DA ($RT1^{avl}$) were used as donors and male Lewis ($RT1^{l}$) as recipients. This strain combination represents a severe mismatch. Heterotopic abdominal heart transplantation was performed with standard microvascular techniques. Graft function was assessed by daily palpation. Cessation of palpable heart beat was defined as rejection. Medications were administered once daily according to Table 1. Cyclosporin A was dissolved in olive oil, 10 mg/ml, and given p.o. under light ether anesthesia via a small orograstric tube or deep i.m. Donor pretreatment was performed by injecting a single dose of 20 mg/kg CyA i.v. 2 h preoperatively. The grafts from these donors were additionally flushed with 10 ml Ringer's solution containing 50 mg/liter CyA, Acetylsalicylate was administered subcutaneously.

Results (Table 1)

Syngeneic control animals survived for more than 500 days. Allogeneic controls with no treatment rejected their grafts uniformly with a mean survival time of 6.2 ± 0.7 days in the BN-LEW animals and 6.1 ± 0.7 days in the DA-LEW combination. CyA, 15 mg/kg, given orally for 14 days resulted in rejection of all grafts within 20 days after discontinuation of therapy. The i.m. administration of

Microsurgical Models in Rats
for Transplantation Research
Eds. A. Thiede et al.

Table 1. Median survival times for eight groups of cyclosporin A-treated rats, BN and DA to LEW combinations

Rat strain combination	Medication: Dose	Medication: Time schedule (days)	Median survival time (days ± SD)
LEW-LEW ($n=6$)	None		>500
BN-LEW ($n=10$)	None		6.2 ± 0.7
DA-LEW ($n=10$)	None		6.1 ± 0.7
BN-LEW ($n=5$)	CyA 15 mg/kg orally	1–14	27.2 ± 3.4
DA-LEW ($n=7$)			28.0 ± 3.3
BN-LEW ($n=5$)	CyA 10 mg/kg i.m.	0, 1, 3, 5, 10	20.2 ± 7.5
BN-LEW ($n=11$)	CyA 10 mg/kg i.m.	0, 1, 2, 3, 4, 5, 6, 7	31.1 ± 3.7
BN-LEW ($n=8$)	CyA 10 mg/kg i.m.	0, 1, 2, 3, 4, 6, 8, 11, 14	57.4 ± 4.2
BN-LEW ($n=8$)	CyA 10 mg/kg i.m.	−1, 0, 1, 2, 3, 4, 6, 8, 11, 14	65.2 ± 6.4
BN-LEW ($n=7$)	CyA + donor pretreatment	0, 1, 2, 3, 4, 6, 8, 11, 14	82.6 ± 7.3
BN-LEW ($n=8$)	CyA + acetylsalicylate	0, 1, 2, 3, 4, 6, 8, 11, 14	>120

10 mg/kg given on postoperative days 0, 1, 3, 5, and 10 prolonged graft survival only moderately. After a 7-day course of 10 mg/kg CyA graft survival increased to 32 days. In this series signs of cyclosporin A toxicity were evident. A schedule for longer administration of a reduced rate resulted in a mean survival time of 57 days. Only slight improvement was obtained by recipient pretreatment beginning with CyA on the day before operation. On the contrary donor pretreatment prolonged graft survival to 82 days. The combination of CyA and acetylsalicylate led to survival for more than 120 days. Histology of a graft at day 120 showed no signs of rejection (Fig. 1).

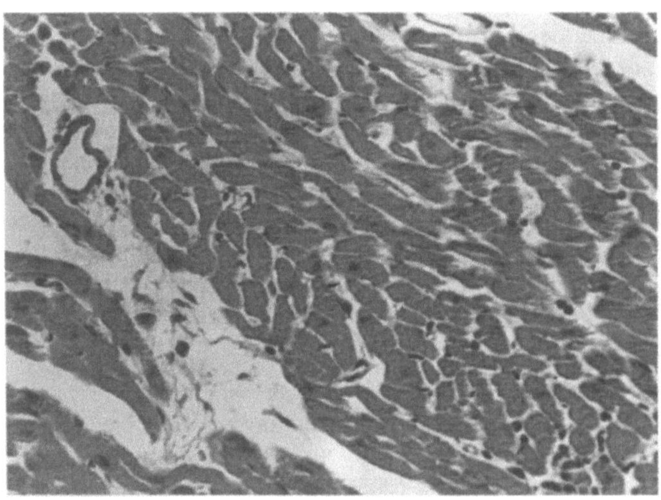

Fig. 1. Histology of a heart after temporary treatment with CyA and ASA 120 days postoperatively

Discussion

Cyclosporin A has a strong immunosuppressive effect on cardiac allografts in severely mismatched rats. However, in the BN-LEW and DA-LEW systems we were not able to obtain indefinite graft survival using the CyA schedules reported by Kavahara et al. (1980) or Nagao et al. (1982). CyA, 10 mg/kg i.m. given daily for a week, resulted in illness from CyA intoxication in nearly all LEW rats. This is in accordance with the findings of Leimenstoll et al. (this volume), who demonstrated a very high cumulation of CyA after daily, i.m. administration. Our strategy of gradually decreasing CyA after an initial treatment of only 5 days resulted in superior graft survival and less side effects from medication. Surprising is that donor and graft pretreatment with CyA led to improved survival, a finding that had been reported from kidney transplantation by Rucker et al. (1982). Indefinite graft survival, however, was only obtained by the combination of CyA and acetylsalicylate. Salicylates have a strong immunosuppressive effect which has been shown by Jamieson et al. (1979) and very recently by Shaw (1983). In our laboratory, graft survival in BN-grafted LEW recipients increased from 6 to 21 days with a daily dose of 200 mg/kg acetylsalicylate (Konertz et al. 1980).

The combination of ASA and CyA might have clinical impact by substituting steroids. The mode of action, however, is still unclear and might range from nonspecific additive possibly prostaglandin-mediated effects to a direct vascular mode of action or competition between both drugs on plasma-protein-binding sites.

References

Engemann R, Ullrichs K, Thiede A, et al. (1982) The value of a physiological liver transplantation model in rats: induction of specific graft tolerance in fully allogeneic strain combination. Transplantation 33:566

Jamieson SW, Burton NA, Reitz BR, Stinson EB (1979) Survival of heart allografts in rats treated with azathroprine and sodium salicylate. Lancet 1:130–131

Kavahara K, Sutherland DER, Rynasiewicz JJ, Najarian JS (1980) Prolongation of heterotopic cardiac allografts in rats by CYA. Surgery 88:594–600

Konertz W, Beck C, Kielholz J, et al. (1980) Prolonged survival of cardiac allografts treated with Acetylsalicylate and Dipyridamole. In: Lambotte (ed) 13th ESSR Congress, Abstr no 404. Karger, Basel

Konertz W, Engemann R, Deltz E, et al. (1984) Organ related differences in graft survival after temporary treatment with Cyclosporin A. Heart Transplant (in press)

Nagao T, White DJG, Calne RY (1982) Kinetics of unresponsiveness induced by a short course of CYA. Transplantation 33:31–35

Rucker J, Toledo Perayra LH, McKenzie GH, Gorden DA (1982) Improvement of kidney transplant survival after graft pretreatment with Cyclosporin A. Transplantation 34:356–359

Shaw JFL (1983) Prolongation of rat cardiac allograft survival by treatment with Prostacyclin or Aspirin during acute rejection. Transplantation 35:526–529

Monitoring of Rejection and Absorption Studies After Small-Bowel Transplantation in the Dog Model

S. Nordgren, Z. Cohen, G. R. Greenberg, and *B. Langer*

Survival following total small-bowel allotransplantation in the dog is greatly prolonged with the use of cyclosporin (Craddock et al. 1983). However, allograft rejection remains a major problem. Therefore, before clinical intestinal transplantation can be undertaken, reliable methods for monitoring graft integrity must be developed. This is important for early therapeutic intervention and host survival.

The aim of the present investigation was to study the intestinal allograft using histological and physiological modalities. The specific objective was to correlate histological findings in biopsies from the transplant with the absorptive capacity of the graft in order to develop a physiological marker of rejection.

Method

Animals and Operative Technique

Mongrel dogs were used. Total small-intestinal autotransplantation was performed in 4 dogs and allotransplantation in 15 dogs. The small bowel, from the ligament of Treitz to the ileocecal valve, was resected on its vascular pedicle. The vascular bed of the graft was perfusd with 1000 ml Ringer's lactate at 4 °C and stored for 60 min. The small bowel was resected from the recipient. Vascular anastomoses were performed between the superior mesenteric artery and the abdominal aorta, and between the superior mesenteric vein and the inferior vena cava. A short segment of the proximal and the distal end of the graft was isolated and closed at one end. The open end was brought through the abdominal wall as a mucous fistula. End-to-end anastomoses were then performed between the recipient duodenum and the proximal end of the graft, and between the recipient terminal ileum and the distal end of the graft. With this technique, the proximal and the distal ends of the graft were accessible from the skin despite the orthotopic position of the major part of the graft.

Immunosuppression

All allotransplanted dogs received cyclosporin, 7.5 mg/kg per day for 14 days. At that time the immunosuppression was discontinued and the dogs were allowed to reject the allograft.

Microsurgical Models in Rats
for Transplantation Research
Eds. A. Thiede et al.

Histological Assessment

Biopsies were obtained from the distal pouch daily from day 14 in allotransplanted dogs and weekly from autotransplanted dogs. The biopsies were fixed in buffered formalin and stained routinely with hematoxylin and eosin. The diagnosis of rejection was based on the demonstration of blunted intestinal villi and the occurrence of an increased inflammatory infiltrate in the lamina propria and the submucosa (Lossing et al. 1982).

Absorption of Cyclosporin

Autotransplanted dogs were tested weekly for 3 weeks. Allotransplanted dogs were tested weekly from day 14. A test dose of cyclosporin in olive oil, 25 mg/kg, was given p.o. to the dogs. Blood samples were drawn at 0, 6, and 24 h after administration of cyclosporin. Plasma was separated and cyclosporin levels were determined using a radioimmunoassay technique (RIA), as described by Sandoz Limited, Basel, Switzerland.

Glucose Uptake

Two-microcurie [^{14}C]glucose (13 µmol/liter) in isotonic saline was instilled into the jejunal pouch. Blood samples were drawn at 0, 5, 10, 15, 30, and 60 min after instillation of the isotope. Plasma radioactivity was determined using a liquid scintillation counter. The test was performed at 3-weekly intervals in the autotransplanted dogs and weekly from day 14 in surviving allotransplanted dogs.

Results

Survival

All four autotransplanted dogs recovered well from the surgical procedure and have remained well for 30–60 weeks. Six allotransplanted dogs died at or before day 6 following transplantation. These dogs were considered technical failures. Three dogs died during ongoing cyclosporin treatment of an ill-defined illness but with normal graft mucosa. The above-mentioned nine dogs were excluded from the study. The remaining six dogs all survived past day 14, when the immunosuppression was discontinued. An additional survival of 12 ± 4.5 days following day 14 was noted in these six dogs, all of which died of rejection.

Absorption of Cyclosporin (Fig. 1)

Plasma cyclosporin levels of 634 ± 126 µg/liter (mean ± SE) were reached in the autotransplanted dogs 6 h following administration of the drug. At 24 h after administration, plasma cyclosporin was 91.8 ± 19.5 µg/liter. In the allotransplanted dogs with normal graft histology, the plasma cyclosporine level 6 h following the oral administration of the drug was 610 ± 81 µg/liter over trough level. Twenty-four hours following the oral dose, the absolute plasma cyclosporin level was 118 ± 76 µg/liter. In allotransplanted dogs undergoing rejection, as diagnosed from simultaneously obtained biopsies from the terminal ileum, the increase in plasma cyclosporin 6 h following the oral dose was 276 ± 107 µg/liter over trough

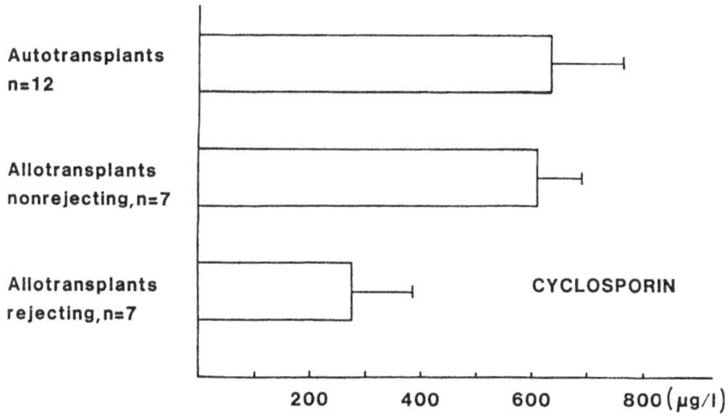

Fig. 1. Increase in cyclosporin concentration 6 h following an oral dose (25 mg/kg in olive oil). Note the significant decrease in dogs undergoing rejection

level. Twenty-four hours after the oral dose in dogs undergoing rejection, absolute plasma cyclosporin was 86 ± 24 µg/liter. The difference between plasma cyclosporin levels at 6 h in dogs with normal graft mucosa and in dogs undergoing rejection is statistically significant ($P < 0.001$).

Uptake of Glucose (Fig. 2)

In autotransplanted dogs, mean plasma radioactivity reached a maximum 5 min following instillation of the isotope at 1233 ± 49 cpm. The plasma radioactivity then decreased in a logarithmic fashion with a half-life of approximately 45 min. In allotransplanted dogs with normal histology in mucosal biopsies, mean radioactivity reached a maximum 15 min following instillation of the isotope at 1133 ± 108 cpm. In allotransplanted dogs undergoing rejection, as diagnosed from mucosal histology, mean plasma radioactivity reached a maximum 15 min following instillation of the isotope at 682 ± 175 cpm. This radioactivity level is significantly lower than the mean radioactivity found in allotransplanted dogs with normal graft histology ($P < 0.01$).

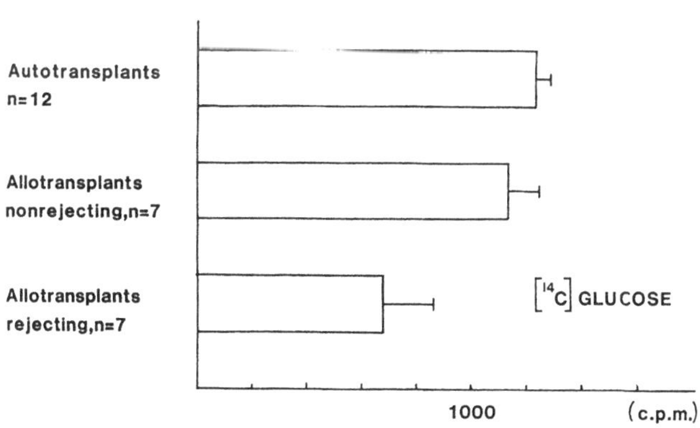

Fig. 2. Diagram showing peak plasma radioactivity following instillation of labeled glucose in the jejunal isolated pouch. Note the significantly decreased uptake in dogs undergoing rejection

Discussion

Uptake of [^{14}C]glucose from transplanted small intestine has been investigated by Cohen et al. (1969) and by Holmes et al. (1971). When short-segment allografts were tested (Cohen et al. 1969), a good correlation between histology and glucose uptake was documented. However, when glucose uptake was studied from long Thiry-Vella allograft loops (Holmes et al. 1971), the correlation with histological findings was poor. It appears that a high glucose concentration, as used by Holmes et al. (1971), allows passive diffusion of glucose across an impaired mucosal barrier. This results in an increased glucose uptake in dogs undergoing rejection. Nonspecific damage of the mucosal barrier, such as caused by infection, also appears to increase the glucose uptake (Holmes et al. 1971). A low glucose concentration as used by Cohen et al. (1969) and in this study, does not exclude passive diffusion of glucose but makes this less significant due to the concentration gradient. The mechanism underlying the decreased uptake of radiolabeled glucose in dogs undergoing rejection has not been investigated.

The absorption of cyclosporin in olive oil was found to be of the same magnitude in autotransplanted dogs as in allotransplanted dogs with normal graft histology. However, this absorption is still significantly lower than in normal, nontransplanted, dogs (Mackenzie et al. 1982). In dogs undergoing rejection, there is a considerable decrease in cyclosporin absorption. Therefore, in intestinal transplantation, administration of cyclosporin by the oral route is hazardous if the graft integrity is not continuously monitored or if plasma levels of cyclosporin are not followed.

In the present study, a good correlation was found between histological findings in the terminal end of the small-bowel transplant, the absorptive capacity of the incontinuity graft, and the glucose uptake in the proximal end of the graft.

In clinical intestinal transplantation, exposure of one end of the graft will be of importance for monitoring purposes. This can be constructed as an end ileostomy, an isolated pouch, or a Roux-en-Y stoma as described by Fortner (Fortner et al. 1972). This "window" to the graft can then be used for histological and physiological monitoring.

References

Cohen WB, Hardy MA, Quint J, State D (1969) Absorptive function in canine jejunal autografts and allografts. Surgery 65:440–446

Craddock G, Nordgren S, Reznick R, Gilas T, Lossing A, Cohen Z, Stiller C, Cullen J, Langer B (1983) Small bowel transplantation in the dog using Cyclosporine. Transplantation 35:284–288

Fortner JG, Sichuk G, Stephen D, Litwin SD, Beattie EJ Jr (1972) Immunological responses to an intestinal allograft with HLA-identical donor-recipient. Transplantation 74:531–535

Holmes JT, Yeh SDJ, Winawer SJ, Kawano N, Fortner JG (1971) Absorption studies in canine jejunal allografts. Ann Surg 174:101–108

Lossing A, Nordgren S, Cohen Z, Cullen C, Craddock G, Langer B (1982) Histologic monitoring of rejection in small intestinal transplantation. Transplant Proc 14:643–645

Mackenzie RD, Nordgren S, Lossing A, Craddock G, Cohen Z, Stiller C, Langer B (1982) Cyclosporine absorption in canine small intestinal transplantation. Transplant Proc 14:646–648

Migration of Donor Passenger Cells During Rat Cardiac and Intestinal Allograft Rejection*

M. A. Hardy, Ch. Iga, K. Satake, N. Kuromoto, and *S. F. Oluwole*

Introduction

Many studies have shown that passenger leukocytes in the donor organ play an important role in the sensitization leading to allograft rejection (Lafferty et al. 1975; Snell 1957; Steinmuller 1967; Summerlin 1973; Talmage et al. 1976). Depletion of passenger leukocytes, prior to transplantation, from endocrine organs by culture techniques (Lacy et al. 1979; Lafferty et al. 1976; Sollinger et al. 1977) and treatments of the donor animal or organ with cytotoxic agents (Freeman et al. 1971; Kyger and Salyer 1973; Steinmuller et al. 1971) have resulted in prolonged allograft survival. Guttmann et al. (1973) and Zincke et al. (1978) have demonstrated that cadaver donor pretreatment with methylprednisolone and cyclophosphamide can lead to improved human renal allograft survival. These results were explained by the hypothesis that the number of passenger leukocytes in the graft (Oluwole et al. 1980) was reduced or that the immunogenicity of the donor cells was altered (Van der Linden et al. 1981). Despite these findings, the relationship between the homing patterns of donor passenger leukocytes in the host and the subsequent host immune response is not clearly defined.

Our previous investigations indicate that depletion of passenger leukocytes from rat cardiac allografts by donor pretreatment with perfusion preservation, antilymphocyte globulin, sublethal total body irradiation, and cyclophosphamide could be effectively studied using indium-111 (^{111}In)-labeled lymphocytes (Oluwole et al. 1980). We also recently reported that ^{111}In-labeled leukocytes and platelets are useful in the early detection of rat cardiac allograft rejection (Oluwole et al. 1981). In the present study ^{111}In-labeled cells were used to define further the cellular dynamics of host immune response during allograft rejection, by focusing on the migration patterns of donor passenger leukocytes during rejection of either rat cardiac or intestinal allografts.

Materials and Methods

Animals. Adult male Lewis (RT1^{l}) and ACI (RT1 k51a) rats were purchased from Microbiological Associates, Walkersville, Maryland. The Lewis rats were used as

* This study was supported by NIH grants HL 14799 and AM 30468

Microsurgical Models in Rats
for Transplantation Research
Eds. A. Thiede et al.

recipients of either heterotopic cardiac or intestinal allografts as well as donors of syngeneic grafts in controls, while ACI rats were used as donors of allografts.

Radioisotope. Indium-111 oxine solution was purchased from Medi-Physics Inc., Emeryville, California. It has a half-life of 67 h and gamma photon emissions at 173 KeV (89%) and 247 KeV (93%).

Heart Transplantation. Adult Lewis rats served as recipients of heterotopic abdominal cardiac allografts from ACI rats or syngeneic grafts from Lewis rats. The donor and recipient rats were anesthetized with ether and intraperitoneal chloral hydrate. Transplantation was performed using the technique of Ono and Lindsey (1969). The donor aorta and the donor pulmonary artery were anastomosed end-to-side to the recipient's abdominal aorta and vena cava, respectively. The technical failure rate was less than 2%. Cardiac survival was assessed by daily palpation, and rejection was considered complete upon cessation of palpable cardiac beat. This was confirmed by histological studies of cardiac grafts.

Intestinal Transplantation. Intestinal transplantation was performed by the modification of the method of Deltz, described elsewhere in this book. The donor intestine was gently flushed with tepid Ringer's lactate while the aorta was flushed with saline solution until the effluent was clear in the vena cava. Carrel patch grafts were used for venous and arterial anastomoses to the aorta and vena cava. Two mucus fistulae were performed and exited in the flanks of the animal, where they could be inspected daily and biopsied as necessary. The operative success rate was 82% and isografts survived indefinitely. The allograft survival in the ACI to Lewis strain combination, as determined by microscopic examination of the allografts, was 6.8 ± 0.48 days.

Cell Separation and Labeling with In-111. We have previously described the methods of isolation and labeling of leukocytes, lymphocytes, and platelets (Oluwole et al. 1981). The labeling efficiency with ^{111}In was 80%–95% for lymphocytes and leukocytes, while it was only 35%–50% for platelets. The viability of the labeled leukocytes was greater than 95% as determined by trypan-blue dye exclusion.

The radioactivity of the tissues excised for scintillation counting was measured in a well-type scintillation counter. After the subtraction of the background levels, the radioactivity in a given specimen (organ) was expressed as a percentage of the total dose injected per gram of wet tissue.

Experimental Design. The Lewis recipients of syngeneic or ACI donor allografts were divided into two major experimental groups. One group consisted of animals used for the study of migration patterns of donor passenger leukocytes from cardiac isografts or allografts. Another group consisted of animals receiving intestinal allografts or syngeneic grafts. The allogeneic donor rats (ACI) or the syngeneic donor rats (Lewis) were infused with 2×10^7 ACI or Lewis-labeled viable leukocytes (15–20 μCi ^{111}In oxine), respectively. The donor rats were killed 24 h later and the heart and the small intestine with the mesentery and MLNs were excised for transplantation into unmodified Lewis recipients. The blood in the isolated donor hearts or the intestines was flushed out through the aorta with 2-ml ice-cold Ringer's lactate solution until the effluent was clear of cells, prior to transplantation. The control groups consisted of Lewis recipients of syngeneic intestinal or cardiac grafts. Four to six Lewis recipients from both the experimental and control groups were killed serially at 1, 3, and 6 h (only 6 h for intestine), then daily from 1 to 6 days after transplantation. The isograft, allograft, native heart, spleen, mesenteric lymph nodes (MLN), liver, cervical lymph nodes (CLN), thymus, and bone marrow (BM) were excised for scintillation counting.

Results

Migration of Passenger Leukocytes out of Cardiac Grafts. Indium-111 concentration in cardiac allografts excised on days 1–6 after transplantation decreased rapidly at 1 h postgrafting (Fig. 1). The depletion of labeled passenger leukocytes following this initial sharp fall became gradual between days 1 and 4, at which time there was a second rapid fall in the isotope count in the allograft. Absence of any further rise in ^{111}In activity in the allograft suggests that the labeled cells which migrate from the heart do not return to the graft. A similar pattern was seen in syngeneic cardiac isografts.

Kinetics of Migration of Donor Passenger Leukocytes to Recipients' Lymphoid Organs. Radio-labeled passenger leukocytes in the cardiac allograft migrated to the spleen and MLN of the recipient within 6 h of allografting and maximum accumulation in the spleen occurred on day 4 after transplantation. The accumulation of radio-labeled passenger leukocytes in the MLN was characterized by two peaks of activity, one at 24 h and a greater one at 5 days after allografting

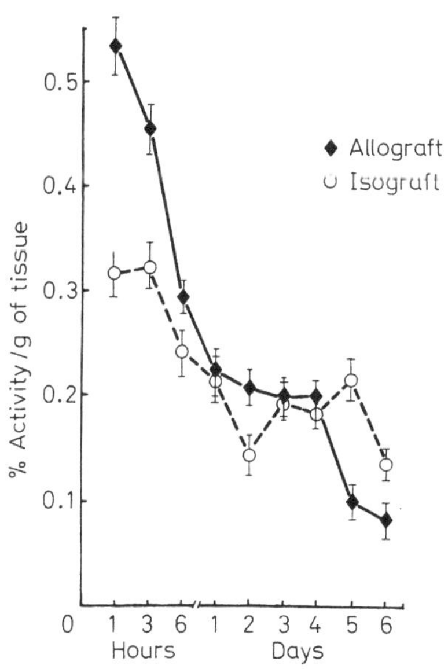

Fig. 1. Depletion of ^{111}In-labeled passenger leukocytes from cardiac allografts (ACI to Lewis) and isografts (Lewis to Lewis), following intra-abdominal transplantation at different time periods

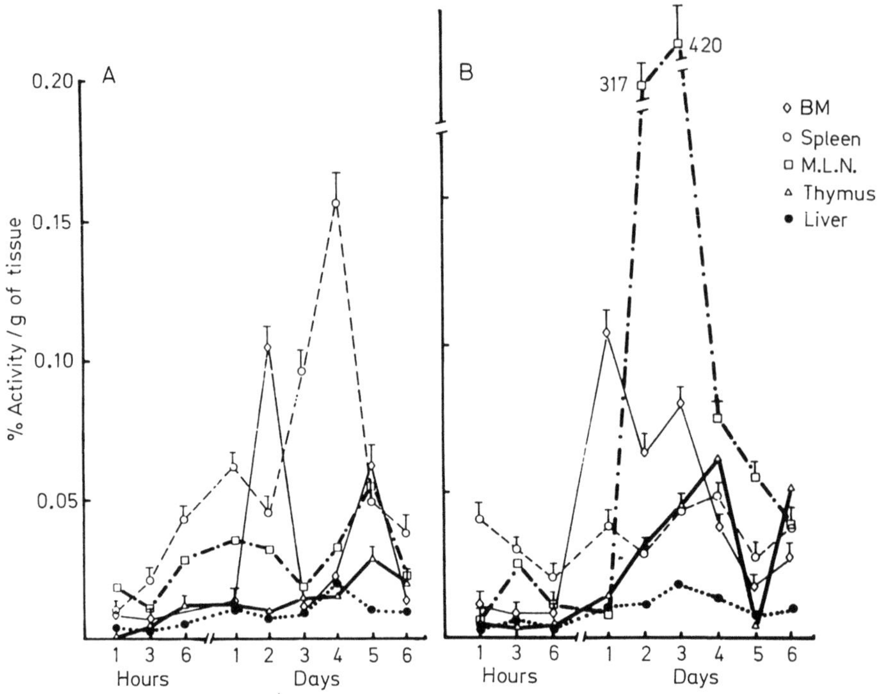

Fig. 2. *A* Migration patterns of ^{111}In-labeled donor passenger leukocytes (transferred on the heart transplant) to the various lymphoid organs in Lewis recipients of ACI cardiac allografts. *B* Migration patterns of ^{111}In-labeled donor passenger leukocytes to the various lymphoid organs in Lewis recipients of syngeneic cardiac grafts

(Fig. 2A). There was no significant accumulation of labeled cells in the thymus or liver. There was a rapid rise of activity in the BM on day 2 after transplantation and again on day 5 after transplantation.

Figure 2B shows the biodistribution of labeled passenger leukocytes from syngeneic cardiac grafts to the recipient's lymphoid organs. The pattern of migration of syngeneic passenger leukocytes to the spleen consistently differed from that of allogeneic passenger leukocytes in that there was minimal accumulation of labeled cells in the spleen at 6 h after transplantation. The accumulation of labeled cells in the MLN of the syngeneic recipient was significantly greater than in the allografts. Similarly, there was a significant accumulation of syngeneic ^{111}In-labeled passenger leukocytes in the BM and thymus in the syngeneic (control) groups. The accumulation of labeled passenger leukocytes in the liver of both syngeneic and allogeneic recipients was minimal and similar.

Migration of Passenger Leukocytes out of Intestinal Grafts. Indium-111 concentration of "passenger" donor lymphocytes in intestinal allografts excised at 6 h to 6 days after allografting decreased rapidly by day 3 after transplantation, following which there was no apparent return of the cells to the graft. In the syngeneic grafts, the maximum decrease in passenger lymphocytes also occurred by day

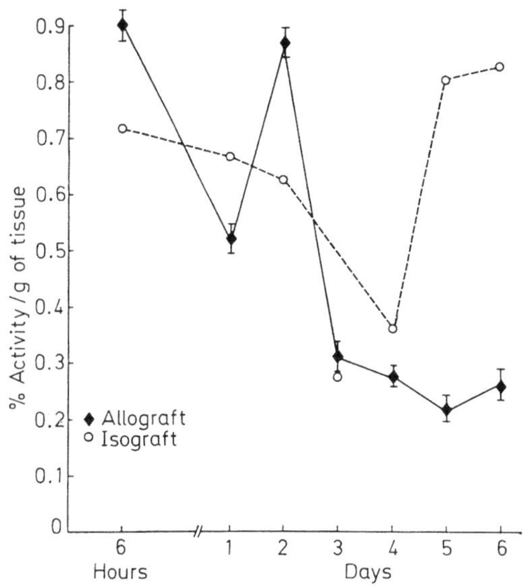

Fig. 3. Depletion of ^{111}In-labeled passenger leukocytes from intestinal allografts (ACI to Lewis) and isografts (Lewis to Lewis) including the graft mesentery, following intra-abdominal transplantation (analogous to Fig. 1)

3 after grafting, but another rise in accumulation of radioactivity 5 days after transplantation suggests significant recirculation of the labeled cells to the isograft. This is shown in Fig. 3. The difference in passenger leukocyte migration between cardiac and small-intestinal allografts may be due to the initially higher population of donor passenger lymphocytes in the small intestine, as compared with the cardiac allograft. Whether other factors influence this migration, particularly in the early phase after transplantation, remains uncertain. Examination of the passenger leukocyte migration from isolated mesentery and its lymph nodes in syngeneic or allogeneic donors suggests that the majority of the donor "passenger" cells originate in the mesenteric lymph nodes of the mesentery, rather than in the small intestine. It is important to note that no recirculation of labeled passenger cells was seen in allografts compared with isografts, where recirculation occurred between days 4 and 5 following transplantation. It appears that the major contribution, in terms of donor passenger leukocytes, originates from the grafted mesentery and mesenteric lymph nodes, rather than from the small intestine itself.

Kinetics of Migration of Small-Intestine Donor Passenger Leukocytes to Recipient's Lymphoid Organs. The results of this study are summarized in Fig. 4 and are preliminary in nature. The migration of radiolabeled leukocytes from the intestinal grafts to the spleen showed a similar pattern, regardless of whether the passenger cells originated in an allograft or an isograft. The first peak of activity occurred on the 2nd day after transplantation, both in isografts and allografts and the second peak occurred on the 6th day after transplantation. Although there was a gradual rise in splenic activity early after transplantation, no significant difference was observed between allografts and isografts. This differs significantly from the findings in cardiac allografts.

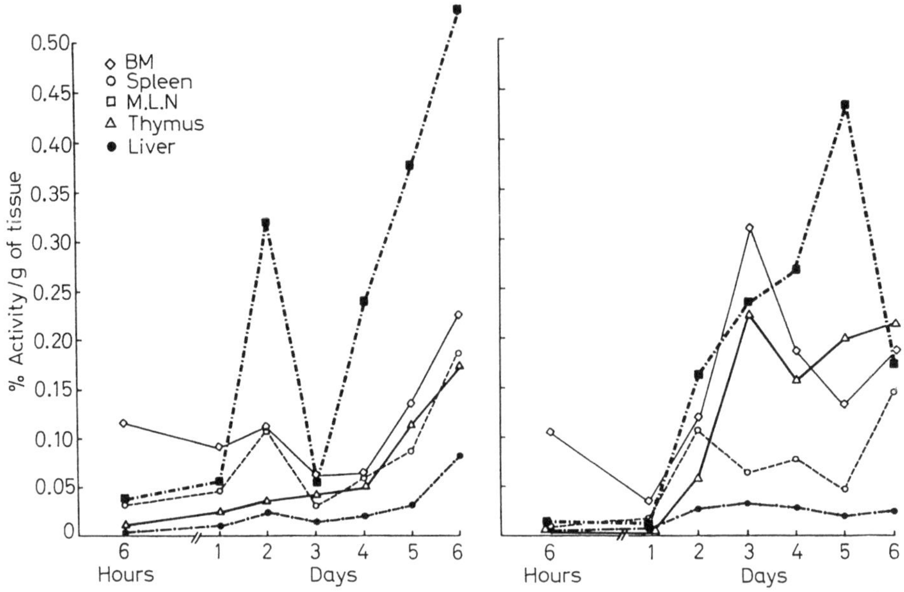

Fig. 4. *A* Migration patterns of ^{111}In-labeled donor passenger leukocytes (transferred on the intestinal and mesenteric ACI allograft) to the various lymphoid organs of Lewis recipients. *B* Migration pattern of radiolabeled donor passenger leukocytes to the various lymphoid organs in Lewis recipients of syngeneic intestinal transplants

The migration pattern of radiolabeled cells from small-intestinal allografts and isografts to the cervical and mesenteric lymph nodes of the host appear to be relatively nonspecific. In the allograft situation, the mesenteric lymph nodes were of the donor strain, and there the migration was markedly decreased as compared with the isograft control. In isografted controls, there was clear evidence of recirculation, to both the cervical and mesenteric lymph nodes of the passenger cells, as evidenced by several peaks of activity detected.

The migration pattern of donor cells to the thymus of the host was striking in the allografted animals and was not previously seen with cardiac allografts. The migration of labeled passenger cells to the bone marrow was similar in intestinal transplantation to that seen in cardiac allografting. There was no significant accumulation of labeled cells in the liver during the initial phase of intestinal transplantation.

Discussion

Acute destruction of allografts is mediated by a complex cellular immune response interacting in both host and graft (Billingham et al. 1955; Gowans 1965; Hall et al. 1978; Strom et al. 1975, 1977; Tilney et al. 1976; Von Willebrand et al. 1979). The afferent limb of allograft rejection, which leads to host sensitization, depends on passenger leukocytes, soluble graft antigens, and antigenic differences between the cell surface receptors of the host and donor (Najarian and Ascher 1979). A knowledge of the dynamics of the migration patterns of the donor passenger leukocytes to the host lymphoid organs and the correlation of such results with the in vivo redistribution of the hosts' cells should lead to a better understanding of the immune mechanisms responsible for acute allograft rejection. We have studied this relationship between the passenger leukocyte homing pattern and the resultant sequence of events in the host of either syngeneic or allogeneic grafts (particularly cardiac grafts) by using ^{111}In oxine labeled cells. Indium-111, with a half-life of 67 h, is suitable for cell labeling and tracing biological processes in vivo. Cells labeled with indium-111 retain their biological activity and remain viable without alteration of function (Thakur et al. 1976). The dose of indium-111 utilized in the present study is small, safe, and nontoxic to the cells or the host (Chisholm et al. 1979).

The study of homing kinetics of mobile passenger leukocytes into recipient lymphoid tissue shows that nearly 50% of the interstitial ^{111}In-labeled cells carried by the cardiac allografts migrated to the host lymphoid organs within 6 h of transplantation. This was followed by a persistent but slow depletion of the labeled cells from the graft until 4 days after transplantation, when a second sharp fall in the allograft isotope count occurred. These changes do not appear to be due to recirculation. In the case of cardiac allografts, the spleen and MLN accumulated the majority of the radioactivity released from the graft. This is consistent with the findings of Van Schilfgaarde et al. (1979) in the rat kidney and heart models. In the case of intestinal allografts, the initial depletion of interstitial cells was much less striking and was apparent only on day 3 after transplantation. There was also a clear-cut recirculation of donor cells in the isografts. The splenic accumulation in intestinal allografts was only very gradual, did not differ strikingly from that seen in isografts, but contrasted sharply with a steep rise in splenic accumulation of radioactivity in cardiac allografts. In the case of small-intestine transplantation, we have shown in preliminary studies that there is little difference in passenger leukocyte migration to the lymphoid organs of the host regardless of whether the graft is an iso- or an allograft. We further demonstrated that the majority of the labeled cells that migrate out of the intestinal graft originate in the mesenteric lymph nodes and not in the intestine itself.

It appears that omit there is a rapid exit of interstitial cells from the allografted heart (within the first 6 h after transplantation), but not from an intestinal transplant. The passenger cells appear to migrate maximally to the spleen and to the regional lymph nodes in the case of cardiac allografts, while in small-intestinal transplantation a surprisingly large number of donor passenger cells migrate to the hosts' thymus and only a relatively small proportion migrate to the spleen and the cervical lymph nodes. In cardiac allografts there may be a transfer of labeled cells from the spleen to the lymph nodes as suggested by a second peak of activity seen in the MLN 5 days after allografting; this may also be true in the case of intestinal transplantation but may be due to recirculation of passenger leukocyte cells and nonspecific migration to the lymph nodes in the cervical area.

The fact that the ^{111}In-labeled is not transferable between cells in vivo strongly suggests that the originally labeled cells in the donor animal migrate out of the graft to the spleen and the MLN and subsequently may also migrate from the spleen to the MLN, perhaps in order to stimulate expansion of sensitized cells. In the case of intestinal transplantation this type of traffic is not as obvious as it is in cardiac allografts, while the traffic to the thymus appears extraordinarily large.

The immediate accumulation of released radiolabeled donor passenger leukocytes in the hosts' central lymphoid organs again suggests that donor passenger cells play an important role in the sensitization of the host to the donor antigens. In the case of cardiac allografts, the splenic accumulation supports the hypothesis that this organ is important in the initial development of effector cells that migrate to the graft to destroy it. In the case of intestinal allografts the delay in splenic accumulation and the large initial accumulation of radiolabeled cells in the thymus may be related more to the phenomenon of GVH, rather than HVG, as in the case of the cardiac transplants. The initial changes in the spleen in the case of cardiac allografts support the hypothesis that the migration of passenger leukocytes to the host lymphoid organs precedes the cellular events in these organs in initiating graft rejection. Such a hypothesis is supported by numerous studies which indicate that the regional lymph nodes and the spleen are the major loci of the early cytotoxic cellular immune response (Billingham et al. 1955; Strom et al. 1977; Tilney et al. 1976; von Willebrand et al. 1979). Whether the different migration pattern seen in intestinal allografts is related to a GVH reaction must await further studies, particularly in the appropriate F_1 hybrid strain combinations. This finding may be related to the fact that the primary source of donor lymphocytes in this model is the mesenteric lymph nodes, which are transplanted with the intestine, rather than interstitial intestinal cells. The issue of whether this is related to the initiation of GVH in small-intestinal transplantation may have major clinical importance in the future and needs further investigations.

In summary, we postulate that the events that lead to graft rejection start with the release of donor antigens into the circulation of the host. The donor antigens, represented here by labeled passenger leukocytes, migrate to, and elicit, proliferative reaction in the central lymphoid organs, particularly in the spleen and MLN, which in turn produce effector cells that home to the graft and destroy it, as we have previously described (Oluwole et al. 1981). Our previous dynamic studies of allograft rejection and the present study of passenger-cell migration confirm previous reports based on biopsy of rejecting allografts and the characterization of the infiltrating cells (Strom et al. 1977; Tilney et al. 1976; von Willebrand et al. 1979), Hayry et al. 1979). The present experiments further suggest that ^{111}In-labeled cells, with their high labeling efficiency, offer an excellent approach for the continued study of the cellular dynamics of graft rejection with particular emphasis on small-intestinal transplantation and the study of the phenomenon of graft versus host reaction and host versus graft reaction.

Acknowledgments. The authors gratefully acknowledge the excellent technical assistance of Margie Prasek and Jitka Mucha, as well as constant encouragement and advice by Dr. Rashid Fawwaz of the Department of Radiology, Columbia University College of Physicians & Surgeons. The secretarial assistance of Renee Grobtuch is deeply appreciated.

References

Billingham RE, Brent L, Medawar PB (1955) Quantitative studies on tissue transplantation immunity. II. The origin, strength and duration of actively and adoptively acquired immunity. Proc R Soc Lond [Biol] 143:58–80
Chisholm PM, Danpure HJ, Healey G, Osman S (1979) Cell damage resulting from the labeling of rat lymphocytes and HeLa S3 cells with In-111 oxine. J Nucl Med 20:1308–1311
Freeman JS, Chamberlain EC, Reemtsma K, Steinmuller D (1971) Prolongation of rat heart allografts by donor pretreatment with immunosuppressive agents. Transplant Proc 3:580–582
Gowans JL (1965) Role of lymphocytes in the destruction of homografts. Br Med Bull 21:106–110
Guttmann RD, Beaudoin JG, Morehouse DD (1973) Reduction of immunogenicity of human cadaver renal allografts by donor pretreatment. Transplant Proc 5:663–665
Hall BM, Dorsch S, Roser B (1978) The cellular basis of allograft rejection in vivo. II. The nature of memory cells mediating second set heart graft rejection. J Exp Med 148:890–902
Hayry P, Soots A, Willebrand EV, Wiktorowicz K (1979) Composition, subclass distribution and preliminary analysis of the functions of host inflammatory cells infiltrating renal allografts during rejection. Transplant Proc 11:785–789
Kyger ER, Salyer KE (1973) The role of donor passenger leukocytes in rat skin allograft rejection. Transplantation 16:537–543
Lacy PE, Davie JM, Finke EH (1979) Prolongation of islet allograft survival following in vitro culture (24 degrees C) and a single injection of ALS. Science 204:312–313
Lafferty KJ, Cooley MA, Woolnough JR, Walker KZ (1975) Thyroid allograft immunogenicity is reduced after a period in organ culture. Science 188:259
Lafferty KJ, Bootes A, Dart G, Talmage DW (1976) Effect of organ culture on the survival of thyroid allografts in mice. Transplantation 22:138–149
Najarian JS, Ascher NL (1979) Causes and management of rejection. Transplant Proc 11:11–16
Oluwole S, Hardy MA, Wang T, Satake K, Todd G, Reemtsma K, Nowygrod R (1980) Donor pretreatment: rat heart allograft survival and measurement of passenger leukocyte depletion with Indium-111. Transplantation 30:31–33
Oluwole S, Wang T, Fawwaz R, Satake K, Nowygrod R, Reemtsma K, Hardy MA (1981) Use of Indium-111-labeled cells in measurement of cellular dynamics of experimental cardiac allograft rejection. Transplantation 31:51–55
Ono K, Lindsey ES (1969) Improved technique of heart transplantation in rats. J Thorac Cardiovasc Surg 57:225–229
Snell GD (1957) The homograft reaction. Annu Rev Microbiol 11:439–458
Sollinger HW, Burkholder PM, Rasmus WR, Bach FH (1977) Prolonged survival of xenografts after organ transplantation. Surgery 81:74–79
Steinmuller D (1967) Immunization with skin isografts taken from tolerant mice. Science 158:127–129
Steinmuller D, Warden G, Coleman M, Lofgreen J, Reemtsma K, Stuart F, Garrick T, Holter A, Lynch A (1971) Prolonged survival of rat heart and kidney allografts irradiated in vivo. Transplantation 12:153–156
Strom TB, Tilney NL, Carpenter CB, Busch GJ (1975) Identity and cytotoxic capacity of cells infiltrating renal allografts. N Engl J Med 292:1257–1263
Strom TB, Tilney NL, Paradysz JM, Bancewicz J, Carpenter CB (1977) Cellular components of allograft rejection: identity, specificity and cytotoxic function of cells infiltrating acutely rejecting allografts. J Immunol 118:2020–2026
Summerlin WT (1973) Allogeneic transplantation of organ cultures of adult human skin. Clin Immunol Immunopathol 1:372–384
Talmage DW, Dart G, Radovich J, Lafferty KJ (1976) Activation of transplant immunity: effect of donor leukocytes and thyroid allograft rejection. Science 191:385–388

Thakur ML, Welch MJ, Joist JH, Coleman RE (1976) Indium-111 labeled platelets: studies on preparation and evaluation of in vitro and in vivo functions. Thromb Res 9:345–357

Tilney NL, Strom TB, MacPherson G, Carpenter CB (1976) Studies on infiltrating host cells harvested from acutely rejecting rat cardiac allografts. Surgery 79:209–217

Van der Linden CJ, Buurman WA, Vegt PA, Greep JM, Jeekel J (1981) A study on the mechanism of donor pretreatment: effect of procarbazine hydrochloride and methyl-prednisolone on immunocompetent cells. Transplantation 32:24–28

Van Schilfgaarde R, Hermans P, van Breda Vriesman PJC (1979) Influence of recirculating lymphocytes on the rejection of first-set rat organ allografts. Transplant Proc 11:1485–1488

Von Willebrand E, Soots A, Hayry P (1979) In situ effector mechanisms in rat kidney allograft rejection. II. Heterogeneity of the effector cells in the inflammatory infiltrate vs. that in the spleen of the recipient rat. Cell Immunol 46:327–336

Zincke H, Woods JE, Khan AU, Holley KE, Leary F (1978) Immunological donor pretreatment in combination with pulsatile preservation in cadaveric renal transplantation. Transplantation 26:207–211

Rejection and Graft Versus Host Reaction Following Small-Bowel Transplantation and Methods of Circumventing Them

E. Deltz, K. Ulrichs, and *A. Thiede*

The small-intestine graft is characterized by a large amount of epithelial surface and a great mass of lymphoid tissue. Both contribute to the elicitation of the rejection reaction. Since with this lymphatic tissue a large quantity of immunocompetent cells is transplanted, which can react against the recipient's antigens, in addition to the rejection reaction a graft versus host reaction (GVHR) may occur.

A general problem in small-bowel transplantation, as in other types of organ grafting, is the rejection reaction. Its dependence on the immunogenetic difference between donor and recipient and the amount of transplanted tissue must be investigated. The special problem in small-intestine transplantation – GVHR – requires detailed investigation in order to characterize it and to clarify its dependence on the degree of histoincompatibility between donor and recipient and the amount of lymphatic tissue transplanted. The third problem is: How can rejection and GVHR be circumvented? Using the experimental model of accessory small-bowel transplantation in the rat (Deltz, this volume, p. 51), we tried to contribute to the elucidation of the questions arising from these three problems.

Rejection Reaction

Bearing in mind the requirement for clinical small-bowel transplantation, namely, a compatible graft of a suitable length (Hardy et al., this volume, p. 337; Stauffer 1983), two major questions arise concerning the rejection reaction.

1. What influence does donor-recipient histoincompatibility have on graft survival and recipient survival?
2. How does recipient survival depend on the length of the small-bowel graft?

In our experiments the rejection reaction was investigated in a fully allogeneic system (donor F344, $RT1^{l}$ or BN $RT1^{n}$; and recipient LEW, $RT1^{l}$). Though in part overlapped by GVHR, a clear pattern of rejection can be shown in a graft (Deltz 1984).

Macroscopically the graft undergoes severe alterations caused by rejection within 5 days postoperatively. The thickened bowel loops are reddened and covered with fibrin. Histologically, progressive damage to the intestinal wall due

Microsurgical Models in Rats
for Transplantation Research
Eds. A. Thiede et al.

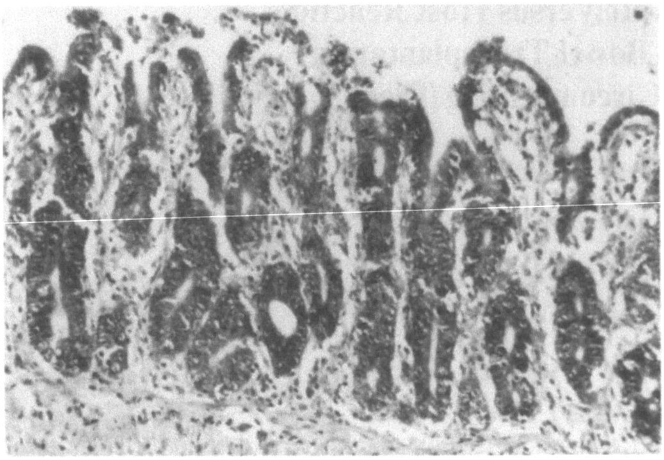

Fig. 1. Rejection reaction in an allogeneic small-bowel graft by infiltration of the lamina propria by mononuclear cells and polymorphocytes

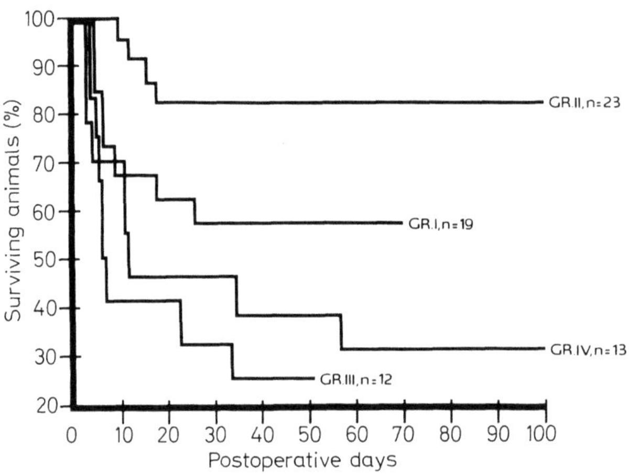

Fig. 2. Survival curves after fully allogeneic small-bowel transplantation

to infiltration by mononuclear cells and polymorphocytes appears (Fig. 1). At the same time it can be seen that the mucosal surface is sloughed off. The recipient dies of perforation of abscesses formed in the graft and of peritonitis.

The survival curves (Fig. 2) of recipients after transplantation of fully allogeneic grafts demonstrate the significance of histoincompatibility between donor and recipient. The first statement that can be made is: In a donor-recipient combination in which the major histocompatibility complex (MHC) differs only in the *C* region and in non-MHC-coded antigens (F344→LEW, GR I, GR II) the survival rate is higher than in a combination in which the donor and recipient differ in MHC- and non-MHC-coded antigens (BN→LEW, GR III, GR IV). The second result of these experiments is: The survival rate is higher after transplantation of half a small bowel (GR II, GR IV) than after transplantation of an entire small intestine (GR I, GR III).

The questions concerning rejection can be answered as follows: The strength of the rejection reaction depends directly on:

1. The degree of histoincompatibility between donor and recipient
2. The length of the graft

Graft Versus Host Reaction

The specific problem of small-bowel transplantation, GVHR, leads to the following questions:

1. What is the morphological and functional correlative of GVHR in the graft?
2. Which lymphoid and nonlymphoid organs of the recipient are involved in GVHR, thus becoming target organs of GVHR?
3. Do the characteristics of GVHR show a specific time course?
4. Does the strength of GVHR depend on the degree of histoincompatibility between donor and recipient?
5. Does the amount of grafted lymphoid tissue influence the strength of GVHR?

For investigations of GHVR a semiallogeneic donor-recipient combination is most suited, since only in this situation does a unidirectional GVHR occur. Small-bowel grafts from parental strain animals were transplanted (F344, $RT1^l$; BN, $RT1^n$) into F_1-hybrid recipient [F344 × CAP, ($RT1^c$); BN × LEW, ($RT1^l$)]. About 14 days after the operation the recipients show the characteristic clinical features of wasting disease.

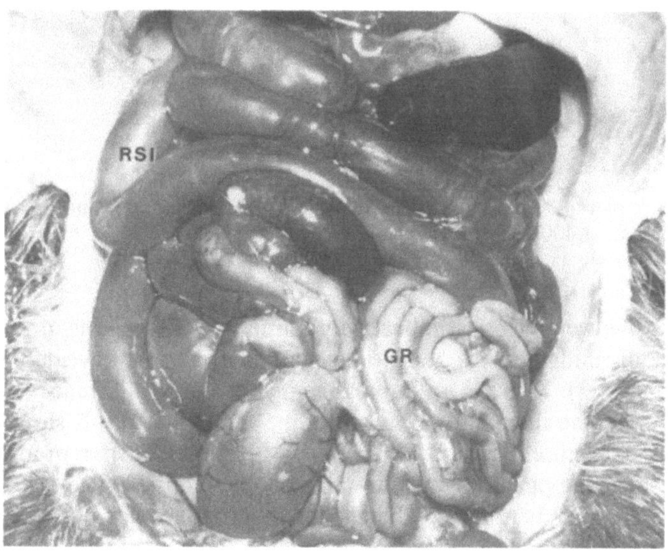

Fig. 3. Autopsy in GVH disease 14 days after semiallogeneic small-bowel transplantation. *GR*, graft; *RSI*, recipient's small intestine

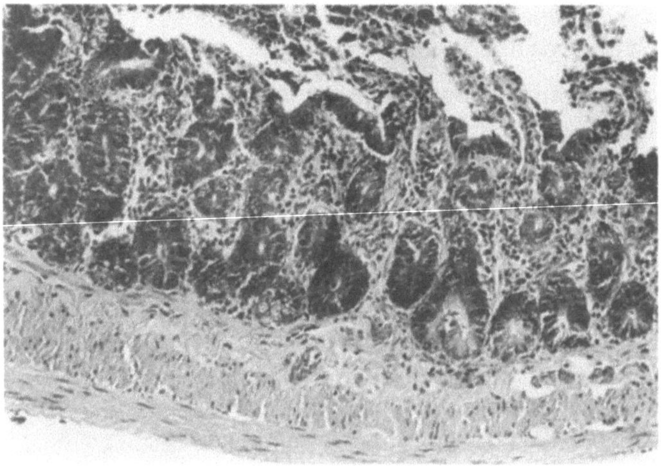

Fig. 4. Severe erosive enteritis of recipient's small bowel in GVH disease. ×160

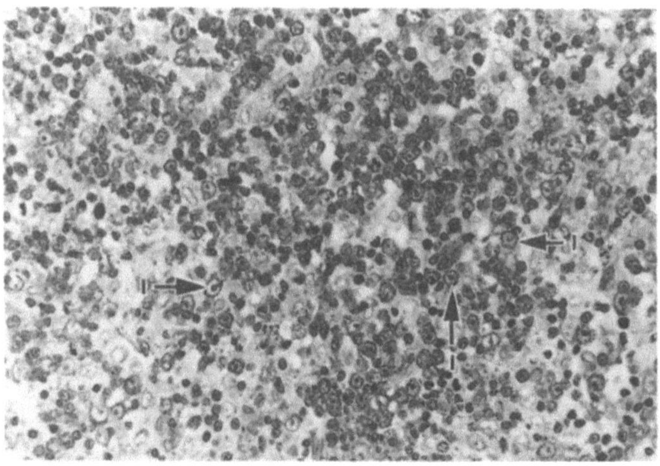

Fig. 5. Histological picture of a mesenteric lymph node of the graft showing immunological stimulation. *I*, immunoblast. ×400

They reveal progressive weight loss and heavy inflammatory alterations of the integument especially around the eyes, nose, and feet. The autopsy revealed a characteristic pattern (Fig. 3). There is an impressive macroscopic difference between the graft (GR) and the recipient's small intestine (RSI). The graft shows no alterations, whereas the small bowel of the recipient appears transparent and puffed up with gas. The same pattern is seen in microscopic pictures. The histological appearance of the graft is normal, whereas the mucosal surface of the recipient's small bowel (Fig. 4) shows severe erosive enteritis with loss of the villi. The lamina propria is infiltrated by numerous immunoblasts and activated lymphoid cells.

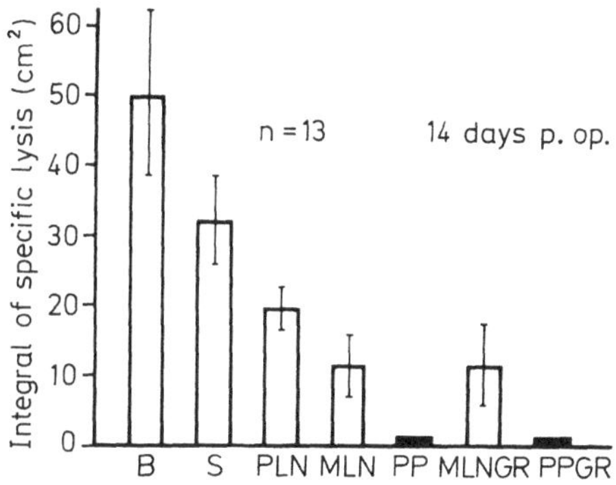

Fig. 6. Anti-host cytotoxic reactivity in the lymphatic compartments of the graft (*MLNGR*, mesenteric lymph nodes of the graft; *PPGR*, Peyer's patches of the graft) and recipient (*B*, blood; *S*, spleen; *PLN*, peripheral lymph nodes; *MLN*, mesenteric lymph nodes; *PP*, Peyer's patches)

The mesenteric lymph nodes of the graft are greatly enlarged. The histological picture (Fig. 5) shows immunoblasts and a proliferation of epitheloid cells in the paracortical area, representing the immunological stimulation against donor antigens in the graft. The spleen of the recipient shows a marked enlargement, which is reflected by the rapid increase in the spleen index up to 12.0 (normal value 1.0) within the first 2 weeks after transplantation. Histologically, the spleen of the recipient shows marked proliferation of immunocompetent cells in the periarteriolar sheath with partial destruction of the normal histological architecture. Another lymphatic system which is involved in GVHR is the peripheral lymph nodes. Like the other lymphatic organs they show distinct enlargement and strong immunological stimulation as can be seen in the appearance of immunoblasts.

Briefly, GVHR following small-bowel transplantation is characterized by proliferation of immunocompetent cells mainly in the mesenteric lymph nodes of the graft which spreads to the recipient's lymphatic organs. The main target organ of the recipient which is attacked by stimulated cells is the recipient's small bowel. All of these clinical features can be summarized as GVH disease, which in the case of strong expression of GVHR leads to the death of the recipients.

The functional results of anti-host T-cell reactivity measured by a microcytotoxicity assay (MCA) (Takasugi and Klein 1970) correspond (Fig. 6) with these morphological findings. Fourteen days after transplantation all lymphatic compartments except Peyer's patches show marked anti-host cytotoxicity. It is higher in the spleen than in the peripheral lymph nodes. The maximum of GVHR cytotoxic activity can be found in the recipient's blood.

This anti-host activity of T cells and the morphological findings show a typical time dependence with maximum expression between days 15 and 25 after transplantation. Then T-cell reactivity decreases, according to regression and fibrosis found in the lymphatic tissues.

The survival curves (Fig. 7) depict the dependence of GVHR on the degree of histoincompatibility between donor and recipient and on the amount of grafted tissue. In the donor-recipient combination in which the difference between the histocompatibility antigens is greater (GR XIII, GR XIV), the survival rate is lower than in combinations with a minor histoincompatibility barrier (GR XI, Gr XII).

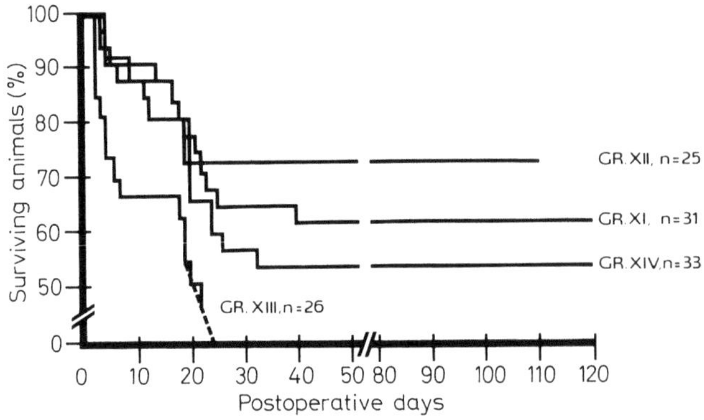

Fig. 7. Survival curves after semiallogeneic small-bowel transplantation: F344→ (F344 × CAP) F_1 (GR XI, XII); BN→ (BN × LEW) F_1 (GR XIII, XIV)

The influence of the amount of grafted lymphoid tissue is revealed by a comparison of transplantation of half a small bowel (GR XII and GR XIV) and grafting of an entire small bowel (GR XI and GR XIII) within one donor-recipient combination. Transplantation of an entire small bowel leads to more pronounced expression of GVHR and death from GVH disease than grafting of half a small intestine. In the case of BN to LEW semiallogeneic transplantation (GR XIII) all recipients succumb to GVHR disease within 22 days.

These results can answer the questions on GVHR.

1. The morphological correlative of GVHR in the graft is the proliferation of immunocompetent cells in the T-dependent areas of the mesenteric lymph nodes. This immunological stimulation corresponds to its functional correlative: antihost cytotoxic activity in this lymphatic compartment.
2. The lymphatic organs of the recipient (lymph nodes, spleen) are involved in GVHR. The most important target organ of GVHR is the small bowel of the recipient.
3. The characteristics of GVHR show a typical time dependence with maximum expression between 15 and 25 days after transplantation.
4. The strength of GVHR depends directly on the degree of histoincompatibility between donor and recipient.
5. The amount of grafted lymphoid tissue influences the strength of GVHR.

Manipulation of Rejection and GVHR

What possibilities are there of circumventing both rejection and GVHR? We investigated the treatment of rejection with immunosuppressive agents as well as modalities for circumventing GVHR by donor treatment and recipient treatment with cyclosporin A (CyA).

Concerning the manipulation of the rejection reaction and GVHR, the following questions arise:

1. Can the small-bowel graft be preserved by immunosuppressive treatment?
2. By which type of donor or graft treatment can GVHR be avoided?
3. Can GVHR be circumvented by recipient treatment?

Cyclosporin A has also proved its immunosuppressive properties in small-intestinal transplantation. An AS ($RT1^l$) to LEW graft, as described in the dog model (Craddock et al. 1983), is usually rejected not later than 8 days postoperatively, thus demonstrating the strong effect of non-MHC antigens in the rejection reaction. After continuous treatment of the recipients with 15 mg CyA/kg body weight/day, all of these grafts are preserved for 6 weeks. After cessation of CyA treatment, however, rejection occurs.

Manipulation of GVHR

In the plast 2 years we have focused our interest on the manipulation of GVHR by donor irradiation with 950 rad prior to the transplantation, or microsurgical removal, of the mesenteric lymph nodes of the graft or CyA treatment of the recipients at a dose of 15 mg/kg body weight given orally for 14 days after transplantation. These experiments were performed in the BN to (BN × LEW) F_1-recipient combination (Deltz et al. 1983). With these methods an impressive increase in the survival rates could be achieved.

After CyA treatment of the recipients a survival rate of 71% (20/28 animals) resulted, and after donor irradiation 77% (20/26) of the animals survived, compared with zero in the control group without treatment. The greatest degree of avoidance of GVHR and GVH disease was achieved by removing the mesenteric lymph nodes of the graft, thus proving these lymph nodes to be most relevant for initiating and maintaining GVHR. This increase in the survival rates is reflected by in vitro findings and the morphological picture. Compared with the control group, a striking decrease in anti-host T-cell reactivity in all lymphatic compartments was evident after CyA treatment both 20 and 120 days after transplantation. The most outstanding feature of flourishing GVHR, the high anti-host cytotoxic activity in the blood of small-bowel recipients, decreases to zero as a result of treatment. The morphological data are in complete accordance with these functional results. The mesenteric lymph nodes of the graft showed no signs of immunological stimulation, nor did the spleen of the recipients reveal a GVHR picture.

A similar pattern of functional data was found after donor irradiation and mesenteric lymph node removal. After irradiation moderate residual activity was found in the spleen, but it decreased to zero 120 days after operation. In accordance with the high survival rate of these animals (80%, 23/26), the removal of the mesenteric lymph nodes of the graft resulted in the most significant reduction of the anti-host T-cell reactivity that represents GVHR 20 days as well as 120 days postoperatively. These results can be summarized:

1. The small-bowel graft can be preserved by immunosuppressive treatment (e.g., CyA).

2. By donor irradiation and removal of the mesenteric lymph nodes of the graft GVHR can be avoided.
3. GVHR can be circumvented by recipient treatment with CyA.

The results of our investigations may help to elucidate the complex immunological reactions induced by small-bowel transplantation either in rejection or in graft versus host reactions. The knowledge of the mechanisms of both reactions makes manipulation feasible, thus providing an indispensable prerequisite for clinical small-bowel transplantation.

References

Craddock GN, et al. (1983) Small bowel transplantation in the dog using Cyclosporine. Transplantation 35:284–288

Deltz E (1984) Die allogene Dünndarmtransplantation. Zuckschwerdt, Munich

Deltz E, et al. (1983) Verteilungsmuster und Kinetik zytotoxischer T-Lymphozyten bei florider und supprimierter GVHR nach Dünndarmtransplantation. Langenbecks Arch Chir [Suppl] 151–155

Stauffer UG (1983) Die Dünndarm-Transplantation – Gegenwärtiger Stand und mögliche Zukunft. Z Kinderchir 38:286–289

Takasugi M, Klein E (1970) A microassay for cell-mediated immunity. Transplantation 9:219–227

Immunological Reactions Induced by Graft Versus Host Reaction Following Small-Bowel Transplantation

K. Ulrichs and *E. Deltz*

As a biological model for the analysis of graft versus host reaction (GVHR), caused by continuously supplied immunocompetent cells of a lymphoid-tissue-rich organ, small-intestine (SI) transplantation was performed in rats.

Material and Methods

Animals. Rats of the inbred strains LEW ($RT1^l$), BN ($RT1^n$), F344 ($RT1^{lv1}$), and CAP ($RT1^c$) and their F_1 progeny were used; recipients were aged 6 months and the graft donors were aged 3–4 months.

Technique of SI-Transplantation. The technique of small-intestine transplantation has been described in detail by Deltz (this volume, p. 51).

Test Procedure for Cell-Mediated In Vitro Reactivity. Cell-mediated in vitro reactivity of the various lymphatic compartments of the recipient (blood, spleen, peripheral lymph nodes, mesenteric lymph nodes, Peyer's patches) and of the graft (mesenteric lymph nodes, Peyer's patches) was tested at different times (14–30, 50 or 90–110 days) postoperatively using the microcytotoxicity assay (MCA) (Takasugi and Klein 1970). Embryonic fibroblasts of parental strains and a third party strain served as target cells.

Experimental Groups. Four experimental groups were formed, grafting either total or half SI of parental origin into F_1 hybrids: (I) 1/1 F344→ (F344 × CAP) F_1, (II) 1/2 F344→ (F344 × CAP) F_1, (III) 1/1 BN→ (LEW × BN) F_1, and (IV) 1/2 BN→ (LEW × BN) F_1.

Results

Mortality Rates. Graft versus host reaction induced mortality of semiallogeneic graft recipients in the four experimental groups is shown in Fig. 1. As can be seen, mortality of SI recipients clearly depends on (a) the kind of strain combination used and (b) the amount of lymphatic tissue grafted (total or half small intestine).

Microsurgical Models in Rats
for Transplantation Research
Eds. A. Thiede et al.

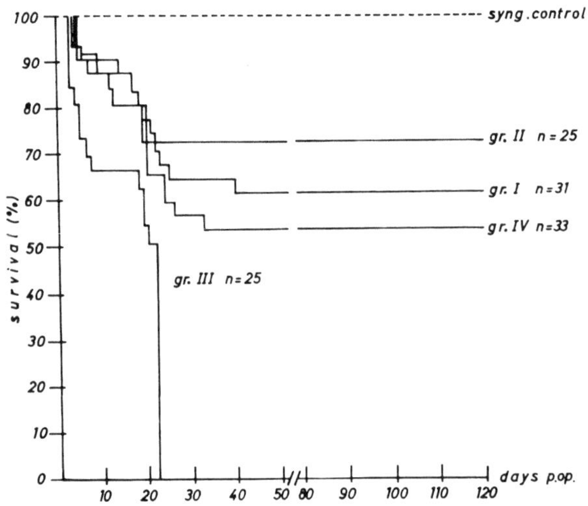

Fig. 1. Survival of small-intestine graft recipients

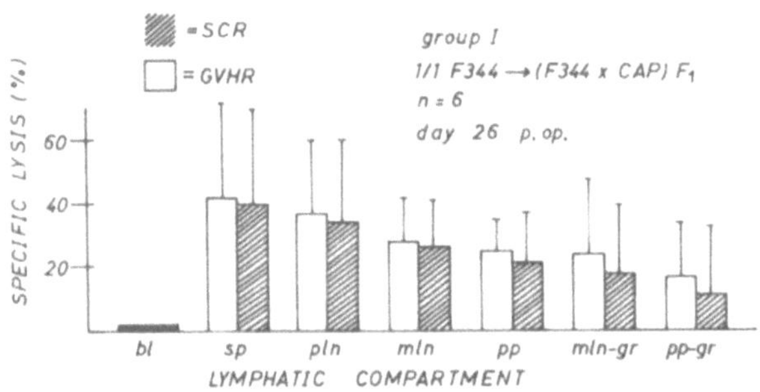

Fig. 2. Cell-mediated in vitro reactivity of the various lymphatic compartments of the recipient (*bl*, blood; *sp*, spleen; *pln*, peripheral lymph nodes; *mln*, mesenteric lymph nodes; *pp*, Peyer's patches) and of the graft (*mln-gr*, mesenteric lymph nodes; *pp-gr*, Peyer's patches)

Characterization of GVHR. For measuring GVHR, embryonic fibroblasts from the parental strain which did not provide the graft were used as targets. Figures 2–4 demonstrate cell compartment, strain combination, and serum factor dependence of in vitro measurable GVHR. With regard to cell compartment dependence, two main patterns have been observed: (a) lack of reactivity in the transport compartment blood combined with reactivity in the various lymphatic tissue compartments, interestingly including donor and recipient Peyer's patches; (b) remarkable reactivity in the transport compartment blood combined with reactivity in the various lymphatic tissues, excluding donor and recipient Peyer's patches. These strongly differing patterns appear to be correlated with the mortality rates in the strain combinations studied, insofar as GVH reactivity in blood is observed in animals which die within the following week. On the other hand, long-term survivors without blood compartment reactivity contain factors in their serum which are able to inhibit cell-mediated reactivity (Fig. 4).

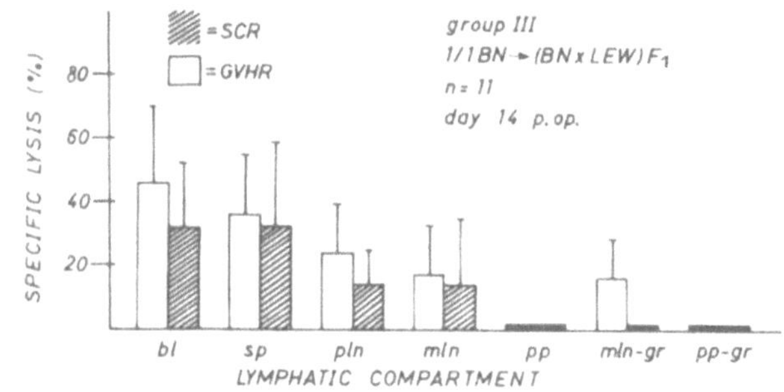

Fig. 3. Cell-mediated in vitro reactivity of the various lymphatic compartments of the recipient (*bl*, blood; *sp*, spleen; *pln*, peripheral lymph nodes; *mln*, mesenteric lymph nodes; *pp*, Peyer's patches) and of the graft (*mln-gr*, mesenteric lymph nodes; *pp-gr*, Peyer's patches) in group III

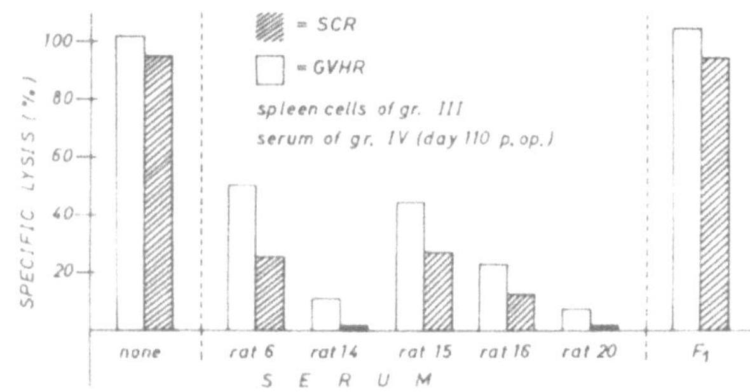

Fig. 4. Inhibition of cell-mediated in vitro reactivity by serum

Characterization of "Self-Constituent Reactivity" (SCR). For measuring SCR, embryonic fibroblasts from the parental strain which provided the graft were used as target cells. As is evident from Figs. 2–4 the observed SCR follows the same pattern of appearance as that of GVHR, i.e., it shows cell compartment, strain combination, and serum factor dependence. However, though dependent on GVHR, it follows independent kinetics, when measuring both reactivities on days 25, 50, and 110 postoperatively (data not shown in figures).

Further Characterization of Reactive Cells and Targets. Reactive spleen cells were shown to be SAL sensitive, that is sensitive against specified antilymphocyte serum, which is exhaustively macrophage absorbed (Müller-Ruchholtz et al. 1976). Furthermore, no reactivity was observed when using third party target fibroblasts; that is, both GVHR and SCR appeared to be specific.

Discussion

Two strain combinations strongly differing in GVHR-induced mortality have been investigated. From the findings in Fig. 1, we conclude that mortality differences between the two strain combinations are not very impressive when grafting only half SI, but quite impressive when grafting total SI. Strong differences occur even within one strain combination when grafting half versus total SI. Considering that the F344/CAP combination is characterized by three but the LEW/BN combination by five MHC mismatches, this difference in histoincompatibility, obviously fully effective only in full-length grafts, could be regarded as a reasonable explanation for the differences in mortality.

The most interesting finding concerning GVHR after SI transplantation, namely, its compartmentalization, enables the hypothesis to be put forward that lymphocytes may become more aggressive to the host, the better they are able to move freely in the blood. Once rats have survived the GVHR crisis between days 10 and 20, cytotoxic effector cells are missing in the transport compartment blood and appear to be "locked in" into the lymphoid tissue compartments. A similar "locking-in phenomenon" (nonreactivity in blood versus reactivity in the splenic compartment) was observed several years ago in rats specifically tolerating skin allografts (Bröcker et al. 1975). Additionally to this cell-mediated phenomenon, serum factors become demonstrable which block cytotoxic effector cells in vitro. How far these two phenomena can be correlated, and which the nature of the serum factor(s) is, awaits clarification.

Autoantibodies, elicited by GVHR, were observed some years ago by Gleichmann and Gleichmann (1976). On the other hand, an in vitro activation of F_1-hybrid cytotoxic T-lymphocytes specific for self H-2 has been described by the Cudkowicz group (Nakano et al. 1981). The findings described here may be considered in this context and even regarded as a connecting link in that in vivo (rather than under the more or less artificial in vitro conditions) developing cell-mediated autocytotoxicity (rather than autoantibody production) has been observed. Concerning mortality, when SCR as well as GVHR can be demonstrated in the transport compartment, the question must be raised whether it is SCR rather than GVHR which kills the graft recipients.

A number of questions call for further investigation: (a) What is the nature of the autoreactive cell? (b) Where does the autoreactive cell originate from (Nakano et al. 1981 found the F_1-hybrid cell to be reactive)? (c) What is the meaning of the independent kinetics of GVHR and SCR? (d) How far are both reactivities determined by the continuous supply of immunocompetent cells after SI transplantation?

References

Bröcker EB, Schienle W, Müller-Ruchholtz W (1975) Time- and compartment-dependent variation of lymphocytotoxicity in sensitized animals. Pathol Microbiol 38:35

Gleichmann E, Gleichmann H (1976) Graft-versus-host reaction: a pathogenic principle for the development of drug allergy, autoimmunity, and malignant lymphoma in non-chimeric individuals. Hypothesis. Z Krebsforsch 85:91–109

Müller-Ruchholtz W, Wottge HU, Müller-Hermelink HK (1976) Bone marrow transplantation in rats across strong histocompatibility barriers by selective elimination of lymphoid cells in donor marrow. Transplant Proc 8:537–541

Nakano K, Nakamura I, Cudkowicz G (1981) Generation of F_1 hybrid cytotoxic T lymphocytes specific for self H-2. Nature 289:559–563

Takasugi M, Klein E (1970) A microassay for cell-mediated immunity. Transplantation 9:219–227

Combined Pancreas and Kidney Transplantation in the Rat

M. S. Nolan, N. J. Lindsey, N. P. Ingram, A. Herold,
D. N. Salter, and *M. Fox*

Introduction

Most centres that have applied pancreatic transplantation for the treatment of Type-I insulin-dependent diabetes mellitus have performed simultaneous renal grafting (Dubernard et al. 1981; Baumgartner et al. 1983; McMaster 1983). The combined procedure currently carries a higher success rate than pancreas transplantation when performed alone (Sutherland 1983). Although experimental models do exist in large mammals for simultaneous pancreaticorenal transplantation (Severyn et al. 1982), no satisfactory microsurgical model has been described. We report a microsurgical model for combined pancreaticorenal transplantation. This permits the use of highly inbred rat strain with their concomitant advantages.

Materials and Methods

Donor Operation

Female AGUS × WAG ($RT1^l \times RT1^u$) F_1-hybrid rats (150–220 g) were always used as donors and all vessel ligation was performed using 7-0 silk, unless otherwise stated. Under ether anaesthesia, following a midline incision the hepatic artery was ligated and divided and the portal vein cleared of connective tissue. The pancreas was separated from the stomach by ligation and division of the short gastric and left gastric vessels. A segment of aorta, bearing the coeliac, superior mesenteric and right renal arteries was mobilised. Following displacement of the abdominal viscera to the left-hand side a 1-cm length of infrarenal vena cava was mobilised and the right adrenal artery arising from the right renal artery was ligated and divided. The right kidney and ureter were freed and the viscera displaced to the right-hand side, allowing ligation and division of the middle colic and superior mesenteric vessels. The common bile duct was ligated at the duodenum and care was taken to preserve the pancreaticoduodenal arcade during separation of the pancreas and duodenum. Following isolation of the pancreas, the proximal common bile duct was either cannulated at the level of the liver hilum (2 fine gauge, FG; intravenous, i.v.; cannula outside diameter, OD 0.75 mm) and

Microsurgical Models in Rats
for Transplantation Research
Eds. A. Thiede et al.

the pancreatic duct system was obliterated using 0.05–0.08 ml latex (polyisoprene).

Malaysian Rubber Producers, Research Association, Brickendonbury, Herts, England (McMaster 1983), or alternatively simply transected without ligation at the same position. Splenic vessels at the hilum of the spleen were ligated but not divided, the spleen proving useful as a handle to minimise graft trauma. The donor rat was subsequently heparinised (100 IU heparin sodium/ml in 5 ml 0.9% w/v saline) via the inferior vena cava, and approximately 5 min later the proximal aorta above the coeliac artery was occluded using a Scoville-Lewis microvascular clip. The suprarenal vena cava was then ligated and divided, while the infrarenal vena cava was transected. The combined pancreaticorenal graft was removed by transection of the portal vein, distal and proximal aorta and ureter, flush with the donor bladder. Both organs were immersed in 5 ml cold (4 °C) modified Collins C2 solution (Collins et al. 1969) and flushed clear of blood using 2 ml of the same medium via the aortic segment. The portal vein was identified and a 3-mm-long 6-FG, 2.10-mm-OD (Portex Ltd., Hythe, Kent, England) polythylene cuff was slipped over the vein, the ends of which were everted over the cuff and secured using a circumferential 7-0 silk ligature. A further 0.2 ml cold perfusate was gently flushed through the pancreas to ensure that twisting of the portal vein had not occurred during cuff preparation. The organ preparation was then placed in 45 ml of the same medium and refrigerated (2°–4 °C) until transplantation.

Recipient Operation

Diabetic male (AGUS × WAG) F_1-hybrid rats were used as recipients. Diabetes was indueced by i.p. injection of streptozotocin (50 mg/kg body weight) and recipients were used 5–7 days later if serum glucose levels >22 mmol/litre, associated with glycosuria and loss of body weight. The technique of combined transplantation is shown diagrammatically in Fig. 1. Under ether anaesthesia, following a midline incision, the left renal vessel were mobilised from each other. The procedure often necessitated ligation and dissection of the left adrenal artery and testicular veins. Similarly, 1.5-cm lengths of infrarenal abdominal aorta and vena cava were mobilised. A vessel approximator was applied across the aorta and the latter transected between the clamp. A 3-mm-long 5-FG (OD, 1.64 mm) polyethylene cuff was slipped over the proximal end of the transected recipient aorta and a 3-mm-long 4-FG (OD, 1.34 mm) polyethylene cuff was slipped over the distal end of the transected recipient aorta. The transected ends of the aorta were everted over the cuffs and secured using circumferential 6-0 silk ligatures. The organ preparation was removed from cold perfusate and orientated so to align the donor and recipient's aorta. The donor aorta was interposed between the transected end of the recipient aorta using a cuff anastomosis as previously described (Nolan et al. 1982). The renal, coeliac and superior mesenteric arteries were occluded independently, and the vessel approximator removed restoring blood flow to the recipient's pelvis and lower limbs. The left renal artery was ligated and the left renal vein was ligated as close to the hilum of the kidney as possible. The vein was occluded at its junction with the inferior vena cava and a small incision made in its upper surface. The cuff of the donor portal vein was inserted into the veno-

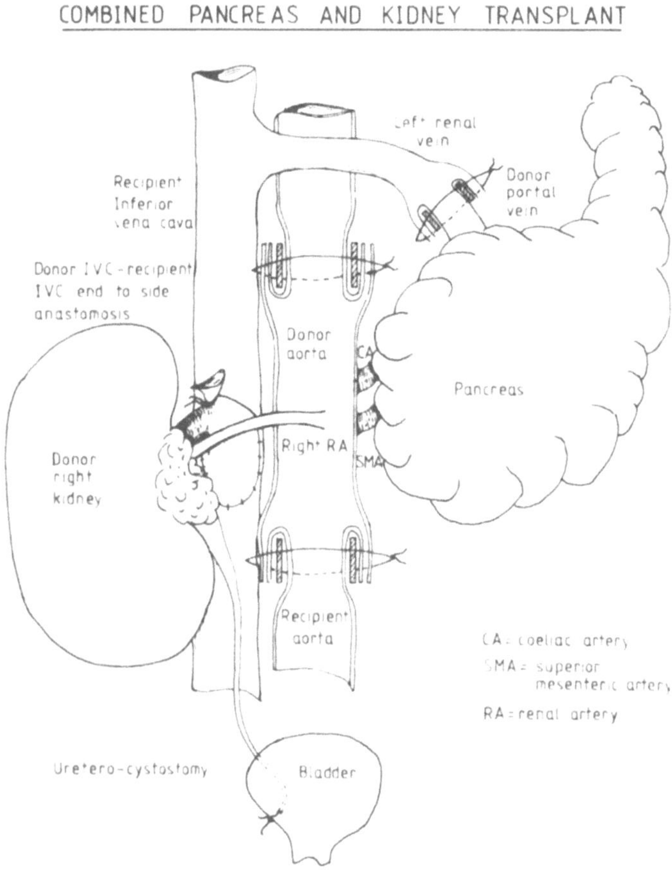

Fig. 1. Diagrammatic representation of the technique for combined pancreaticorenal transplantation in the rat

tomy, securing the anastomosis with a circumferential (7-0) silk ligature. The venous clamp together with the clips occluding the coeliac and superior mesenteric arteries was released, restoring pancreatic graft blood flow. Following occlusion of the recipient inferior vena cava, a venotomy was performed and the vessel rinsed with 0.9% saline. For renal venous drainage, end-to-side donor to recipient caval anastomosis was performed using eight to ten interrupted 10-0 monofilament nylon sutures and the anastomosis surrounded with surgical gauze. The proximal and distal recipient caval clips were released, followed by removal of the renal arterial clip.

Ureteric drainage was accomplished by ureterocystostomy. This was performed through a stab incision in the recipient bladder and the donor ureter was secured inside the bladder to the posterior wall using a 7-0 silk ligature. In cases where pancreatic duct to ureter anastomosis was performed, the proximal common bile duct of the graft was anastomosed end-to-end (four interrupted, 10.0 monofilament sutures) to the recipient left ureter. Immediately postoperatively all recipients received 4 ml human plasma protein fraction i.v. at 37 °C and 25 mg Cefuroxime sodium (Glaxo Laboratories, Greenford, Middlesex, England). Contralateral nephrectomy was performed usually 4–5 days postoperatively.

Blood samples were taken for measurement of random serum glucose on the first postoperative day and subsequently for 4 days after contralateral nephrectomy, then at 50 and 100 days postoperatively. For animals surviving beyond 100 days oral glucose tolerance tests (OGTTs: 2.5 g glucose/kg body weight) and intravenous glucose tolerance tests (IVGTT: 1 g glucose/kg body weight) were performed at 115 and 120 days postoperatively, respectively. IVGTT *K* values (percentage decline of serum glucose per minute) were calculated according to the method of Moorhouse et al. (1964). Some animals underwent removal of the pancreas graft immediately following the IVGTT and the recurrence/nonrecurrence of diabetes was noted over three consecutive days thereafter, after which time the transplanted kidney was removed. Both transplanted organs were fixed in formal saline for routine light microscopy examination.

Results

A total of 44 isografts have been performed, the first 14 of which enabled the development of a satisfactory technique. Of the subsequent 30 transplants 6 rats died within 24 h (internal haemorrhage) and a further 2 rats died on the 3rd and 5th postoperative days from undetermined causes. The remaining 22 animals (15 with latex-obliterated pancreas grafts and 7 with the pancreatic duct anastomosed

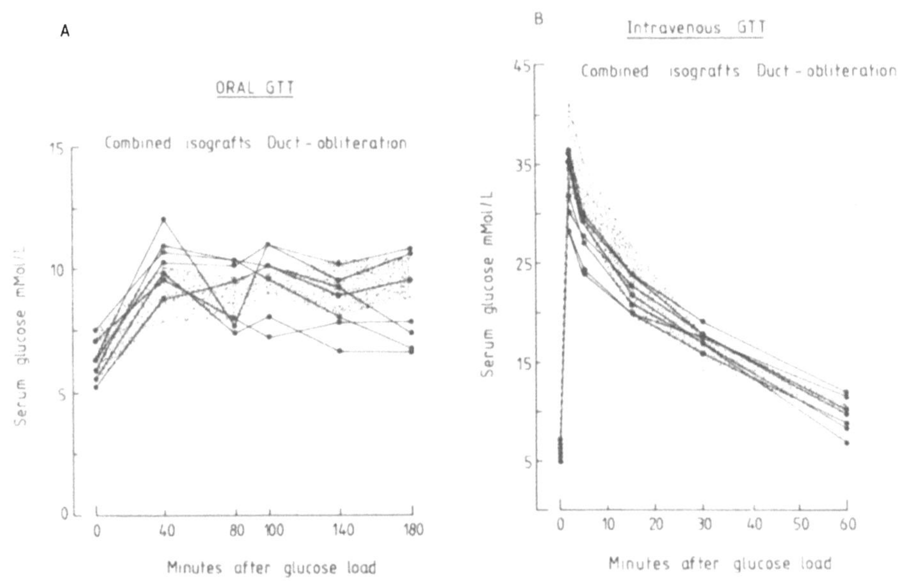

Fig. 2. A Results of OGTTs at 115 days postoperatively for seven recipients of combined isografts with latex obliteration of the pancreas graft. **B** Results of IVGTTs at 120 days for the same animals. Included in both *A* and *B* are the mean ±1 SD glucose values (*stippled area*) for six normal non-transplanted control rats

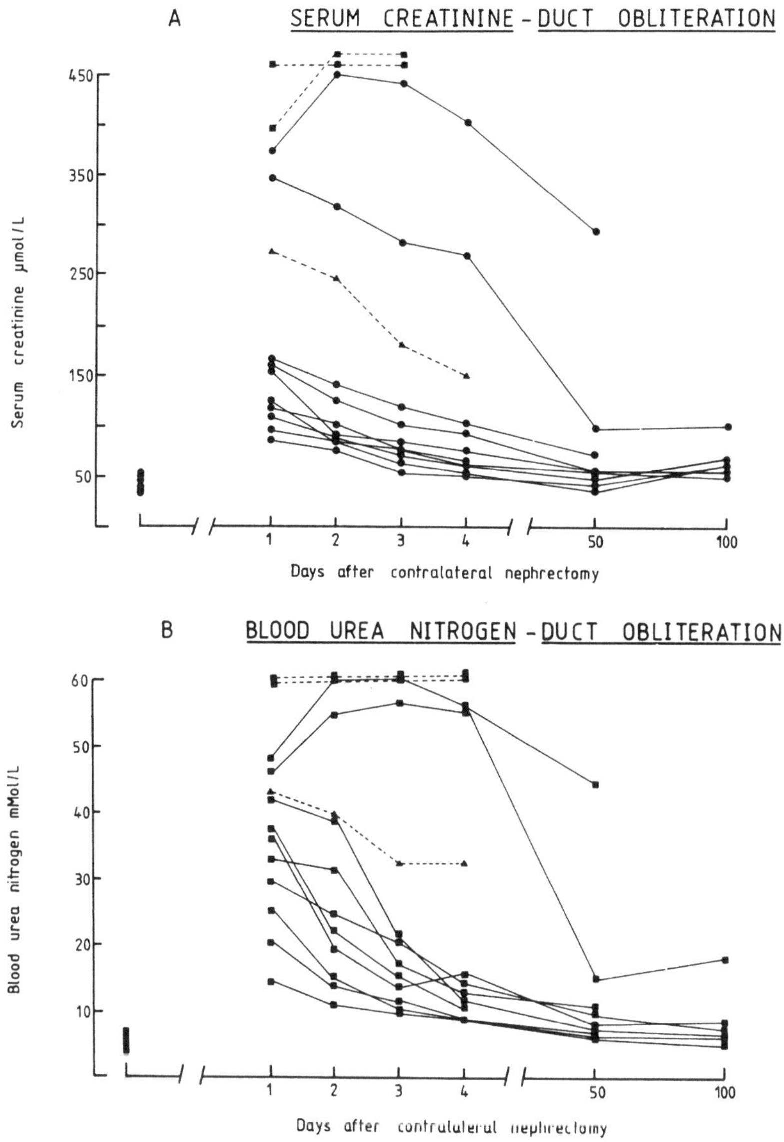

Fig. 3. A Serum creatinine levels prior to the induction of diabetes and following contralateral nephrectomy for recipients of combined pancreaticorenal isografts with latex obliteration of the pancreatic ducts. **B** Blood urea nitrogen levels at the same time intervals for the same animals. **A, B** ■--------■, death from uraemia; ▲--------▲, death from infection

to the recipient ureter) were considered technically successful. Of these, 15 (68%) (10 with latex-obliterated pancreas grafts and 5 with the pancreatic duct anastomosed to the recipient ureter) have survived beyond 50 days, 7 of which (all with latex-obliterated pancreas grafts) have survived beyond 100 days. The remaining rats have not yet completed the 100-day postoperative course at the time of writing. Total pancreas and renal cold ischaemia for the 15 survivors was 147 ± 28 min, pancreatic warm ischaemia was 10 ± 1 min and renal warm ischaemia was 34 ± 9 min. Total recipient operative time was 121 ± 27 min. The group of 22 technically successful combined isografts all showed immediate reversal of their diabetes and no animal died or was killed due to recurrence of diabetes. The 15 long-term survivors have all remained normoglycaemic and aglycosuric and have shown a normal weight gain. The results of OGTT for the seven long-term survivors with latex-obliterated ducts are shown in Fig. 2 A. All were normal when compared with a group of normal non-transplanted control animals. Similarly the results of IVGTTs (Fig. 2 B) for the same animals at 120 days showed no difference in the *K* values when compared with normal control animals; *K*, 1.93 ± 0.4%/min and *K*, 1.99 ± 0.19%/min respectively (mean ± SD, $P > 0.05$, using Student's *t* test).

Temporary uraemia was always seen immediately following contralateral nephrectomy and this correlated directly with the degree of renal warm ischaemia. However, renal function within normal limits was evident by 50 days postoperatively as shown by serum creatinine and blood urea nitrogen levels (Fig. 3 A, B, respectively). In four animals (all with latex-obliterated pancreas grafts) removal of the graft's pancreas was performed immediately following the IVGTT. All animals reverted to the diabetic state within 24 h and remained diabetic over the three consecutive days, indicating that a spontaneous recovery of the recipient's own beta-cell function had not occurred. Of the seven animals dying before 50 days that were considered technically successful, five deaths were due to uraemia (three rats – all with latex-obliterated pancreas grafts – died from excessive renal warm ischaemia of > 55 min, days 6, 7, and 8 postoperatively; and two rats – one with pancreatic duct to ureter anastomosis and one with a latex-obliterated graft – died from uraemia caused by ureteric obstruction, days 7 and 10 postoperatively). A further animal with pancreatic duct to ureter anastomosis was killed on the 7th postoperative day due to leakage of the ureterocystostomy and one animal (with a latex-obliterated pancreas graft) died on the 14th postoperative day due to infection.

The histological appearance of latex-obliterated pancreas grafts removed at 120 days postoperatively showed a total disruption of exocrine tissue but intact islets of Langerhans. Latex was identifiable within the pancreatic ducts by a giant-cell reaction. Transplanted kidneys removed at 123 days postoperatively in most cases appeared normal. Occasional cases showed mild tubular dilatation and focal chronic interstitial inflammation.

Discussion

We have previously used cuff anastomoses for isolated pancreas transplantation in the rat (Nolan et al. 1982) and the addition of only one conventional suture anastomosis – for renal venous drainage – has now enabled extension of this technique to simultaneous pancreaticorenal transplantation.

Cuff techniques as originally described by Nitze (1897) have been used with increasing frequency for vascular anastomoses in small animals. Although initially employed in the rabbit (Heron 1970; Green 1973), Heron (1971) first reported anastomotic techniques in the rat. Since this time cuff techniques have been used in the rat for orthotopic liver transplantation (Kamada and Calne 1979; Miyata et al. 1980) and heterotopic heart (Shaw 1983) and pancreas grafting (Brynger et al. 1983).

In the current experimental technique for combined pancreaticorenal transplantation we have not encountered arterial or venous thrombosis or anastomotic leakage due to the use of cuff anastomoses, and only slight intimal vascular proliferation has occurred at the anastomotic site.

The technique is reliable and reproducible and technically simplifies microvascular surgery in a small-animal model. For ureteric drainage of the donor kidney we initially performed ureteroureterostomy, employing a polyethylene stent, and although this technique minimized the operative time it was abandoned as obstruction occurred in some cases. End-to-end anastomoses using four interrupted 10-0 monofilament sutures was also used during development of the technique, but this prolonged substantially the operative time. Implantation of the donor ureter into the recipient bladder was selected as a compromise (Fisher and Lee 1965) and using this technique ureteric obstruction has still occurred in two cases – both of which were due to the donor ureter coming out of the recipient bladder because of being inadequately secured.

The histological appearance of latex-obliterated pancreas grafts at 120 days postoperatively was very similar to that observed at 1 month postoperatively (Nolan et al. 1983). The interstitial inflammation observed in transplanted kidneys at the same time period is probably attributable to the technique of ureteric drainage used and was always associated with a mild degree of hydronephrosis.

To our knowledge the only other report of simultaneous pancreaticorenal grafting in the rat was performed en bloc with liver, spleen and duodenum (Maki et al. 1977). However, biochemical data were not presented and function of either transplanted organ could not be assessed as the rats were not diabetic and contralateral nephrectomy was not performed.

Acknowledgment. We wish to thank the Sheffield Area Kidney Association for financial aid.

References

Baumgartner D, Largiader F, Uhlschmid G, Binswanger U (1983) Rejection episodes in recipients of simultaneous pancreas and kidney transplants. Transplant Proc 15:1330–1331

Brynger H, Mjoornstedt L, Olausson M (1980) Heterotopic grafting of the pancreas to the neck in the rat – an experimental model. Transplant Proc 12:148–149

Collins GM, Bravo-Shugarman M, Terasaki P (1969) Kidney preservation for transportation. Initial perfusion and 30 hours ice storage. Lancet 2:1219–1222

Dubernard JM, Traeger J, Touraine JL, Malik MC, Martin X, Devonec M (1981) Patterns of renal and pancreatic rejection in double grafted patients. Transplant Proc 13:305–307

Fisher B, Lee S (1965) Microvascular surgical techniques in research with special reference to renal transplantation in the rat. Surgery 58:904–914

Green CJ (1973) Rabbit renal autografts as an organ preservation model. Lab Anim 7:1–11
Heron I (1970) Kidney transplantation in the rabbit. A new method. Acta Pathol Microbiol Scand 78:90–95
Heron I (1971) A technique for accessory cervical heart transplantation in rabbits and rats. Acta Pathol Microbiol Scand 79:366–372
Kamada N, Calne RY (1979) Orthotopic liver transplantation in the rat. Technique using cuff for portal vein anastomosis and biliary drainage. Transplantation 28:47–50
Maki T, Sakai A, Petirossi O, Kountz S (1977) En bloc transplantation of the liver, pancreas, duodenum, spleen and kidney in the rat. Transplantation 24:256–262
McMaster P (1983) Pancreas transplants. Hos Update 9:369–374
Miyata M, Fischer JM, Fuhs M, Isselhard W, Kasai Y (1980) A simple method of orthotopic liver transplantation in the rat. Cuff technique for three vascular anastomoses. Transplantation 30:335–338
Moorhouse JA, Grahame GR, Rosen NJ (1964) Relationship between intravenous glucose tolerance and the fasting blood glucose level in healthy and in diabetic subjects. J Clin Endocrinol 24:145–159
Nitze M (1897) Gefäßnaht(discussion). Zentralbl Chir 24:1042
Nolan MS, Lindsey NJ, Savas CP, Slater DN, Fox M (1982) Pancreatic transplantation in the rat: a simplified method using aortic interposition and cuff techniques. Transplantation 33:327–329
Nolan MS, Lindsey NJ, Savas CP, Herold A, Slater DN, Beck S, Fox M (1983) Pancreatic transplantation in the rat. Long term study following different methods of management of exocrine drainage. Transplantation 36:26–33
Severyn N, Olsen L, Miller J, Kayriakides G, Rabinovitch A, Flaa C, Mintz D (1982) Studies on the survival of simultaneous canine renal and segmental pancreatic allografts. Transplantation 33:606–612
Shaw JFL (1983) Comparison of the effects of acetylsalicyclic acid and sodium salycyclate on prolongation of rat cardiac allograft survival and on inhibition of rat platelet aggregation. Transplantation 36:33–36
Sutherland DER (1983) Current status of pancreas transplantation: Registry statistics and overview. Transplant Proc 15:1303–1307

Immunological Characteristics of Vascularized Pancreas Transplantation *

J. Klempnauer, B. Steiniger, K. Wonigeit, and *E. Günther*

Introduction

Transplantation of the vascularized pancreas may not only correct the deranged intermediary metabolism of diabetes mellitus, but may also prevent, halt, or to some degree even reverse the vascular changes associated with diabetes. The results of clinical pancreas transplantation, however, have generally been disappointing (Sutherland 1983). Unlike in renal and cardiac transplantation long-term allograft survival has only rarely been accomplished. Besides technical problems mainly related to the management of the exocrine secretion, this has been ascribed to the fact that the pancreas may be more susceptible to the immune assault or may be more immunogenic than other primarily vascularized organ transplants. In an attempt to investigate possible immunological characteristics of pancreas transplantation an immunogenetic analysis of the effect of different classes of histocompatibility antigens was performed in a model of pancreas whole organ grafting in streptozotocin diabetic rats. By using donor recipient strain combinations which differed either at major histocompatibility complex (MHC) or non-MHC antigens or both it was possible to investigate the relative contribution of MHC and non-MHC histocompatibility antigens to rejection of vascularized pancreas allografts. Furthermore, the differential effect of class I and class II MHC alloantigens could be assessed by using intra-MHC recombinant rat strains. The individual histocompatibility antigens differed markedly in their capacity to induce rejection of pancreas allografts. Moreover, striking differences could be demonstrated of the effect of various classes of histocompatibility antigens in pancreas transplantation when compared with data obtained with heart or kidney transplants in comparable strain combinations.

Materials and Methods

Organization of the Rat MHC. The rat MHC consists of several closely linked gene loci. The RT1.*A* region codes of class I antigens and the RT1.*B* region codes for the expression of class II MHC antigens. The RT1.*C* region represents a fur-

* This work was supported by Stiftung Volkswagenwerk AZ I/38 510

Microsurgical Models in Rats
for Transplantation Research
Eds. A. Thiede et al.

ther class I locus and closely resembles the Qa antigens of the mouse. Recombinant rat strains which separate a RT1.*D* region within the RT1.*B* complex were not included in this study. Histocompatibility antigens not encoded within the MHC are summarized as non-MHC alloantigens (Günther, this volume, p. 83).

Animals. In order to assess the differential effect of MHC and non-MHC histocompatibility antigens in pancreas transplantation, rats of the strains AVN, LEW, and LEW.1A were used. LEW.1A rats express the MHC of AVN on the genetic background of LEW. Appropriate donor recipient strain combinations differed either at MHC or non-MHC alloantigens or both. Taking advantage of the congenic rat strains LEW.1A and LEW.1U and their intra-MHC recombinants LEW.1AR2, LEW.1WR1, and LEW.1WR2, it was possible to investigate the relative importance of the individual MHC gene regions. In appropriate strain combinations donor and recipient differed either at their entire MHC or of the RT1.*A*, RT1.*B*, and RT1.*C* regions only.

Technique of Pancreas Transplantation and Definition of Rejection. Pancreas whole organ transplantation was performed in streptozotocin diabetic rats. By careful preservation of the pancreaticoduodenal arcade it was possible to transplant the entire islet-cell mass of a single donor pancreas. The exocrine secretion was suppressed by ligating the common bile duct at the liver hilum and at its entrance into the duodenum. The pancreas was isolated on a segment of the aorta giving off the celiac axis and the superior mesenteric artery.

The venous outflow was provided by the portal vein. Heterotopic intraperitoneal transplantation was performed by means of end-to-side anastomoses of the donor aorta and portal vein to the recipient infrarenal aorta and vena cava. Prospective recipients of pancreas transplants were rendered diabetic by a single i.v. dose of 55 mg/kg streptozotocin. Diabetes was confirmed by repetitive demonstration of blood glucose levels above 18 mmol/liter. Transplant function was monitored by daily determination of the nonfasting blood glucose levels. Rejection was defined by recurrence of hyperglycemia above 14 mmol/liter on at least two consecutive days. The functional demonstration of rejection was always confirmed by histological examination of the rejected organ.

Results

Isogenic Controls. Following pancreas whole organ transplantation in streptozotocin diabetic LEW recipients (*n*, 10), normoglycemia was established within 6 h after transplantation and maintained throughout the observation period of 100 days. Graft removal on day 100 was followed by prompt recurrence of diabetes. Duct ligation caused complete diasappearance of acinar cells, proliferation of centroacinar cells, and neoformation of small pancreatic ducts while the islets remained grossly normal (Steiniger et al. 1984).

Effect of MHC and Non-MHC Alloantigens. MHC-incompatible LEW.1A pancreas grafts were rejected by LEW recipients after a mean of 13 days. Additional incompatibility at non-MHC alloantigens in the strain combination AVN→LEW

Table 1. Effect of MHC and non-MHC alloantigens in vascularized pancreas allotransplantation

Donor	RT1	Recipient	RT1	Histocompatibility barrier	Rejection times (days)
LEW.1A	*a*	LEW	*l*	MHC	10, 12, 14, 14, 15, 17
AVN	*a*	LEW.1A	*a*	non-MHC	13, 15, 16, 18, 18
AVN	*a*	LEW	*l*	MHC plus non-MHC	8, 9, 10, 10, 11, 12

shortened allograft survival significantly to a mean of 10 days. Unexpectedly, all non-MHC-incompatible pancreas allografts in the strain combination AVN→LEW.1A were uniformly rejected in a rapid fashion. The rejection times did not differ significantly from those of MHC-incompatible grafts only (Table 1).

Effect of Class I and Class II MHC Alloantigens. Fully MHC-incompatible pancreas grafts in the strain combination LEW.1A→LEW.1U were rejected more rapidly than grafts incompatible at individual RT1 regions only (Table 2). The individual MHC gene regions differed markedly in their capacity to induce rejection of vascularized pancreas allografts. Incompatibility at either class I or class II antigens was sufficient to induce pancreas rejection. Class I antigens encoded for by the RT1.*A* region had a stronger effect than class II antigens coded for by the RT1.*B* region. Following transplantation across a RT1.*A* barrier rejection occurred after a mean of 14 days compared with 27 days when transplanted across a RT1.*B* region incompatibility. Only one out of six RT1.*C* region disparate grafts was rejected on day 26; the remaining animals remained normoglycemic for 100 days and promptly became hyperglycemic upon graft removal. In all instances the functional demonstration of rejection was confirmed by histological examination. The histological pattern of rejection did not differ according to the degree of histoincompatibility. Rejected organs were characterized by dense cellular infiltration, massive fibrosis, and disintegration of the islets (Steiniger et al. 1984).

Table 2. Effect of class I and class II MHC alloantigens in vascularized pancreas allotransplantation

Donor	RT1			Recipient	RT1			Histocompatibility barrier	Rjection times (days)
	A	*B*	*C*		*A*	*B*	*C*		
LEW.1A	*a*	*a*	*a*	LEW.1U	*u*	*u*	*u*	Full RT1	9, 9, 9, 9, 9, 10, 10
LEW.1A	*a*	*a*	*a*	LEW.1WR2	*u*	*a*	*a*	RT1.*A*	11, 11, 14, 14, 16, 16
LEW.1WR2	*u*	*a*	*a*	LEW.1WR1	*u*	*u*	*a*	RT1.*B*	16, 21, 29, 31, 31, 33
LEW.1A	*a*	*a*	*a*	LEW.1AR2	*a*	*a*	*u*	RT.1*C*	26, >100×5
LEW.1A	*a*	*a*	*a*	LEW.1WR1	*u*	*u*	*a*	RT1.*A* plus RT1.*B*	8, 8, 8, 9, 10, 10

Discussion

The effect of various classes of histocompatibility antigens in pancreas rejection was studied in a model of pancreas whole organ transplantation in congenic and recombinant rat strains. In streptozotocin diabetic recipients graft function was monitored by daily blood glucose measurements and rejection was defined by recurrence of hyperglycemia. Relapse of diabetes is a very sensitive and reliable functional parameter of pancreas rejection. The functional demonstration of rejection was always confirmed by histological examination. False-positive results, i.e., recurrence of hyperglycemia without rejection or false-negative results, i.e., rejection without recurrence of hyperglycemia, were not observed in this study. Furthermore, prompt recurrence of diabetes upon graft removal in nonrejecting strain combinations clearly indicated that normoglycemia was entirely due to the presence of the graft.

The development of genetically defined inbred and congenic strains of rodents has facilitated a genetic analysis of the biological significance of MHC gene products. Histocompatibility antigens encoded by the MHC play a dominating role in eliciting rejection responses. This study confirms the role of major histocompatibility antigens also in transplantation of the vascularized pancreas. When donor and recipient differed at their entire MHC rapid rejection of all pancreas grafts occurred. Marked differences, however, were noted of the capacity of the individual MHC gene regions to induce rejection. Class I differences alone were sufficient to induce rejection. Compatibility at class II antigens or the RT1.*B* region did not ensure permanent survival of vascularized pancreas allografts. The effect of class II antigens encoded by the RT1.*B* region was less pronounced but they were also able to induce rejection on their own. A cumulative effect of class I and class II antigens was demonstrated when donor and recipient differed at both the RT1.*A* and RT1.*B* regions. No in vivo effect of the RT1.*C* region gene products could be demonstrated when transplantation was performed across an isolated RT1.*C* region barrier or in combination with the RT1.*A* and RT1.*B* regions.

A similar effect of class I antigens was noted in transplantation of isolated mouse islets across H-2K and H-2K plus D disparities (Marrow et al. 1983). Lack of a stimulus from class II alloantigens did also not ensure permanent islet allograft survival. Unlike in vascularized pancreas transplantation, however, islets were only rarely rejected when transplanted across a class II (H-2I) incompatibility only.

The pancreas is a composite organ consisting of two functionally and morphologically distinct components, the exocrine and the endrocrine part. Transplantation is indicated only for endocrine insufficiency. In vascularized pancreas transplantation, however, the islets are inevitably transplanted along with the exocrine tissue. Possible differences of the expression of histocompatibility antigens might therefore be very relevant for the immunogenicity of vascularized pancreas grafts compared with isolated islets. Islets of Langerhans have been shown to express class I antigens while exocrine cells were largely negative (Hart et al. 1983). Class II antigens have been demonstrated only on interstitial dendritic cells both within the islets and the exocrine pancreas but not on the B cells themselves (Hart et al. 1983; Steiniger et al. 1985). There is strong evidence that class I antigens have a dominating effect in rejection of both vascularized pancreas and islet transplantation. Their strong effect in pancreas transplantation is in contrast to findings obtained in identical strain combinations with heart (Klempnauer et al., this volume, p. 121) and kidney (Paris and Günther 1980) transplantation which were not rejected when transplanted across an isolated RT1.*A* barrier. An explanation for these organ-specific differences of the effect of RT1.A^a gene products is difficult to find but may be related to differences of the antigenic structure of the RT1.A^a molecule (Diamond et al. 1983).

The relative contribution of MHC and non-MHC alloantigens to rejection of vascularized pancreas allografts was studied in donor-recipient strain combinations differing at MHC or non-MHC alloantigens or both. Non-MHC histocompatibility antigens were capable of inducing rejection responses which were as strong as those caused by MHC antigens. In sharp contrast heart allografts were not rejected when transplanted across the same non-MHC incompatibility (Klempnauer et al., this volume, p. 121, 1984). This phenomenon is not a peculiarity of the strain combination used, since the strong effect of non-MHC alloantigens could be demonstrated in a total of six different strain combinations sharing the RT1 haplotypes *a*, *l*, and *u* (Klempnauer et al. 1983a). The strong effect of non-MHC alloantigens may have clinical implications with regard to the value of prospective HLA matching in vascularized pancreas transplantation, since even in the case of MHC identity strong rejection responses have to be anticipated caused by non-MHC alloantigens. Furthermore, this phenomenon provides a possible explanation for the observation that cyclosporin A immunosuppressive proved less effective in clinical pancreas transplantation than in cardiac or renal allografting. This correlates with experimental findings that in fully allogenic strain combinations using identical short-term cyclosporin A immunosuppression it has been difficult to suppress rejection of pancreas allografts, while heart and kidney allografts enjoyed permanent survival (Klempnauer et al. 1982, 1983b). It is crucial to determine whether the effect of non-MHC alloantigens in pancreas transplantation reflects the cumulative effect of multiple non-MHC antigens or whether it is caused by an alloantigenic system expressed on the pancreas but not on heart or kidney. In any case it is important to come to an estimate of the number of systems involved by the analysis of segregating populations.

The final point concerns the localization of the functionally important non-MHC alloantigens within the pancreas. If the relevant non-MHC antigens are situated exclusively on the exocrine part of the pancreas, the effect of non-MHC antigens may be overcome by transplantation of pure islets of Langerhans. If the functionally significant non-MHC antigens are located on the islets themselves, the strong effect of non-MHC histocompatibility antigens can also be anticipated in islet transplantation and is inheritant to all forms of endocrine pancreatic transplantation.

Acknowledgment. We gratefully appreciate the excellent technical assistance of Margot Giesecke, Anne Franke, and Petra Peschel.

References

Diamond AG, Butcher GW, Howard JC (1983) The antigencity and structure of RT1A. Transplant Proc 15:1554–1556

Hart DNJ, Newton MR, Reece-Smith H, Fabre JW, Morris PJ (1983) Major histocompatibility complex antigens in the rat pancreas, isolated pancreatic islets, thyroid and adrenal. Transplantation 36:431–435

Klempnauer J, Kasahara K, White DJG, Calne RY (1982) Differential susceptibility to Cyclosporin A immunosuppressive of fully allogenic vascularized heart and pancreas transplants in the rat. Eur Surg Res 14:128–129

Klempnauer J, Wonigeit K, Günther E, Pichlmayr R (1983a) Pancreas whole organ transplantation in the rat. Evidence for a strong effect of non-MHC incompatibilities. Transplant Proc 15:1649–1651

Klempnauer J, Wagner E, Wonigeit K, Pichlmayr R (1983b) Pancreas and kidney allograft rejection responds differently to cyclosporine A immunosuppressive. Transplant Proc 15:3001–3003

Klempnauer J, Hoins L, Steiniger B, Günther E, Wonigeit K, Pichlmayr R (1984) Evidence for a differential importance of MHC and non-MHC alloantigens in pancreas and heart transplantation in the rat. Transplant Proc 16:778–780

Marrow CE, Sutherland DER, Steffes MW, Kaufman D, Najarian JS, Bach FH (1983) Differences in susceptibility to rejection of mouse pancreatic islet allografts disparate for class I or class II major histocompatibility antigens. J Surg Res 34:358–366

Paris A, Günther E (1980) Kidney grafting between rats which carry recombinant major histocompatibility haplotypes. Immunogenetics 10:205–209

Steiniger B, Klempnauer J, Brüsch U, Wonigeit K (1984) Histology of rejection in rat pancreas allografts with suppressed or preserved exocrine function. Transplant Proc 16:783–784

Steiniger B, Klempnauer J, Wonigeit K (1985b) Phenotype and histological distribution of interstitial dendritic cells in the rat pancreas liver, heart and kidney. Transplantation 38:169–175

Sutherland DER (1983) Pancreas transplantation: registry statistics and overview. Transplant Proc 15:1303–1307

Is Preserved Exocrine Secretion Mandatory for Optimal Endocrine Function in Vascularized Pancreas Transplantation?

U. Brüsch, J. Klempnauer, and *B. Steiniger*

In vascularized pancreas transplantation the exocrine tissue is inevitably transplanted along with the islets. However, transplantation of the pancreas is indicated for endocrine insufficiency only. A great variety of methods have been used in the management of exocrine secretion, both in clinical and experimental transplantation (Sutherland 1981 a, b). Despite the numerous different surgical methods employed, there are only two principal procedures: transplantation with preserved and with suppressed exocrine secretion.

In the rat we compared the functional and histological outcome of these two principal techniques. The exocrine secretion was preserved by pancreaticoduodenal transplantation, and suppression of the exocrine secretion was accomplished by duct ligation.

Material and Methods

Animals. Isogeneic pancreas whole organ transplantation was performed in male LEW rats, 2 months old, weighing 200–250 g. The graft recipients were rendered diabetic by a single i.v. dose of 55 mg/kg streptozotocin. Diabetes was confirmed by repetitive blood glucose levels above 18 mmol/liter. Transplantation was performed 4–6 days following the induction of diabetes. A total of 44 rates received a pancreas transplant with either suppressed (SUP, $n = 22$) or preserved (PRES, $n = 22$) exocrine secretion. Age and sex-matched diabetic (D, $n = 12$) and normal (N, $n = 7$) rats served as controls.

Monitoring. Body weight and nonfasting blood glucose levels were measured daily. Intravenous glucose tolerance tests (i.v. GTT, 1 g glucose/kg) were performed on days 10 and 100 after transplantation. The k values were calculated according to Lundbaek (1962). Food and water consumption, urine output, and production of feces were measured in metabolic cages shortly after the i.v. GTTs.

Anesthesia. Fentanyl-fluanison (Hypnorm, Janssen) was used for the donor operation and the i.v. GTTs, while the recipient operation was performed with ketamine (KetaVet, Parke-Davis) anesthesia.

Microsurgical Models in Rats
for Transplantation Research
Eds. A. Thiede et al.

Donor Operation. Preservation of the exocrine secretion was accomplished by pancreaticoduodenal transplantation, originally described by Lee et al. (1972). Ligation of the common bile duct both at the liver hilum and its entrance into the duodenum was performed in order to induce suppression of the exocrine secretion. Whole organ grafting was possible by preserving the pancreaticoduodenal arcade as blood supply for the head of the pancreas. With both forms of pancreas transplantation the arterial blood supply was provided by a segment of the aorta, giving off the celiac axis and the superior mesenteric artery. The portal vein was used for venous drainage. In both groups the pancreas was flushed and stored in cold saline solution (0.9%). Cold ischemic time was below 2 h.

Recipient Operation. The aorta and vena porta of the graft were anastomosed end-to-side to the infrarenal aorta and vena cava of the recipients using overrunning sutures. In pancreaticoduodenal transplantation and end-to-side duodenojejunostomy was performed.

Graft Histology. For histological examination in both experimental transplanted groups ten rats were killed consecutively during the first 10 days. Four rats on day 28 and eight rats on day 100 post-transplant underwent surgical graft removal. All pancreas grafts were fixed in Bouin's solution, processed for paraffin histology, and stained with hematoxylin and eosin. The presence of insulin was demonstrated by an indirect immunoperoxidase procedure.

Statistics. The Wilcoxon rank sum test was used for statistical evaluation. The limit of significance was set to $P \leqq 0.01$.

Results

In streptozotocin diabetic graft recipients pancreas whole organ transplantation established normoglycemia within 6 h after transplantation. The perioperative mortality was less than 5%. During the observation period of 100 days there was no significant difference in the nonfasting blood glucose levels of rats receiving a pancreas with preserved (5.6 ± 0.6 mmol/liter, mean $\pm$ SD) or suppressed (5.8 ± 0.9 mmol/liter) exocrine secretion. These values did not differ from normal controls (5.2 ± 0.7 mmol/liter). A normal growth curve and clinical observation revealed no difference compared with unmodified controls. Graft removal was always followed by prompt recurrence of hyperglycemia. Streptozotocin induced a permanent and stable diabetes, with polyphagia, polyuria, and polydipsia as the main clinical signs. The metabolic features with elevated nonfasting blood glucose levels above 18 mmol/liter and impaired k values below 1.0 min^{-1} remained unchanged during the whole observation period. In all the parameters tested there was a highly significant difference ($P < 0.005$) of diabetic compared with both normal and transplanted rats (Table 1, Fig. 1). No statistically significant differences were found in recipients of pancreas grafts with preserved or suppressed exocrine function. None of the pancreas-grafted rats differed from normal controls either on day 10 or on day 100 (Table 1, Fig. 1). A decrease in food and water consumption was a general feature in all groups.

Table 1. Metabolic studies performed 100 days after pancreas transplantation ($\bar{x} \pm SD$)

	Normal controls $n=7$	Diabetic controls $n=12$	Suppressed exocrine function $n=8$	Preserved exocrine function $n=8$
Water intake (ml/24 h)	19.0± 5.1	128.7±14.4	20.8± 2.7	20.8± 5.1
Food intake (g/24 h)	17.7± 3.4	34.6± 2.3	17.4± 2.8	14.2± 1.4
Urine output (ml/24 h)	10.2± 1.9	113.2±14.8	10.6± 2.2	13.6± 4.8
Feces (g/24 h)[a]	4.9± 2.0	13.4± 2.6	5.8± 1.3	5.4± 1.6
Weight gain (%/100 days)	66.2±12.1	9.8±14.3	70.3±18.0	79.9±17.5

[a] Dry weight

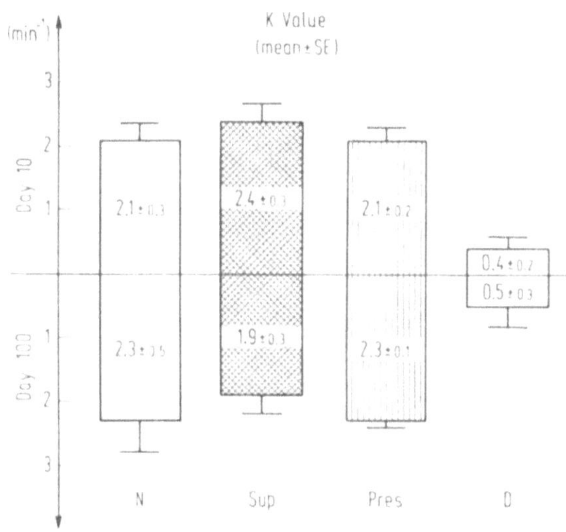

Fig. 1. *k* Values estimated 10 and 100 days following pancreas whole organ transplantation. *Pres*, preserved exocrine secretion; *Sup*, suppressed exocrine secretion; *N*, normal controls; *D*, diabetic controls

Histological Results. Within 5 days after duct ligation the acinar cells became atrophic and totally disappeared. There was a proliferation of centroacinar cells and neoformation of small pancreatic ducts. The atrophy of the exocrine pancreas was accompanied by interstitial-cell infiltration of macrophages and lymphocytes which persisted up to 100 days after transplantation. These histological alterations caused by duct ligation were mostly completed by day 10. The majority of the islets of Langerhans were well preserved and fibrosis was not a striking histological feature. Normal staining of insulin was demonstrated in all of these islets. Following pancreaticoduodenal transplantation both macroscopic and histological architecture of the transplanted pancreas were perfectly preserved. Both the exocrine pancreas and the islets of Langerhans remained unchanged.

Discussion

Adequate management of the exocrine secretion is the main technical problem in vascularized pancreas transplantation. It is controversial whether preservation of the exocrine secretion is mandatory for optimal endocrine graft function. In a model of pancreas whole organ isotransplantation in streptozotocin-diabetic rats a functional and histological assessment of pancreas grafting with preserved or suppressed exocrine secretion was performed. In both experimental groups the entire islet-cell mass of a single donor was transplanted. Prompt recurrence of diabetes after transplant removal indicated that the normoglycemia was entirely due to the presence of the graft. Irrespective of the method used pancreas transplantation completely ameliorated the streptozotocin-induced diabetes mellitus. Despite fundamental differences in histological graft appearance no statistically significant difference of endocrine function was noted of transplants with preserved or suppressed exocrine secretion. There was some reduction of the *k* values from day 10 to day 100 in rats receiving duct-ligated isografts; yet these remained within the normal range (Fig. 1). However, in long-term experiments we could not detect a significant reduction of the *k* values within 9 months after transplantation, representing half of the life span of the rat. In conclusion, in the rat the preserved exocrine secretion is not mandatory for optimal endocrine graft function of vascularized pancreas transplants. This statement is in accordance with the data reported by Orloff et al. (1975), Brekke et al. (1980), and Martin et al. (1980), who also detected no deterioration of endocrine graft function following transplantation of duct-ligated and duct-occluded pancreas whole organ grafts.

Our results seem to contradict those of Nolan et al. (1983), who reported impaired *k* values 6 and 9 months after transplantation of segmental duct-ligated isografts. This may be related to the higher functional reserve in whole organ than in segmental organ transplantation. Due to the redundancy of islet-cell mass, deterioration of endocrine function will occur only when the number of functioning islets drops below a critical level.

We would not detect an adverse effect of suppressed exocrine secretion on the quality of endocrine function in isografts; however, in allografts the profound morphological alterations induced by duct ligation may have an impact on allograft immunogenicity of vascularized pancreas transplants.

Acknowledgment. We greatly appreciate the excellent technical assistance of Margot Giesecke, Annegret Franke, and Petra Peschel.

References

Brekke IB, Gullesen I, Refsum SB, Flatmark A (1980) Long-term endocrine function of duct ligated pancreas isotransplants in rats. Eur Surg Res 12:167–178

Lee S, Tung KSK, Koopmans H, Chandler JG, Orloff MJ (1972) Pancreaticoduodenal transplantation in the rat. Transplantation 13:421–425

Lundbaek K (1962) Intravenous glucose tolerance as a tool in definition and diagnosis of diabetes mellitus. Br Med J 1:1507–1513

Martin X, Faure JL, Eloy R, Margonari J, Amiel J, Gelet A, Dubernard JM (1980) Long-term survival of pancreatic isografts in rats. Transplant Proc [Suppl 2] 12:126–128
Nolan MS, Lindsey NJ, Savas CP, Herold A, Beck S, Slater DM, Fox M (1983) Pancreatic transplantation in the rat. Transplantation 36:26–33
Orloff MJ, Lee S, Charters AC, Grambort DE, Storck LG, Knox D (1975) Long term studies of pancreas transplantation in experimental diabetes mellitus. Ann Surg 182: 198–205
Sutherland DER (1981 a) Pancreas and islet transplantation. I. Experimental studies. Diabetologia 20:161–185
Sutherland DER (1981 b) Pancreas and islet transplantation. II. Clinical trials. Diabetologia 20:435–450

The Effect of Temporary Cyclosporin A Treatment on the Endocrine Function of Pancreaticoduodenal Allografts in the Rat*

W. Timmermann, T. Schang, and *A. Thiede*

Introduction

In pancreatic transplantation in the rat, cyclosporin A (Ciclosporin, Sandimmune) has not been able to induce tolerance after temporary treatment (Rynasiewicz et al. 1980; Garvey et al. 1980; Klempnauer et al. 1984). In these experiments, pancreatic tissue was transplanted as a duct-ligated segmental graft, a technique which leads to changes in the macroscopic as well as in the microscopic aspect of the organ even in isografts (Bitter-Suermann and Säve-Söderbergh 1978; Shah et al. 1980; Orloff et al. 1975). In this study we wanted to investigate the effect of a short-term cyclosporin A (CyA) treatment on the duration of the endocrine function of allografts transplanted in a technique which preserves the physiological secretion drainage via the donor's duodenum. In these grafts, the morphology remains unchanged (own observation; Orloff et al. 1975), thus minimizing influences on the graft resulting from the technique of transplantation. To elucidate the potency of CyA to prevent rejection of pancreaticoduodenal allografts in dependence on the histocompatibility barrier between donor and recipient, we investigated two allogeneic strain combinations, one with differences on both MHC and non-MHC antigens and one with differences only on non-MHC antigens.

Material and Methods

Adult rats of the inbred strains Lewis (RT1^{l}), DA (RT1^{av1}), and AS (RT1^{l}) were used as donors; Lewis (RT1^{l}) rats always served as recipients. The donor-recipient combinations, the postoperative treatment, and the number of animals in each group are summarized in Table 1. Diabetes in recipients was induced by a single i.v. injection of 55 mg/kg streptozocin (Upjohn) and confirmed when blood glucose levels exceeded 400 mg/dl. The technique for pancreaticoduodenal transplantation was similar to that described by Lee et al. (1972). The whole pancreas and the attached duodenum are transplanted. The oral end of the duodenum is closed, and the aboral end is used for an end-to-side anastomosis to the recipient's first jejunal loop. The arterial blood supply is provided by a segment of the

* Supported by Deutsche Forschungsgemeinschaft, SFB 111/Teilprojekt B 12

Microsurgical Models in Rats
for Transplantation Research
Eds. A. Thiede et al.

Table 1. Groups, numbers, strains, and treatment of animals

Strain combination	Histocompatibility barrier	Group	Treatment	No. of animals
LEW → LEW	Syngeneic	I	–	30
AS (RT1[l]) → LEW (RT1[l])	Allogeneic, MHC compatible, and	II	–	8
	non-MHC incompatible	III	CyA 14 days 15 mg/kg/day	8
DA (RT1[avl]) → LEW (RT1[l])	Allogeneic, MHC incompatible, and	IV	–	12
	non-MHC incompatible	V	CyA 14 days 15 mg/kg/day	8

donor's aorta bearing the celiac axis and the superior mesenteric artery; the venous outflow is provided by the portal vein. Both vessels are anastomosed end-to-side to the recipient's infrarenal abdominal vessels. So the graft is placed heterotopically in the recipient's abdomen, and drainage of the exocrine pancreatic secretion is provided via the donor's duodenum to the recipient's jejunum. Endocrine function of the grafts was determined by daily measurement of blood glucose levels. Cessation of endocrine function was defined by recurrence of hyperglycemia above 200 mg/dl on two successive days. CyA was dissolved in olive oil and given orally at a dose of 15 mg/kg per day for 14 days, starting with the day of operation.

Results

Isografts. All 30 Lewis rats receiving isografts (group I) became normoglycemic the day after the operation and remained so for over 300 days. The grafts could neither macroscopically nor microscopically be distinguished from normal pancreases.

Allografts. The AS→Lewis combination is MHC compatible but non-MHC incompatible. In untreated allografts (group II) recurrence of hyperglycemia can be

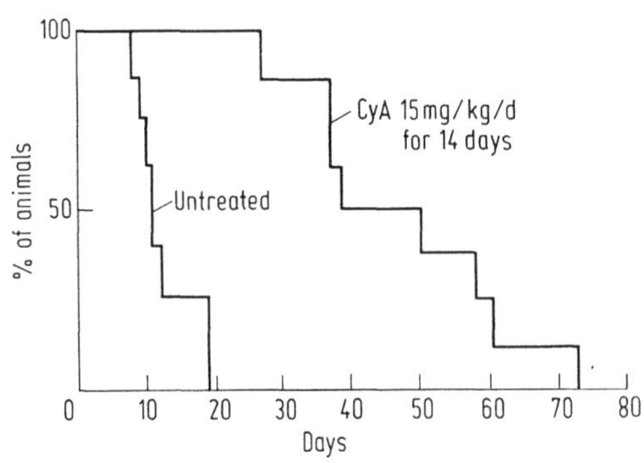

Fig. 1. Duration of endocrine function of pancreaticoduodenal allografts, allogeneic non-MHC-incompatible combination AS-→LEW, without CyA therapy (12.4±4.0 days, $n=8$, group II) and with CyA therapy (47.5±14.4 days, $n=8$, group III)

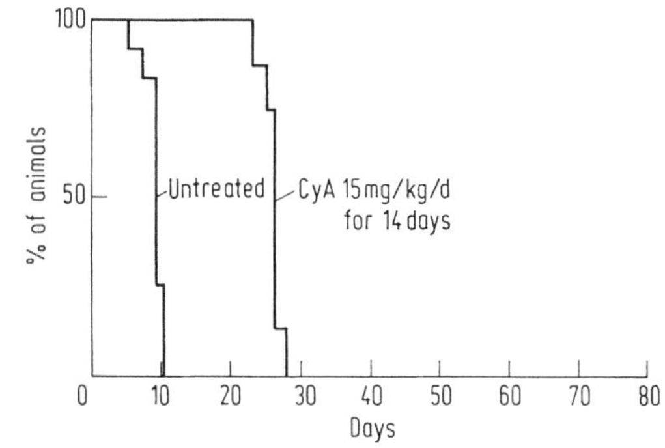

Fig. 2. Duration of endocrine function of pancreaticoduodenal allografts, allogeneic non-MHC- and MHC-incompatible combination DA→LEW, without CyA therapy (7.9±0.9 days, $n=12$, group IV) and with CyA therapy (25.6±1.1 days $n=8$, group V)

observed between 8 and 19 days, the mean being 12.4±4.0 days (n, 8). This period is prolonged after treatment with CyA (group III) up to 27–73 days (n, 8) (Fig. 1).

The second allogeneic combination studied, DA→Lewis, is incompatible in both MHC and non-MHC antigens. In animals who were untreated (group IV), the endocrine function of the allografts ceased between 5 and 9 days (n, 12). CyA treatment (group V) could extend this period up to 23–27 days (n, 8) (Fig. 2). All recipients of allografts survived the period of treatment and rejection and were finally killed for histological examination.

Histology

In both allogeneic combinations treated with CyA the morphological changes in the transplants are basically the same. Beginning with day 4, a specific mononuclear infiltrate can be found in the pancreatic tissue, spreading from the connective tissue between the pancreatic lobules. The islets themselves show no infiltration, but in some islets there seems to be singular cell necrosis. The transplanted duodenum appears to be completely unchanged and without any infiltration. Some days after cessation of therapy with CyA, when the animals are still normoglycemic, the exocrine tissue disappears and only pancreatic ducts remain. The islets are disintegrated by proliferating connective tissue, the duodenum shows mild cellular infiltrations, and the intestinal villi are shorter and bulky. After cessation of endocrine function the pancreas itself is totally destroyed, the duodenum is heavily infiltrated, and the villi have almost disappeared, but there is still a continous layer of undifferentiated endothelium.

Discussion

Pancreaticoduodenal allografts are rejected depending on the histocompatibility barrier between donor and recipient, and non-MHC antigens seem to have a considerable influence on the rejection process. This finding and the time of rejection of untreated allografts both correspond with those found by other investigators for duct-ligated segmental pancreatic grafts (Klempnauer et al. 1983; Rynasiewicz et al. 1980; Garvey et al. 1980).

It has been shown that the applied dose of CyA, 15 mg/kg per day for 14 days, is sufficient to induce long-term graft tolerance in the rat for heart, kidney, and liver allografts which were incompatible with the recipient in MHC and non-MHC antigens (Timmermann et al. 1980; Homan et al. 1980; Engemann et al. 1984). For pancreaticoduodenal grafts, this dose of CyA is able to prevent the rejection response for the time of its application. After withdrawal of the drug, rejection takes place in both the examined strain combinations and the time interval is dependent on the histocompatibility barrier. So, in contrast to other organs in the rat, CyA can significantly influence the rejection response to pancreatic allografts but cannot, at least in the applied dosage, induce tolerance. This has been shown for duct-ligated segmental grafts as well (Garvey et al. 1980; Rynasiewicz et al. 1980).

The transplanted duodenum seems to be rejected in both untreated allografts (Lee et al. 1972) and CyA-treated allografts in a milder or slower way than the pancreas itself, at least regarding the cellular reaction within the graft.

Mononuclear infiltration in the graft, progressive destruction of the exocrine tissue, and finally disintegration and destruction of the islets have to be regarded as the pattern of the rejection process of pancreatic tissue under and after therapy with CyA. Technically induced influences caused by duct ligation might be of the same appearance (Shah et al. 1980; Orloff et al. 1975), but can be excluded because a model with secretion drainage is used.

With regarding to the fact that the duration of endocrine function is the same for pancreaticoduodenal allografts and duct-ligated segmental grafts with and without CyA therapy, it seems that in the rat the technique used for grafting has no major influence on the rejection pattern of pancreatic transplants. The failure of CyA to induce tolerance in pancreatic transplantation in the rat appears to be organ specific.

References

Bitter-Suermann H, Säve-Söderbergh J (1978) The course of pancreas allografts in rats conditioned by spleen allografts. Transplantation 26:28–34

Engemann R, Ulrichs K, Thiede A, Müller-Ruchholtz W, Hamelmann H (1983) Induction of liver graft tolerance in a primarily nontolerant strain combination with temporary treatment of Cyclosporin. Transplant Proc 15 [Suppl 1]:2986–2991

Garvey JFW, McShane P, Poole MD, Millard PR, Morris PJ (1980) The effect of cyclosporin A on experimental pancreas allografts in the rat. Transplant Proc 12:266–269

Homan WP, Fabre JW, Williams KA, Millard PR, Morris PJ (1980) Studies on the immunosuppressive properties of cyclosporin A in rats receiving renal allografts. Transplantation 29:361–366

Klempnauer J, Wonigeit K, Günther E, Pichlmayr R (1983) Pancreas whole organ transplantation in the rat: evidence for a strong effect of non-MHC incompatibilities. Transplant Proc 15:1649–1651

Klempnauer J, Wagner E, Wonigeit K, Pichlmayr R (1983) Pancreas and kidney allograft rejection responds differently to cyclosporine immunosuppression. Transplant Proc 15 [Suppl 1]:3001–3003

Lee S, Tung KSK, Koopmans H, Chandler JG, Orloff MJ (1972) Pancreaticoduodenal transplantation in the rat. Transplantation 13:421–425

Orloff MJ, Lee S, Charters AC, Grambort DE, Storck LG, Knox D (1975) Long term studies of pancreas transplantation in experimental diabetes mellitus. Ann Surg 182:198–206

Rynasiewicz JJ, Sutherland DER, Kawahara K, Gorecki P, Najarian JS (1980) Cyclosporin A prolongation of segmental pancreatic and islet allograft function in rats. Transplant Proc 12:270–274
Shah KH, Bitter-Suermann H, Säve-Söderbergh J (1980) Morphological findings in duct-ligated pancreas grafts in the rat. Transplantation 30:83–89
Timmermann W, Calne RY, White DJG, White D, Thiede A (1980) Die Wirkung von Cyclosporin A und Cytimun auf die Überlebenszeit von RT-1 allogenen Rattenherztransplantaten. Chir Forum Langenb. Arch. Chir (Suppl.) 130–143

The Influence of Exocrine Secretion Management on Graft Survival in Different Pancreatic Transplant Models

D. Baumgartner

Pancreatic transplantation is performed for endocrine replacement therapy in juvenile onset diabetics or insulin-dependent diabetic experimental animals; however, the most difficult problems with pancreatic transplantation relate to the exocrine component of the organ.

Like any other transplanted organ, pancreatic grafts are subject to acute episodes of in situ inflammation most often caused by rejection, exocrine stasis, ischemia, or operative trauma. In the exocrine pancreas these specific stimuli initiate a uniform unspecific reaction characterized by exocrine-cell death and consecutive acute pancreatitis with varying degrees of autodigestion. Thus, pancreatic self-destruction becomes independent of the original challenging agent, and this process often results in permanent structural damage to the organ, functional exocrine and possibly endocrine tissue being replaced by a fibrous scar. The severity of this reaction and its effects on endocrine function and graft survival vary from species to species, making evaluation of new techniques of transplantation, organ preservation, immunosuppression, and above all, management of exocrine secretions in various animal models and species mandatory before these innovations are transferred to clinical pancreatic transplantation. In this paper the results of our experimental studies on exocrine secretion management in dogs will be compared with clinical findings.

Effect of Pancreatic Duct Occlusion on Pancreatic Segments In Situ

Studies in Dogs. In 45 mongrel dogs the head of the pancreas was removed. The ductal system of the pancreatic tail was occluded with either neoprene or prolamine, allowed to drain freely into the peritoneal cavity, or anastomosed to the left ureter after a left nephrectomy. Details of surgical technique have been described elsewhere (Baumgartner et al. 1981 a, 1983 a). Histological examination of biopsies obtained at 24 h–3 months showed an unchanged pancreatic architecture in pancreases with exocrine drainage to the urinary system. Leaving the duct open for free drainage to the peritoneal cavity results in spontaneous duct occlusion within 1–2 weeks and histologies of this group resemble those of duct-occluded pancreases where there were signs of pancreatitis initially, followed by replacement of the exocrine tissue by a fibrous tissue. Intravenous glucose tolerance

Microsurgical Models in Rats
for Transplantation Research
Eds. A. Thiede et al.

Table 1. Intravenous glucose tolerance test (0.5 g/kg glucose) performed in dogs 1 week after partial pancreatectomy and duct occlusion or drainage. *K* values and serum insulin (mean ± SE)

	n	*K* 1–60 (%)	Basal insulin (μU/ml)	Peak insulin (μU/ml)
Duct occlusion				
Neoprene	8	1.1±0.1	3±1	14±3
Prolamine	7	1.6±0.3	7±2	27±7
Duct drainage				
Peritoneum (open duct)	23	1.2±0.2	6±1	32±4
Ureter	7	1.8±0.1	5±1	28±5

tests (IVGTTs) (0.5 g/kg glucose) were performed at 1 week postoperatively. *K* values from all glucose samples in the groups obtained between 1 and 60 min after injection and basal and peak insulin values are given in Table 1. Mean *K* value was highest in dogs with patent pancreaticoureterostomy; however, the difference to dogs with prolamine-occluded pancreases was not statistically significant. In a similar study in the same canine system the nature of postduct occlusion pancreatitis was assessed more closely by serial biopsies and differentiation of hematological cells in fine-needle aspirates from the pancreatic segments. Techniques have been described in more detail elsewhere (Baumgartner et al. 1983 b). In addition to the above-mentioned histological changes, it was found that this type of pancreatitis is characterized by a significant influx of lymphocytes in varying stages of activation and lymphoblasts and that morphological differentiation of these infiltrates occurring at 14–120 days postoperatively from those observed in acute rejection is virtually impossible.

Experience in Man. Our experience with occlusion of the pancreatic duct in patients undergoing pancreatic resection is limited to two cases where the ductal system was occluded on reoperation for pancreatic fistulae. In both cases pancreatic secretions stopped and glucose profiles over 24 h remained as preoperatively.

Comment. Although duct occlusion results in severe changes of pancreatic architecture with near total destruction of the exocrine component and moderate morphological alteration of the islets, the endocrine function is well preserved and no significant short-term advantage of duct drainage can be demonstrated.

Effect of Pancreatic Duct Occlusion in Segmental Pancreas Transplants

Long-Term Studies in Dog Autografts. In our previously described model of segmental intraperitoneal autotransplantation pancreatectomy of the head and right limb of the pancreas was delayed for 12 months to prevent malnutrition due to exocrine insufficiency in 18 animals (Baumgartner et al. 1981 b). In four dogs the duct had been injected with neoprene, in six it had been left open to obliterate spontaneously, and in eight a pancreaticoureterostomy had been performed. One

Table 2. Intravenous glucose tolerance test (0.5 g/kg glucose) at 13 months in dogs with segmental autografts 4 weeks after delayed removal of head and right limb of the pancreas. See text for details

	n	Survivors at 1 year	*K* 1–60[a] (%)	Basal insulin (μU/ml)	Peak insulin (μU/ml)
Open duct	6	4	1.5±0.3	9±1	30±16
Neoprene-occluded duct	4	4	1.1±0.3	10±3	26± 9
Duct drainage to ureter	8	4	1.2±0.3	9±3	24±11

[a] Mean±SE

Table 3. Intravenous glucose tolerance test (0.5 g/kg glucose) at 1 week and 12–17 months in dogs with segmental autografts with open ducts. *K* values and serum insulin (mean±SE)

	n	*K* 1–60 (%)	Basal insulin (μU/ml)	Peak insulin (μU/ml)
1 week	7	1.3±0.1	7±1	29±4
12–17 months	5	1.3±0.2	7±2	19±6

graft with an open duct was lost from arterial thrombosis, and one animal from this group died of pneumonia. Three animals of the pancreaticoureterostomy group died with pancreatic abscesses and one of graft vein thrombosis. Four animals in each group were available for evaluation by IVGTT at 13 months. As the results given in Table 2 show, there was no difference in *K* values or insulin response between the groups. Histology of occluded or open duct pancreases was similar to that of corresponding pancreatic segments in situ; however, in dogs with pancreaticoureterostomy there was a certain amount of fibrosis even in the presence of a patent anastomosis. In another experiment the long-term effect of pancreatic fibrosis on endocrine function was studied by comparing in autografts with open duct and immediate pancreatectomy IVGTTs at 1 week of seven consecutively transplanted dogs with those of five dogs surviving for 12–17 months. As shown in Table 3 there was again no difference between the groups; however, it must be taken into account that the five long-term survivors represent a highly selected population, all dogs with poor graft function having died in the interval.

Experience with Intraperitoneal Segmental Pancreatic Allografts with Prolamine-Occluded Duct in 15 type I Diabetics

As published in detail elsewhere (Baumgartner and Lagiadèr 1984), nine of these patients had excellent initial graft function. Primary function was demonstrable but inadequate in two patients, in one due to rapidly deteriorating myocardial function with a fatal outcome after 36 h and in the other due to a cold ischemia time of 23 h. In this latter patient endocrine function recovered after several days.

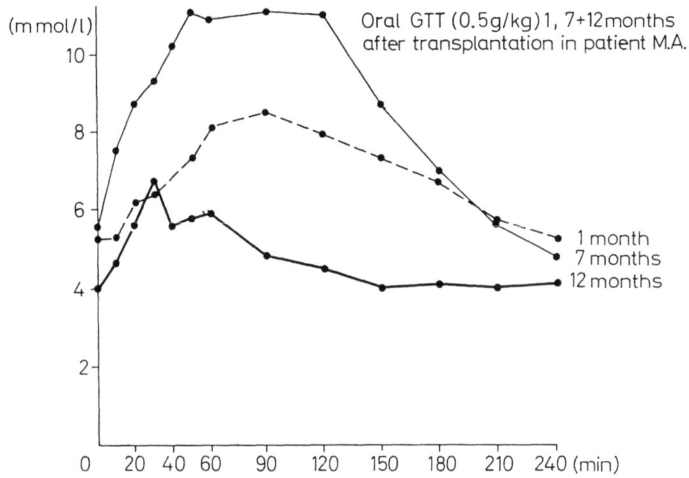

Fig. 1. Plasma glucose in serial oral GTT in patient M.A. after segmental intraperitoneal pancreas transplantation with prolamine-occluded duct. Glucose load at 1 and 12 months: 0.5 g/kg. At 7 months 0.5 g/kg after heavy breakfast

In the late course in this patient there were recurrent exocrine fistulae, and endocrine function slowly deteriorated over a 10-month period. Histological examination of the graft removed at this point showed signs of ongoing inflammation and a picture reminiscent of chronic sclerosing pancreatitis. In this patient, as well as in the four patients with primarily nonfunctioning transplants graft failure was not attributable to duct occlusion. In the absence of additional factors that induce prolonged inflammation in the pancreas such as ischemia, excessive operative trauma, infection, or rejection, duct occlusion does not seem to result in progressive deterioration of endocrine function. Metabolic control has remained stable in our two patients with functioning grafts for over 30 months; this is also illustrated by IVGTTs at 1 month and 1 year in one of these patients (Fig. 1).

Comments. These experimental and clinical data show that postduct occlusion pancreatitis and fibrosis do not result in progressive loss of endocrine function and therefore do not jeopardize graft survival. Furthermore, they illustrate that pancreatitis and fibrosis are unspecific reactions of the graft to various stimuli and that factors such as operative trauma and risk of infection should also be taken into consideration in the evaluation of techniques for exocrine secretion management.

Discussion

The main difficulty in the evolution of pancreatic transplantation has been to combine the requirements of metabolic efficiency with those of the safety of the procedure. The originally practiced extraperitoneal whole organ or segmental transplantation with enteric drainage proved to be effective but extremely dangerous. In consequence the exocrine pancreas was totally eliminated and islet suspensions or pancreatic microfragments were transplanted safely but without metabolic effect. Intraperitoneal placement of segmental grafts with polymer-occluded duct was subsequently shown to be a safe and effective procedure, making pancreatic transplantation clinically feasible again. However, graft survival remained significantly lower than that of other organs. This and the gross histological alterations found in organs with occluded duct have triggered renewed interest in enteric or ureteral drainage procedures for intraperitoneal transplantation. The data presented here suggest that histological changes induced by duct occlusion are of no consequence for endocrine function. Similar findings have been reported for pancreatic segments in situ in patients undergoing partial pancreatectomy for tumors or chronic pancreatitis (Di Carlo et al. 1984, Brekke et al. 1984) and for patients after segmental pancreas transplantation with neoprene-occluded duct (Pozza et al. 1984). Transplantation of a maximal islet mass is desirable since rejection processes are focal in nature and a large functional reserve may therefore prolong graft survival. It is important to realize, however, that duct occlusion is not the only mechanism inducing pancreatitis and it is quite possible that the theoretical advantages of duct drainage are offset by increased operative trauma and a higher infection rate. The international pancreas and islet transplantation registry statistics (Sutherland 1984) show survival of a higher proportion of grafts with enteric drainage than of duct-occluded grafts; however, the mean follow-up time in grafts with enteric drainage is significantly shorter than that of grafts with duct occlusion. If only transplants performed in 1983 are considered, survival statistics are similar for both groups, and at the University of Minnesota, where both techniques are practiced, there is also no difference in graft survival between duct-occluded grafts and grafts with enteric drainage (Sutherland 1983, personal communication). One-year graft survival with both techniques is still not higher than 30%, this indicates that the ideal technique of pancreatic transplantation remains to be found. From the data available at the present time, we concluded that the theoretical advantages of duct drainage in clinical pancreatic transplantation do not outweigh the risks of this procedure and we therefore adhere to our described technique of segmental intraperitoneal transplantation with polymer-occluded duct.

References

Baumgartner D, Largiadèr F (1984) Simultaneous renal and intraperitoneal segmental pancreatic transplantation: the Zurich experience. World J Surg 8:267–269

Baumgartner D, Hansen SL, Sutherland DER (1981) Endocrine function and histology of pancreatic segments after duct occlusion with neoprene or ethibloc. Eur Surg Res 13:33–334

Baumgartner D, Sutherland DER, Heil J, Kyriakides GK, Najarian JS (1981 b) Long term canine segmental pancreas transplants with the duct left oben, neoprene injected duct pancreaticoureterostomy. Transplant Proc 13:812–814

Baumgartner D, Illig R, Sutherland DER (1983 a) Effect of venous drainage to the vena cava and denervation on endocrine function of pancreatic segments in dogs. Transplant Proc 16:769–772

Baumgartner D. Burger H, Rubeli M, Mona D (1983 b) Changes in pancreatic aspiration cytology induced by duct obliteration. In: Proc of the 2nd international workshop on renal transplant aspiration cytology. Wichtig, Milan, p 167–172

Brekke IB, Berga A, Heen L, Flatmark A (1984) Pancreatic endocrine function after duct occlusion. Transplant Proc 16:739–740
Di Carlo V, Chiesa R, Pontiroli AE, et al. (1984) Use of neoprene for the block of the exocrine pancreatic secretion: clinical and metabolic evaluation in man. Transplant Proc 16:737–738
Pozza P, Secchi A, Poutiroli AE, Bosi E, et al. (1984) Influence of steroid administration on the endocrine function in neoprene-injected segmental pancreas allotransplantation. Transplant Proc 16:707–708
Sutherland DER (1983) International pancreas and islet transplantation registry statistics: Transplant Proc 15:2597–2601

Investigations of Glucose Tolerance Following Pancreatic Transplantation in Rats with Streptozotocin-Induced Diabetes

R. Meister and *K. Paul*

In addition to problems of immune tolerance and the technique of transplantation one must pose the question of how a transplanted pancreas will function under conditions of daily life. In order to investigate this problem we initiated a trial in an animal model. Two questions were of particular interest:

1. Would pancreas transplantation in diabetic recipient animals normalize fasting and stimulated hyperglycemia, the probable cause of vascular damage in diabetics?
2. What are the basal values for insulin, glucagon, and glucose and the secretion kinetics of insulin following various stimulation tests.

Material and Methods

Male inbred Lewis rats (170–350 g) were used. Induction of diabetes was accomplished by administration of streptozotocin (65 mg/kg). Pancreaticoduodenal grafts were transplanted using the technique developed by Sun Lee (San Diego). The following tests were used to evaluate glucose regulation:

1. Intravenous glucose tolerance test (IVGTT): 0.354 g glucose/250 g bodyweight, serial blood samples at −2, 5, 10, 20, 60, and 100 min
2. Maximal stimulation of beta cells (MSB): 0.365 g/250 g body weight glucose intragastrally, 30 min later tolbutamide (6 mg/250 g) and glucagon (4 µg/250 g) intravenously (samples at −2, 30, 45, 60, 75, and 90 min)
3. Oral glucose tolerance test (OGTT): 0.73 g glucose/250 g body weight intragastrally with blood samples at −2, 15, 30, 45, 90, and 120 min

All tests were performed using general anesthesia (sodium pentobarbital) in laparotomized animals, with serial blood samples taken from the vena cava. Insulin and glucagon levels were determined by established radioimmunoassays, and glucose with the oxidase method.

Three groups of animals ($n = 6$) were compared in each test:

1. Sham-operated normal controls (N = I)
2. Sham-operated diabetic controls (D = II)
3. Diabetic animals with pancreatic transplants (PTP = III)

Microsurgical Models in Rats
for Transplantation Research
Eds. A. Thiede et al.

Table 1. Intravenous glucose tolerance test

		−5 0	5	10	20	60	100 min
Glucose (mg/dl)	I	92/70–122 (6)	547/447–593 (6)	487/428–545 (6)	427/177–305 (6)	270/175–305 (6)	202/116–301 (6)
	II	267/105–354** (6)	779/646–871** (6)	776/688–874** (6)	699/633–814** (6)	632/523–729** (6)	581/506–813** (5)
	III	100/70–133[b] (6)	628/503–829* (6)	546/488–694* (6)	508/443–647*[b] (6)	345/140–519[b] (6)	335/67–454[b] (6)
Insulin (μU/ml)	I	21/18–47 (6)	45/20–77 (6)	43/21–51 (6)	46/33–54 (6)	63/50–81 (6)	44/30–59 (6)
	II	3/2–4[a] (7)	3/2–3[a] (5)	3/2–3[a] (6)	3/2–3[a] (6)	3/2–4[a] (6)	3/2–3[a] (5)
	III	11/4–20[b] (6)	25/12–114[b] (6)	42/15–152[b] (6)	65/24–348[b] (6)	66/28–273[b] (6)	50/45–62[b] (5)

*, $P<0.05$; **, $P<0.05$; [a] $P<0.001$ vs. I; [b] $P<0.01$ vs. II

Table 2. Maximal stimulation of beta cells

		−2 min	30 min	45 min	60 min	75 min	90 min
Glucose (mg/dl)	I	83.8/80–102** (9)	172.0/146–181 (9)	231.0/175–277 (9)	272.0/115–325 (9)	321.0/249–439 (9)	343.0/306–464 (9)
	II	142.0/124–303 (7)	405.0/313–509 (7)	413.0/356–469 (7)	435.0/357–516 (7)	439.0/377–561 (7)	492.0/371–605 (7)
	III	91.5/90–116 (4)	154.5/71–237 (4)	183.5/131–291 (4)	183.0/137–272 (4)	195.5/134–329 (4)	150.0/113–332 (4)
Insulin (μm/ml)	I	17.0/3.2–30.4 (10)	65.2/21–150.1 (10)	39.0/23.4–217.6 (10)	21.0/12.2–37.1 (9)	17.0/11.4–27.1 (10)	15.7/9.4–30.9 (10)
	II	10.0/9.2–14.1 (7)	11.0/7.8–15.4 (7)	9.8/8.8–13.8 (7)	10.2/9.2–13.5 (7)	10.3/8.0–13.8 (7)	9.8/8.4–14.4 (7)
	III	9.1/2–17.2 (4)	13.2/7.8–43.4 (4)	43.8/33–296.8 (4)	54.2/45.6–223.0 (4)	43.1/28.6–210 (4)	29.7/29.4–441.0 (4)

** median/range; ( ) numbers

Results

Intravenous Glucose Tolerance Test. The *K* values (median/range) reflect complete normalization of the glucose removal by PTP (normal controls 0.94/0.63–1.46 mg/dl; diabetic controls, 0.28/0.08–0.36 mg/dl; PTP, 1.03/0.42–2.32 mg/dl). Basal insulin is lower in the transplanted group but significantly higher than in the diabetic group. After glucose load, integrated insulin in the PTP group is higher than in diabetics and even higher than in the normal controls ($P<0.05$). We can conclude that PTP can normalize fasting blood glucose levels and those induced by intravenous glucose load. This is achieved by way of higher integrated insulin. Islets in vascularized pancreatic grafts react adequately to intravenously given glucose.

Maximal Stimulation of Beta Cells. Here the synthesis, storage, and secretion capacity of islets in the transplanted pancreas was investigated. Basal glucose and insulin levels do not differ in normal controls and the PTP group. However, both vary significantly from the diabetic controls ($P<0.05$). The integrated glucose values are similar for both normal controls and the PTP group, but the diabetic controls show a significant difference when compared with the transplanted diabetic group ($P<0.001$). There is no statistical difference between the levels of integrated insulin in normal controls and in the PTP group, but both show a highly significant difference when compared with diabetic controls ($P<0.001$). We can conclude that the beta cell within the islets of the transplanted pancreas is capable of producing, storing, and liberating insulin following maximal beta-cell stimulation with glucose, glucagon, and tolbutamide.

Oral Glucose Tolerance Test. The OGTT is comparable to the everyday challenge to the islets. Basal glucose levels are slightly elevated in the PTP group when compared with the normal controls, but are significantly different from the diabetic controls. In contrast the integrated glucose levels are similar for both the normal controls and the PTP group, whereas both are significantly different from dia-

Table 3. Oral glucose tolerance test

	Control (6)	Diabetic (8)	Diabetic PTP (5)
Glucose			
Basal (mg/dl)	84/71–101	146/114–436*	118/105–145[c], *
Integrated; area[a]	165/69–216	271/197–346[c]	158/123–208*
Insulin			
Basal (μU/ml)	22/12– 30	10/ 4– 14[c]	20/ 16– 25***
Integrated; area[b]	23/ 0– 32	0.3 – 1– 0.4[c]	62/ 13–127[c], **

( ), number of observations
[a] mg/dl × min × 10^{-2}
[b] μU/ml × min × 10^{-2}
[c] U-test used to determine significance
*, $P<0.05$; **, $P<0.01$; ***, $P<0.001$; versus control in second column and versus diabetic in third column

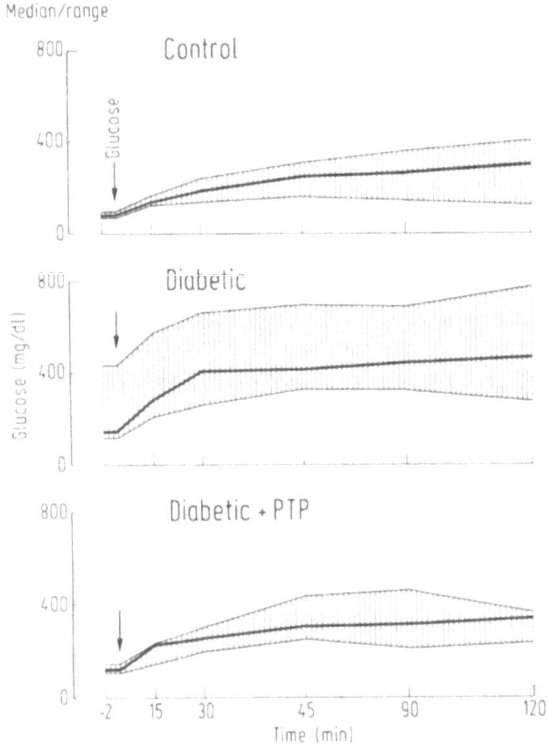

Fig. 1. Effect of pancreas transplantation (*PTP*) on the OGTT in streptozotocin-diabetic rats

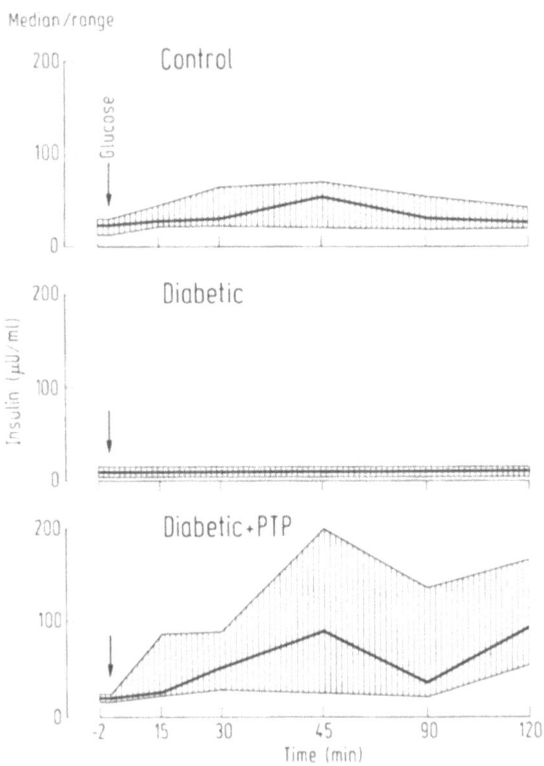

Fig. 2. Effect of pancreas transplantation (*PTP*) on plasma insulin in the OGTT in streptozotocin-diabetic rats

betic controls. Basal insulin levels in the PTP group do not differ from those in the normal group. Again both differ significantly from the diabetic controls. Integrated insulin levels are significantly higher in the PTP group when compared with the normal group and even more so when compared with the diabetic controls. Diabetic hyperglucagonemia (median/range), which in addition to the insulin deficiency is the reason for poor glucose homeostasis in diabetics, is normal (diabetics, 417/271–1130 pg/ml; controls, 139/59–640 pg/ml, $P<0.05$; normalized following PTP: 222/75–550 pg/ml).

We conclude that PTP induces normal basal glucose and insulin values and reduces streptozotocin-induced hyperglucagonemia. PTP shows normal integrated glucose levels following oral stimulation, which signifies sufficient stimulation of insulin in a transplanted pancreas and obviously does not require intact innervation. This normal glucose regulation is achieved by significantly elevated stimulated insulin values (Figs. 1, 2).

Successful Allotransplantation of Islets as Composite Graft in a Strong Rejection Model

G. H. Müller, H. Reece-Smith, P. McShane, and *P. J. Morris*

The tremendous success of transplantation surgery is due to both better surgical management and immunosuppressive therapy. However, surgical treatment of the diabetic has so far not been very successful. This is partially due to the difficulties in keeping the endocrine part of the pancreas – the islets – alive. Immunological reactions such as cell invasion of the graft and nonimmunological factors which are also able to destroy the graft must be avoided. The latter is avoided when the islets alone are transplanted. After excision and fragmentation of the pancreas, islets can be separated from fat, lymphoid tissue, and exocrine tissue using enzymes and gradient centrifugation. Islet isografts prepared in this way have been successfully transplanted under the kidney capsule, in the spleen, and intraportally. However, allotransplants using a strong model (DA to Lewis) were all rejected early. Using the same strain combination, we were able to produce a donor-specific unresponsiveness for hearts and kidneys following a primary auxiliary liver transplantation and short-term cyclosporin A therapy. The question

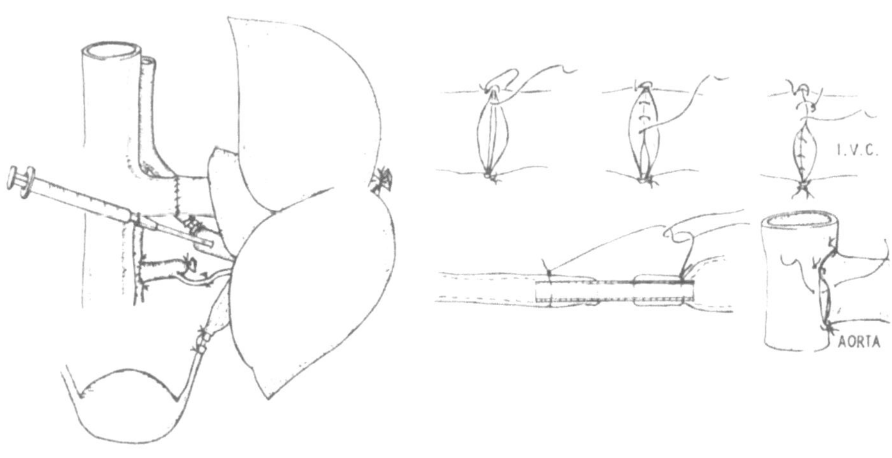

Fig. 1. Liver graft: site of implantation and technique of the various anastomoses: inferior vena cava (*IVC*) to renal vein, celiac artery to aorta, portal vein to renal artery, bile duct to ureter. The islets are introduced into the portal vein

Microsurgical Models in Rats
for Transplantation Research
Eds. A. Thiede et al.

asked was whether that same combination of liver transplantation and CyA would create a tolerance for the apparently more immunogenic islets.

We selected the following protocol: Female Lewis rats, weighing 175–200 g, were made diabetic by injecting 65 mg/kg body weight streptozotocin into a tail vein. Only animals which showed blood sugar levels above 22 mmol/liter twice in 14 days were given transplants. Following a left nephrectomy, a DA liver was implanted using the renal vessels: the renal artery to the portal vein using a stent for the anastomosis and continuous 10-0 sutures for an end-to-end anastomosis of the graft inferior vena cava and the renal vein of the recipient (Fig. 1). The hepatic artery was anastomosed end-to-side to the aorta using an aortic segment. The bile duct was anastomosed to the left ureter, again using a stent. One day later, islets from six donor rats were injected into the arterialized portal vein of the graft. Blood sugar was monitored daily until the 3rd week and then every 3rd day. Cyclosporin A (CyA) was given orally every day for only 14 days, starting with the day of the liver transplant. The day of rejection was taken as being the first of two consecutive days when blood sugar levels were higher than 11 mmol/liter.

The following experiments were carried out:

Group 1: Five diabetic Lewis rats were transplanted with a DA liver and DA islets.
Group 2: Five diabetic Lewis rats were treated additionally with 10 mg/kg CyA for 14 days.
Group 3: Five diabetic Lewis rats transplanted with a DA liver and Brown Norway islets and given CyA treatment.
Group 4. Five diabetic Lewis rats were injected with DA islets intraportally.
Group 5. Five diabetic Lewis rats were injected with DA islets intraportally and were treated with CyA.

Table 1. Results for time of normoglycemia

Group 1:	3, 7, 9, 10, 11	days
Group 2:	15, 19, 26, >100, >100	days
Group 3:	3, 5, 5, 6, 14	days
Group 4:	2, 2, 4, 4, 5	days
Group 5:	3, 5, 6, 7, 8	days

Every animal became normoglycemic, although only for a short interval when without immunosuppression (Table 1). CyA treatment resulted in prolonged islet survival in the group without liver grafts (using the Wilcoxon rank sum test for statistical evaluation), but the results were not significant. Significantly increased islet survival was observed in the group with liver grafts plus islets and further prolongation was noted when cyclosporin A was given. This effect was seen to be liver donor strain specific. Third party islets (BN) were rejected early.

The results again show the characteristic immunological status of the liver as a transplant. They also show that the strong immunogenicity of the endocrine tissue does not constitute a barrier to allotransplantation, and they further show that CyA can be an important tool in the regulation of the immune response.

The results are not immediately transferable to the human model for the following reasons: Firstly, despite the strong rat rejection model, it seems to be easier to produce a state of tolerance in the rat than in humans. Secondly, the experimentally induced diabetes differs from the human diabetes type I or II. Thirdly, islets are still much easier to obtain from rats than from the human pancreas. However, the following can be applied: The recipient's immune response is altered by specific immunosuppression: a tolerance of antigens develops, induced by CyA and treatment of the recipient, or by liver grafting. Both methods show promising results.

The Immunogenicity of Pancreatic Islets *

D. W. R. Gray, H. Reece-Smith, P. McShane, and *P. J. Morris*

The idea of transplanting islets of Langerhans as free grafts has long been attractive because such a procedure should be free of the major complications of whole organ pancreatic transplantation. The introduction of reproducible techniques for isolation of islets in large numbers from the rodent pancreas led to experiments reporting the ease with which islet isografts could be transplanted with reversal of diabetes (Ballinger et al. 1972). Initially it was hoped that islet tissue and other endocrine tissues would be less immunogenic than organs such as the kidney and heart (Lillehei et al. 1970).

An effective and frequently used model of islet transplantation in the rodent has been the injection of islets into the portal vein of a diabetic recipient (diabetes usually being induced by streptozotocin injection) (Kemp et al. 1973). This leads to a rapid fall in blood sugar of the diabetic rat, but in the case of the allograft to an equally rapid rise after rejection 2–5 days later. The rejection times for untreated intraportal islet allografts have been well documented by other centres. Our own figures are shown in Table 1, where islets have been transplanted across a major histocompatibility barrier in the rat. It can be seen that rejection is rapid even where the donor is heterozygous and rejection appears to occur more rapidly than rejection of kidney grafts across the same histocompatibility barriers (Table 2).

Immunosuppression has a marked effect on the survival of renal allografts as shown in Table 2. A similar improvement in survival for islet allografts has been difficult to achieve using a variety of immunosuppressive agents including azathiaprine, prednisolone, cyclophosphamide, antilymphocyte sera and cyclosporin A in varying dosages and combinations. Our own experience (Morris et al. 1980) using intraportal islet transplants in a variety of strain combinations is summarized in Table 3. Our results agree with reports from other centres (Sutherland 1981) that the most effective immunosuppressive agent is antilymphocyte serum and that other methods only produce a substantial increase in islet survival at a dose that is toxic to the recipient.

These results led to the suggestion that pancreatic islets, far from being less immunogenic than other tissues, are in fact more susceptible to immune attack. This

* Supported by a grant from the Medical Research Council of the United Kingdom

Microsurgical Models in Rats
for Transplantation Research
Eds. A. Thiede et al.

Table 1. Survival of intraportal islet allografts without immunosuppression

Donor strain	Recipient strain	Graft survival (days)[a]
Lewis	Lewis	> 100 × 6
Lewis	DA	3, 4, 4, 4, 5
Fl (Lew/DA)	Lewis	4, 4, 4, 4, 5
Fl (Lew/DA)	DA	4, 4, 4, 5, 6
DA	Lewis	2, 3, 3, 4, 4, 5, 6

[a] Rejection designated as a consistent rise in blood sugar to greater than 11 mmol/litre

Table 2. Rejection of rat renal allografts

Donor	Recipent	No.	Treatment	Mean survival (days)
DA	Lewis	19	None	10 ± 0.8
Lewis	DA	8	None	10.25 ± 0.4
DA	Lewis	5	CyA 10 mg/kg	7, > 100 × 4
Lewis	DA	26	CyA 10 mg/kg	8, 10, 15, 17, 18, > 65 × 21

Table 3. Effect of immunosuppression on survival of intraportal islet allografts

Donor strain	Recipient strain	Treatment	No.	Survival (days)	
				Mean	Range
DA	Lewis	Nil	5	5	(2– 5)
DA	Lewis	1 ml ALS × 4	5	26	(22– 32)
DA	Lewis	Lewis anti-DA 500 µl	5	8	(5– 11)
DA	Lewis	CyA 5 mg/kg/day	4	5	(1– 13)
DA	Lewis	CyA 10 mg/kg/day	5	6	(3– 8)
DA	Lewis	CyA 20 mg/kg/day	6	6	(1– 13)
DA	Lewis	CyA 40 mg/kg/day	6	6	(2– 21)
DA/Lew	Lewis	Nil	5	5	(5– 6)
DA/Lew	Lewis	ALS	6	32	(28– 38)
DA/Lew	Lewis	Cyclo. 10 mg/kg/day	5	6	(6– 9)
DA/Lew	Lewis	Lewis anti-DA 500 µl	5	16	(12–360)

Cyclo, cyclophosphamide; CyA, cyclosporin A; ALS, antilymphocyte serum; Lewis anti-DA, enhancing serum

idea was supported by studies comparing survival times of islet allografts to that of whole organ pancreatic allografts in the same strain combinations (Morris et al. 1980; Reckard et al. 1979) and also to the survival times of islet, kidney and heart allografts in the same animal (Reckard et al. 1981). It was suggested that islets were especially sensitive to humoral rejection (Naji et al. 1979).

Is the short survival time of allografted islets actually related to an inherent immunogenicity of islets? There is some evidence that other mechanisms may play a part in the failure of allogeneic islets. The diagnosis of rejection by a rise in blood sugar is obviously a sensitive, and no doubt, earlier method of detecting

Table 4. Effect of transplanting increasing numbers of islets

Donor	Recipient	Treatment	No. of islets	Graft survival (days)
Fl (Lew/DA)	Lewis	Nil	600– 800	4, 4, 4, 4, 5
Fl (Lew/DA)	Lewis	Cyclophosphamide 10 mg/kg/day	600– 800	6, 6, 6, 7, 9
Fl (Lew/DA)	Lewis	Cyclophosphamide 10 mg/kg/day	1500–3000	8, 9, 11, 13, 30, 100

Table 5. Effect of site of implantation on islets

Combination	Site	Survival (days)
Lewis to DA	Liver	3, 4, 4, 4, 5
Lewis to DA	Renal capsule	5, 7, 7, 8, 8, 8, 11, 13, 17

Table 6. Effect of transplanting islets as a vascularized graft

Donor	Intermediate	Recipient	Treatment	Survival (days)
Lewis	Nil	DA	Nil	5, 7, 7, 8, 8, 8, 11, 13
Lewis	Lewis	DA	Nil	6, 7, 8, 8, 8
Lewis	Lewis	DA	CyA 10 mg/kg/day	11, $>100\times5$

rejection of allogeneic islets than is the case with rejection of kidney or skin allografts. This may be related to the insulin reserve of the allogeneic islets. In support of this is the effect of using a larger number of islets to implant (Finch and Morris 1977) when the rejection time is significantly delayed (Table 4).

Another factor may be the site of islet transplantation. We have shown (Reece-Smith et al. 1981, b) that the survival time of allogeneic islets increases if the islets are simply placed under the renal capsule (Table 5). In fact survival times are not dissimilar to that of skin allografts. These data imply that implantation beneath the renal capsule results in some delay of effector mechanisms against the allogeneic islet, perhaps macrophages. This idea is further supported by experiments where the isolated islets are first transplanted under the kidney capsule of a diabetic, syngeneic rat. This rat acts as an intermediate host, and after 48 h the combined kidney and islet graft is transplanted to an allogeneic rat as a vascularized graft (Table 6). Rejection of the islet graft then occurs in a time similar to the renal allograft. Furthermore, immunosuppression with cyclosporin A produces prolonged survival of both the kidney and islets (Reece-Smith et al. 1981). This suggests that the new site of the islets has conferred immunological properties similar to that of the kidney on which the islets have implanted.

As can be seen from the data presented in Table 2, cyclosporin A is highly effective in preventing rejection of renal allografts, at a dose of 10 mg/kg for 14 days, yet has almost no effect on islets at the same dose. It is possible that islets have antigens not present on renal allografts, possibly exposed by the isolation

Table 7. Effect of long-surviving renal allograft on survival of allogeneic islets

Donor kidney strain	Donor islet strain	Recipient strain	Islet Tx site	Islet graft survival (days)
None	Lewis	DA	Intraportal	3, 4, 4, 4, 5
None	Lewis	DA	Renal capsule	5, 6, 7, 7, 7, 8, 8, 8, 8, 11, 13
Lewis (1)	Lewis	DA	Renal capsule	45, 66, 67, 85, >100×3
Lewis (2)	Lewis	DA	Renal capsule	80, >100×5
Lewis	Lewis	DA	Intraportal	15, 90, >100×3
Lewis	PVG	DA	Renal capsule	3, 6, 9, 10, 11
Lewis	BN	DA	Renal capsule	7, 9
None	Lewis	PVG	Renal capsule	5, 6, 6, 7, 7
Lewis	Lewis	PVG	Renal capsule	70, >100×4

Long-surviving kidneys = >75 days after transplantation and CyA 10 mg/kg/day (14 days)

technique. If this were so, allografted islets transplanted into recipients that had already accepted a renal allograft of the same strain as the islets might be expected to reject in the normal time. Previous reports of such experiments using rats with long-surviving renal allografts induced by enhancing sera showed no increase in islet survival (Nash et al. 1978; Perloff et al. 1981).

We performed a similar experiment using rats with long-surviving renal allografts induced by cyclosporin A treatment. Islets of the same strain as the kidney were placed under the renal capsule without further immunosuppression (Reece-Smith et al. 1982). Prolonged survival of the islets was observed, in some cases indefinitely (Table 7). There remained the possibility that the effect was mainly due to transplanting the islet under the renal capsule of a grafted kidney (with loss of lymphatics, etc.) rather than true unresponsiveness. The experiment was repeated, this time comparing islets transplanted to the intraportal site to islets transplanted under the kidney capsule (Gray et al. 1983). Prolonged survival was noted without further immunosuppression. To confirm the effect as real the experiment was repeated using a different strain combination, with the same results. Finally the specificity of the effect was confirmed using unrelated islets from two strains. These experiments imply that there are no tissue-specific islet antigens separate from the antigens carried by the kidney.

In conclusion islets are probably not inherently more immunogenic than other tissues and probably do not carry tissue-specific antigens separate from those of the kidney. However, islets are probably more susceptible to damage if exposed directly to an immune response, either specific or nonspecific.

References

Ballinger WF, Lacy PE (1972) Transplantation of intact pancreatic islets in rats. Surgery 72:175–186

Finch DR, Morris PJ (1977) The effect of increasing islet numbers on survival of pancreatic islet allografts in immunosuppressed diabetic rats. Transplantation 23:104–106

Gray DW, Reece-Smith H, Fairbrother B, McShane P, Morris PJ (1984) Isolated pancreatic islet allografts in rats rendered immunologically unresponsive to renal allografts: the effect of the site of transplantation. Transplantation 37:434–437

Kemp CB, Knight MJ, Scharp DW, Ballinger WF, Lacy PE (1973) Effect of transplantation site on the results of pancreatic islet isografts in diabetic rats. Diabetologia 9:486–491

Lillihei RC, Simmons RL, Najarian JS, et al. (1970) Pancreatico-duodenal allotransplantation: experimental and clinical experience. Ann Surg 172:405–436

Morris PJ, Finch DR, Garvey JF, Poole MD, Millard PR (1980) Suppression of rejection of allogeneic islet tissue in the rat. Diabetes [Suppl] 29:107–112

Naji A, Barker CF, Silvers WK (1979) Relative vulnerability of isolated pancreatic islets, parathyroid, and skin allografts to cellular and humoral immunity. Transplant Proc 11:560–562

Nash JR, Peters M, Bell PR (1978) Studies on the enhancement of rat islet allografts. Transplantation 25:180–181

Perloff LJ, Naji A, Silvers WK, McKearn TJ, Barker CF (1981) Enhancement of whole pancreas and islet allografts. Transplant Proc 13:1423–1427

Reckard CR, Stuart FP, Schulak JA (1979) Immunologic comparisons of isolated pancreatic islet and whole-organ allografts. Transplant Proc 11:563–566

Reckard CR, Stuart FP, Clayman JL, Buckingham F, Schulak JA (1981) Differential susceptibility of segmental and isolated islet allografts of rat pancreas to rejection and enhancement. Transplant Proc 13:819–822

Reece-Smith H, DuToit DF, McShane P, Morris PJ (1981) Prolonged survival of pancreatic islet allografts transplanted beneath the renal capsule. Transplantation 31:305–306

Reece-Smith H, Homan WP, DuToit DF, McShane P, Morris PJ (1981) A technique for transplanting pancreatic islets as a vascularized graft and prevention of rejection with cyclosporin A. Transplantation 31:442–444

Reece-Smith H, Homan WP, McShane P, Morris PJ (1982) Indefinite survival of isolated pancreatic islets in rats rendered immunologically unresponsive to renal allografts. Transplantation 33:452–453

Sutherland DE (1981) Pancreas and islet transplantation. I. Experimental studies. Diabetologia 20:161–185

A New Approach to Experimental Pancreatic Islet Transplantation: Ultraviolet Irradiation *

H. Lau, K. Reemtsma, and *M. A. Hardy*

Introduction

Pharmacological immunosuppression has been relatively ineffective in preventing pancreatic islet allograft rejection. Although antilymphocyte serum (ALS) and antithymocyte serum (ATS) were, on occasion, found to be effective in the prevention of islet allograft rejection (Reckard et al. 1973; Finch and Morris 1977; Nelken et al. 1976), and recent studies with the use of cyclosporin, with or without ALS (Rynasiewicz et al. 1980, 1982), have also increased islet allograft survival in experimental models, no pharmacological immunosuppression has been effective in inducing indefinite acceptance of islet allografts.

Remarkable progress has been made in the past 5 years in the field of islet transplantation with special concentration on islet allografting. The various methods of altering the pancreatic allografts, previously summarized by Barker (1982; Barker et al. 1980), have relied primarily on elimination of passenger leukocytes and/or dendritic cells by organ culture techniques (Lacy et al. 1979 a; Lafferty et al. 1976; Sollinger et al. 1977), on the use of anti-Ia antibodies and complement (Faustman et al. 1981), and, most recently, on the use of ultraviolet irradiation treatment of donor-specific transfusions, or on direct ultraviolet irradiation of donor islets (Lau et al. 1983, 1984).

The effect of UV light from various sources on the immune mechanisms has been studied by many authors, both in vivo and in vitro, and has recently been well summarized by Daynes and Krueger (1983). The ability of UV irradiation to modulate immunological activity may begin by causing alterations in antigen-presenting cell function, and may culminate in the appearance of specific T-suppressor cells. We have performed a series of experiments based on the hypothesis that transplantation antigen, whether it be class I or class II major histocompatibility complex antigen, is not the primary barrier to grafting, and that "passenger" leukocytes in the donor islet tissue, or in the donor-specific blood, may provide the major immunogenic stimulus for the host, which may be altered to a tolerogenic stimulus with the use of UV irradiation. We have looked at the effect of ultraviolet B (UVB) irradiation on rat mitogenic and allogeneic lympho-

* Supported by NIH grant AM 30468 and JDF grant CU 50190201

Microsurgical Models in Rats
for Transplantation Research
Eds. A. Thiede et al.

cyte responses in vitro and we have utilized UVB irradiation of donor-specific blood transfusions as well as direct UV irradiation of islets in relation to rat islet allograft survival in vivo.

Materials and Methods

Animals

All donor and recipient rats were obtained from Microbiological Associates, Inc., Walkersville, Maryland. ACI ($RT1^a$) rats were used as recipients in all experiments and were made diabetic by intravenous injections of 60 mg/kg streptozotocin (courtesy of Dr. Dulin, Upjohn Company). Rats used as islet donors in these experiments included Lewis ($RT1^l$) and Wistar/Furth ($RT1^u$).

Islet Source and Route of Administration

Rat islet isolation was performed by the Lacy technique using collagenase digestion and Ficoll separation (Lacy et al. 1979 b). Four to six donors were used for one diabetic recipient. The islets (1200–1500) were transplanted intraportally into each diabetic recipient after handpicking and counting.

Source of UV Light and Method of Delivery of Irradiation

For all the following experiments the readily available sunlamps (Westinghouse FS-20 and FS-40) were used. Calibration of the UV fluorescent light source was performed using a UV light meter (UV Product, California). In most of the experiments the UV irradiation was UVB. UV irradiation was given to cells suspended in a balanced salt solution in open petri dishes at distance of 4–20 cm from the culture dishes and the irradiance was calibrated each time with the ultraviolet irradiation (UVX) light meter. The amount of UV irradiation depended on whether blood cells, purified lymphocytes, islet cells, or dendritic cells were being irradiated, and titration curves for functional damage for each cell type were obtained. The details of each method are outlined under the specific experiments.

Operative Procedures

All operations in rodents were performed under ether or intraperitoneal chloral hydrate anesthesia. Rejection and function of transplanted islets were monitored by blood glucose levels, initially on a daily basis and subsequently three times per week. Second set and third party heterotopic cardiac allografts were performed in selected animals with prolonged islet allograft survival by standard microsurgical techniques described elsewhere in this book.

Methods for In Vitro Studies of Cellular Mechanisms

Mixed lymphocyte culture (MLC) assay in the rat was done with thoracic-duct-derived lymphocyte responder cells, or with lymphocytes from blood or lymph node cells, against stimulator spleen cells. The cultures were performed in 96 well tissue culture plates with an optimal stimulation time of 5 days. Labeling with tritiated thymidine was done at 18 h prior to cell harvesting and scintillation counting.

The isolation of dendritic cells has been previously described (Mason et al. 1981). Abdominal lymph nodes were removed from rats 6 weeks prior to thoracic duct drainage. Lymph was collected over a 36-h period and the resultant cells were enriched for dendritic cells (DCs) by high-density bovine serum albumin (BSA) centrifugation step. Resultant light-density cells had a population of approximately 70% DCs with a distinct morphological appearance. These cells were used in MLCs with thoracic duct lymphocytes as responder cells.

Results of In Vitro Experiments

The Effect of UVB on Concanavalin A Responses of Mouse and Rat Lymph Node Lymphocytes

We have evaluated the effect of UVB (Westinghouse FS-20 sunlamp) on lymph node lymphocytes obtained from lymph nodes of Lewis rats or C57 Bl/6 mice. The cells were separated by teasing and passing through a #60 that irradiation with UVB at 10 cm for 30 s resulted in no response to 5 µg/ml concanavalin A (conA) of lymphocytes of mouse or rat origin. At the same time, the lymphocytes remained viable as determined by trypan blue exclusion. Higher doses of UVB resulted in no response and no killing, even after 2 min of UVB irradiation. These findings, summarized in Table 1, show that UVB irradiation of lymphocytes, even for 30 s, results in lack or response to a potent mitogen, conA, without apparent decrease in viability of cells, and is consistent with findings previously reported (Morrison et al. 1979; Cripps et al. 1978).

The Effect of Ultraviolet Irradiation (UVB) on Allogeneic MLC

In vitro studies of Lewis peripheral blood lymphocytes obtained from UV-irradiated blood, or from UV-irradiated splenocytes, had no significant stimulation of ACI thoracic duct lymphocytes as compared with Lewis lymphocytes obtained

Table 1. Effect of UVB on concanavalin a responses of mouse and rat lymph node lymphocytes

Dose of UVB (s)	[H^3]Thymidine incorporation of lymphocytes (cpm+SD)	
	Lewis rats	C57 BL/6 mouse
0	57189±2296	32156±3109
30	415± 80	960± 101
60	2314±1506	1054+ 189
90	179± 46	840± 204
120	2038±1912	1124± 198
150	859± 184	620± 89

Table 2. Effect of UV irradiation on the stimulatory activity of lewis and Wistar/Furth leukocytes in MLR

Responder at 5×10^5 cells/well	Stimulator	cpm ± SD[a]
ACI	ACI PBL[b]	465 ± 153
ACI	LEW PBL	5371 ± 543
ACI	LEW PBL with UV	772 ± 102
ACI	ACI splenocytes[c]	625 ± 282
ACI	W/F splenocytes	102404 ± 12305
ACI	W/F splenocytes with UV	709 ± 124
ACI	LEW splenocytes	25474 ± 22419
ACI	LEW splenocytes with UV	802 ± 201

[a] [H^3]Thymidine incorporation after 96-h culture with 16-h thymidine pulse
[b] Purified peripheral blood lymphocytes used as stimulators at 5×10^5 cells/well
[c] X-Irradiated splenocytes at 5×10^5 cells/well

from non-UV-irradiated whole blood or spleen. Similar results were obtained when W/F splenocytes obtained from UV-irradiated spleen-cell suspensions were used as the stimulator cells. The results are summarized in Table 2. In this experiment UV irradiation of cells was performed with two Westinghouse FS-20 lamps at a 10-cm distance from the dishes, with variable times of exposure. As expected, UVB irradiation abolished the mixed lymphocyte response (MLR), as has been previously described by several investigators (Lindahl-Kiessling and Safwenberg 1971; Hayry and Andersson 1976).

The Effect of UV Irradiation on Rat Dendritic Cells as Stimulators in MLR

The exact nature of the "passenger" leukocyte which has been implicated in causing graft rejection is not clear. The rat dendritic cells have been previously demonstrated to be extremely powerful as accessory cells in T-cell proliferation and in causing acute rejection of otherwise "passenger" leukocyte-depleted rat kidneys (Lechler and Batchelor 1982). MLCs were performed using ACI thoracic duct lymphocytes (TDLs) as responders and Lewis rat afferent lymph-derived dendritic cells (DCs) as stimulators. The isolated dendritic cells were gamma irradiated (1600 R), prior to UV irradiation and their use in MLCs. The source of UV irradiation for this experiment was a bank of two Westinghouse FS-20 lamps which have a flux of 1 mW/cm^2 at 310 nm (UVX-radiometer, UV Product, California), measured at 10 cm from the source. A high stimulation index (SI) of greater than 400 was obtained with 10^5 DCs, while when the number of DC stimulators was decreased to 0.125×10^5 the SI remained markedly elevated (greater than 162) when DCs were not UV irradiated. Dendritic cells subjected to UV irradiation at exposures of 800–1000 J/m^2 were completely ineffective as stimulators in the MLC, with a resulting SI of less than 3. Although DCs are extremely powerful allogeneic stimulators as demonstrated in the MLC (Mason et al. 1981; Steinman and Witmer 1978) and in causing graft rejection (Klinkert et al. 1982), they appear to be inactivated as stimulators by UV irradiation, but not by gamma irradiation.

Results of In Vivo Experiments

The Effect ov UV Irradiation on Donor-Type Blood Transfusions as Preconditioning for Allotransplantation

Preliminary results of these experiments have been recently reported and are summarized in Table 3. The ACI rats were made diabetic with streptozotocin and when their diabetes was stable, as measured by fasting blood glucose of greater than 300 mg/dl on consecutive occasions, were used as recipients of islets in all the following groups. Groups I and IA received Lewis or W/F islet allografts intraportally and were not pretreated with any blood transfusions. In the first series of experiments Lewis islets were used as allografts and group II animals received 1 ml Lewis blood that was not UV irradiated at 3, 2, and 1 weeks) prior to Lewis islet transplantation, while group III received the same pretransplant donor-specific blood transfusion protocol, but the blood was irradiated with UVB prior to injection. In group IV, donor-nonspecific blood, i.e., W/F, was given at 3, 2, and 1 week(s) prior to transplantation of Lewis islets, to determine the specificity of this treatment regimen. In all instances, blood was obtained from the donor animals and adjusted with balanced salt solution to a 50% packed cell volume prior to irradiation in open petri dishes and subsequent injection into the diabetic recipients (Lau et al. 1983). Following this experiment, as identical protocol was conducted using W/F isolated islet transplantation into ACI diabetic recipients using donor-specific blood transfusions; the parallel groups are referred to as IA, IIA, and IIIA. The islets transplanted in all the above experiments were fresh, noncultured, and isolated by sedimentation gradient and subsequent handpicking.

In the Lewis to ACI strain combination, islet allografts transplanted into diabetic hosts pretreated with UV-irradiated blood were accepted with full conversion of the diabetic animals to normoglycemia. The allografts continue to survive without any immunosuppression for almost a year. Donor specificity of this treatment was demonstrated by the ineffectiveness of UV-irradiated W/F blood in relation to subsequent Lewis islet transplantation and by the acceptance by fully converted animals of second set Lewis heart allografts and rejection of third party W/F hearts. All control groups rejected the islet allografts between the 3rd and 14th days post islet grafting and became hyperglycemic and polyuric.

Table 3. Islets allograft survival in ACI diabetic host pretreated with blood transfusions

Group	Treatment	Islet donor	n	Mean survival time ± SD (days)
I	None	LEW	5	8.2 ± 2.9
II	LEW blood	LEW	5	8.8 ± 4.1
III	LEW blood and UV	LEW	10	"Indefinite" > 250
IV	LEW blood and UV	W/F	4	7.5 ± 3.0
IA	None	W/F	4	6.5 ± 0.5
IIA	W/F blood	W/F	3	2.0 ± 0.0
IIIA	W/F blood and UV	W/F	6	21.1 ± 12.0

In another experiment, the survival of W/F islets in ACI diabetic recipients transfused with W/F UV-irradiated blood was examined (Table 3). Donor-specific transfusion with UV-irradiated blood (W/F) led to prolonged survival of W/F islet allografts mean survival time of 21.1 ± 12.2 days). This differed from the experiments using Lewis donors in that the allograft survival was not indefinite. It appears, therefore, that the effectiveness of the use of UV-irradiated donor-specific blood transfusions as a method of induction of acceptance of islet allografts may be dependent on the degree of MLC reactivity in the absence of UV irradiation.

The Effect of UV Irradiation on Pancreatic Islet Isograft Survival in Lewis Rats

Encouraged by the results obtained by donor-specific transfusions and UV irradiation in the Lewis to ACI islet transplantation model, we examined the direct effect of UV irradiation on islet isografts prior to using that method in an allograft model. The initial efforts included UV irradiation of islet isografts separated by sedimentation gradient and handpicking as previously described, to determine the appropriate dose of UV irradiation to avoid functional islet damage. The doses of UV irradiation that were chosen were based on the effectiveness of UV irradiation in abrogating the MLC response using rat dendritic cells. Lewis rat islets were UVB irradiated with a dose range of 600–1200 J/m^2. Irradiation of syngeneic islets (1200 per recipient) with 600 or 900 J/m^2 UVB prior to transplantation resulted in permanent conversion to normoglycemia of all syngeneic diabetic recipients, as shown in Table 4. When Lewis rat islets were irradiated with 1000 J/m^2 and subsequently transplanted intraportally into diabetic Lewis recipients, the conversion of diabetic animals to normoglycemia lasted for less than 5 days; islets irradiated with 1100 or 1200 J/m^2 failed to convert the diabetic animals at all. Thus, the UV irradiation dose that can abrogate the proliferative response in the MLC using 10^5 DCs as stimulator cells has no deleterious effect on the in vivo endocrine function of syngeneic islet grafts irradiated with 900 J/m^2.

Table 4. Effect of direct UV irradiation on islet isografts and allografts

Islet treatment (J/m^2)[a]	n	Donor	Recipient	Mean survival time ± SD (days)
600	3	LEW	LEW	Indefinite > 300
900	3	LEW	LEW	Indefinite > 300
1000	3	LEW	LEW	< 5
1200	2	LEW	LEW	< 1
No UV	10	LEW	ACI	6.8 ± 2.7
900	11	LEW	ACI	Indefinite > 250 (8/11)[b]

[a] Islets were handpicked, UV irradiated, and then cultured for 12–24 h at 37 °C in 95% air and 5% CO_2

[b] Three out of eleven allografted animals became hyperglycemic (BS × 200 mg/dl) on days 10, 10, and 18 after grafting. The remaining eight remained normoglycemic for more than 250 days

The Effect of Direct UV Irradiation on Islet Allograft Survival in the Diabetic Rat

The following experiment was performed to determine whether the immunogenicity of allogeneic Lewis islets was reduced and their functional survival prolonged after direct UVB irradiation. Based on our previous experiments described on pp. 206, 210, the isolated, handpicked Lewis rat islets were irradiated with 900 J/m^2 and were subsequently cultured for 12–24 h in 95% air and 5% Co_2 at 37 °C prior to intraportal transplantation into streptozotocin diabetic ACI rats. In preliminary experiments we demonstrated that the use of UV irradiation in this model without culture technique led to no islet allograft prolongation. All the control ACI animals receiving Lewis islets cultured for 24 h at 37 °C rejected their grafts and became diabetic again at 6.8 ± 2.7 days as shown in Table 4. When Lewis islets were UVB irradiated, cultured for 24 h, and then transplanted into diabetic ACI recipients, islet survival was prolonged in 8 of 11 transplanted animals, to more than 250 days, with all 8 animals remaining normoglycemic. These results suggest that treatment of allogeneic rat islets with UV irradiation prior to allotransplantation, at an exposure that is not deleterious to their endocrine function as demonstrated by previous experiments with syngeneic islet transplantation, can reduce the islets' immunogenicity and permits prolonged allograft survival without immunosuppression in most of the animals. In other preliminary studies using W/F islets into ACI diabetic recipients and direct UV irradiation of islets, we did not demonstrate a significant prolongation of islet allograft survival. This appeared to be related to a much higher MLC obtained with W/F to ACI lymphocytes as compared with Lewis to ACI MLCs, done without UV irradiation.

Discussion

Seven different methods have been developed which will prevent rejection of isolated adult islet allografts and, in a few instances, of concordant xenografts (rat to mouse). Five of these methods are based on implied or proven depletion from the donor islets of "passenger" leukocytes (macrophage-like antigen-presenting cells or dendritic cells), while two other methods rely primarily on pretreatment of the recipients with donor blood devoid of, or containing altered Ia-positive, leukocytes. The recent results suggest that the MHC antigen is not the primary barrier to allografting, and that the "passenger" leukocytes carried in the donor tissue provide the major immunogenic stimulus for the host. It therefore makes practical sense, which is also theoretically tenable, to alter tissue immunogenicity by treating the tissue to be grafted, rather than the recipient.

Although the abolishment of the initial recognition of a foreign allograft by the host appears to be critical to successful transplantation, maintenance of a functioning, nonrejecting allograft may depend on the initiation of donor-specific donor T-lymphocytes in the host as we have previously suggested (Lau et al. 1983). Various studies on the immune responsiveness of animals with established islet allografts indicate that a tolerant status can be induced following transplantation of islets devoid of "passenger" leukocytes. This state of unresponsiveness to allogeneic tissue has been demonstrated in other systems by Welsh et al. (1974)

when Ia-negative platelets and RBCs are unable to provoke a primary immune response, and actually attenuate a subsequent challenge with Ia-positive cells. This concept was further elaborated upon by Faustman et al. (1981), where pretreatment of diabetic mice with donor blood depleted of Ia-positive cells may permit subsequent successful transplantation of fresh, untreated allogeneic islets of the blood donor strain. In this study we have further demonstrated a similar phenomenon using UV-irradiated donor-specific blood transfusions used as pretransplant treatment of streptozotocin diabetic rats. Such pretransplant treatment with donor-specific transfusions that have been UVB irradiated permitted indefinite survival of fresh, untreated allogeneic islets of the blood donor strain. It appears, therefore, that immunization with Ia-negative donor blood components induces immunological unresponsiveness to the donor strain in the recipients by mechanisms that have not yet been clearly defined but that we suspect are due to the stimulation of specific suppressor cells. This demonstration of a rapid and simple method of inducing donor-specific unresponsiveness in adult animals, which allows long-term survival of islet allografts, may be a very useful method of inducing maintenance of a functioning, nonrejecting allograft, while the direct UV irradiation of islets may be more important in abolishing the initial recognition of a foreign allograft. It remains to be seen whether the combination of the two methods will be more efficient in obtaining indefinite islet allograft survival in all rodent strain combinations than each method alone.

It may be concluded from our studies that UV irradiation may offer a promising method for reduction of islet immunogenicity and induction of donor-specific unresponsiveness with maintenance of allograft acceptance. It is important to note that direct UVB irradiation, at an appropriate dose, does not alter pancreatic islet endocrine function in isograft experiments and that appropriately treated donor islet tissue can survive indefinitely as an allograft with prompt conversion of the diabetic host to normoglycemia. The possibility of using a combination of pretransplant treatment with UV-irradiated donor blood components and direct UV irradiation of donor islet tissue is presently being explored, as is the use of a brief peritransplant immunosuppresion, particularly with cyclosporin. We feel that this type of approach will abrogate allograft recognition and induce permanent maintenance of islet allografts in the diabetic hosts by reinforcing and potentiating any suppressor mechanism that may appear.

Acknowledgment. We gratefully acknowledge the technical assistance of Mrs. Margaret Prasek and Mrs. Jitka Mucha, as well as the secretarial assistance of Ms. Renee Grobtuch.

References

Barker C (1982) Pancreatic cell transplantation. In Chatterjee SN (ed) Transplantation. PSG Publishing, Littleton, Massachusetts, pp 413–445

Barker C, Naji A, Silvers WK (1980) Immunologic problems in islet transplantation. Diabetes 28(1):86–92

Cripps DJ, Horowityz S, Hang R (1978) Spectrum of ultraviolet radiation on human B and T lymphocyte viability. Clin Exp Dermat 3:43–50

Daynes RA, Krueger G (1983) In Experimental and clinical photoimmunology, vol 2. CRC Press, Boca Raton, Florida
Faustman D, Hauptfeld V, Lacy PE, et al. (1981) Prolongation of murine islet allograft survival by pretreatment of islets with antibody directed to Ia determinants. Proc Natl Acad Sci USA 78:5156, 5159
Finch DRA, Morris PJ (1977) The Effect of increasing islet numbers on survival of pancreatic islets allografts in immunosuppressed diabetic rats. Transplantation 23:104–106
Hayry P, Andersson LC (1976) Generation of T memory cells in one-way mixed lymphocyte culture. Scand J Immuno 5:391–399
Klinkert WEF, LaBadie JH, Bowers WE (1982) Accessory and stimulating properties of dendritic cells and macrophages isolated from various rat tissues. J Exp Med 156:1–19
Lacy PE, Davie JM, Finke EH (1979) Prolongation of islet allograft survival following in vitro culture (24, C) and a simple injection of ALS. Science 204:312–313
Lacy PE, Davie JM, Finke EH (1979)Induction of rejection of succesful allografts of rat islets by donor peritoneal exudate cells. Transplantation 28:415–420
Lafferty KJ, Bootes A, Talmage DW (1976) Effect of organ culture on the survival of thyroid allografts in mice. Transplantation 22:138–149
Lau H, Reemtsma K, Hardy MA (1983) Pancreatic islet allograft prolongation by donor specific blood transfusions treated with ultraviolet irradiation. Science 221:754–756
Lau H, Reemtsma K, Hardy MA (1984) Prolongation of rat islet allograft survival by direct ultraviolet irradiation of the graft. Science 223:607–609
Lechler RBI, Batchelor JR (1982) Restoration of immunogenicity to passenger celldepleted kidney allografts by the addition of donor strain dendritic cells. J Exp Med 155:31–41
Lindahl-Kiessling K, Safwenberg J (1971) Inability of UV-irradiated lymphocytes to stimulate allogeneic cells in mixed lymphocyte culture. Int Arch Allergy 41:670–678
Mason DW, Pugh CW, Webb M (1981) The rat mix lymphocyte reaction: roles of a dendritic cell in intestinal lymph and T-cell subsets defined by monoclonal antibodies. Immunology 44:75–87
Morrison WL, Parrish JR, Block KJ, et al (1979) In vivo effect of VV-B on lymphocyte function. Br J Dermat 101:513–519
Nelken D, Morse SL, Beyer MM, et al (1976) Prolonged survival of allotransplanted islets of Langerhans cells in the rat. Transplantation 22:74
Reckard CR, Zeigler MM, Barker CF (1973) Physiological and Immunological consequences of transplanting isolated pancreatic islets. Surgery 74:91–99
Rynasiewicz JJ, Sutherland DER, Kawahara K, et al (1980) Cyclosporin A prolongation of segmental pancreatic and islet allograft function in rat. Transplant Proc 12:270–274
Rynasiewicz JJ, Sutherland DER, Ferguson RM, et al (1982) Cyclosporin A for immunosuppression; observations in rat, heart, pancreas and islet allograft models and in human renal and pancreas transplantation. Diabetes 31(4):92–108
Sollinger HW, Burkholder PM, Bach FH (1977) Prolonged survival of xenografts after organ culture. Transplant Proc 9:359–365
Steinman RM, Witmer MD (1978) Lymphoid dendritic cells are potent stimulators of the primary mixed leucocyte reaction in mice. Proc Natl Acad Sci USA 75:5132–5136
Welsh KI, Burgos H, Batchelor JR (1977) The immune response to allogeneic rat platelets; AG-B antigens in matrix form lacking. Eur J Immunol 7:267–272

Characterization of Pancreas Islet-Cell Reactivity with Various Immune Sera *

K. Ulrichs

Introduction

In non-MHC as well as in MHC fully allogeneic rat strain combinations, islet grafts are very often as rapidly rejected as skin grafts. Among the many attempts to reduce islet immunogenicity there are such promising studies as the in vitro culture of islet grafts prior to transplantation (Bowen and Lafferty 1980) and the elimination of Ia-bearing cells from murine islets with antibody directed against Ia determinants (Faustman et al. 1981). Little is known about the cell type(s) and number of immunogenic cells in question. Therefore, it was the aim of this study (1) to characterize rat islet cells with sera of known specificity, (2) to try to eliminate putatively immunogenic cells serologically without damaging the insulin-secreting cells, and (3) to test the in vivo relevance of the in vitro observations.

Material and Methods

Animals. Pancreas islet and cell donors were male CAP ($RT1^c$) rats. Graft recipients were LEW ($RT1^l$) rats, made diabetic by a single i.v. injection of 55 mg/kg streptozotocin (Upjohn). Animals were bred and housed at the Dept. of Immunology.

Islet Isolation. Approximately 1500 handpicked clean islets were isolated from ten donor animals using the collagenase technique (Lacy and Kostianovsky 1967).

Islet Dispersion. Fresh islets were enzymatically dispersed into a single-cell suspension with dispase I enzyme (Boehringer, Mannheim) in three to six 15-min steps (each step with 4 ml Ca^{2+}-free, Hepes-buffered Hanks' balanced salt solution (HBSS), pH 7.5, 37 °C, gentle shaking) using a total of 10 000 IU. The supernatants containing single cells were washed twice and collected. The total cell number varied between 10^5 and 10^6 cells per dispersion, depending on the number of islets and the enzyme activity.

* This work was supported by the Deutsche Forschungsgemeinschaft, SFB 111/B12

Microsurgical Models in Rats
for Transplantation Research
Eds. A. Thiede et al.

Tests. Standard immunofluorescence tests were performed in Terasaki microplates with rat lymph node cells, thymocytes, splenic macrophages, and dendritic reticulum cells from the spleen as control cells, and dispersed pancreas islet cells as test cells. Findings were evaluated comparing the fluorescence and phase contrast appearance of the cells. Standard cytotoxicity tests were carried out with the above cells in Cooke microtiter plates using fresh or frozen LEW complement. Standard immunoperoxidase tests were performed with total fresh islets.

Test Sera. Monoclonal antibody (rat to rat) YR5/24 against class II antigens (RT1.*B* region) was commercially obtained (Serotec) and used in 1 : 100 dilutions in the above tests. Monoclonal antibody (mouse to rat) OX6 against class II antigens (RT1.*B* region) was a kind gift from Dr. Wonigeit (Hannover) and used in 1:200 dilutions in the above tests. Rabbit-anti-rat lymphocyte serum (ALS) was prepared, absorbed, and tested as described in detail elsewhere (Müller-Ruchholtz et al. 1976). To prepare specified antilymphocyte serum (SAL), sequential absorptions with erythrocytes (E), peritoneal exudate cells (P), and fetal liver cells were performed. It may be pointed out that each absorption was proven to be serologically exhaustive, maintaining 100% cytotoxicity at dilutions of 1 : 4–1 : 16 in sera which were used only undiluted. Rabbit-anti-rat macrophage serum (AMS) was prepared, absorbed, and tested similar to ALS, using peritoneal exudate cells (10^8) for immunization. To prepare specified antimacrophage serum (SAM), sequential absorptions with erythrocytes, thymocytes (T), lymphocytes (L), and adult liver cells were performed.

Results

Reactivity with Rat Monoclonal Anti-Ia Antibody

As shown in the first row of Table 1, among the freshly suspended islet cells, containing mainly hormone-secreting cells, 30% of the visible macrophages, making 3%–5% of the total cell number, and 50% of the visible lymphocytes, making approximately 5% of the total cell number, reacted positively with YR5/24. However, as also shown, islet dendritic cells (I-DC, approximately 1% of the total cell number) did not react with anti-Ia in contrast to the known reactivity of spleen dendritic control cells. (The antibody reacted clearly positively with 30%–40% of the control lymphocytes and 50% of the control macrophages, thus confirming the supplier's data; data not shown). As to be further seen in Table 1, the antibody was nontoxic to dispersed islet cells when tested in combination with rat complement in cytotoxicity assays. These and the above results have been confirmed using the second monoclonal anti-Ia antibody OX6.

Reactivity with SAL

Exhaustive absorptions with erythrocytes and peritoneal exudate cells yield an antiserum (ALS-EP) which specifically reacts with control lymphocytes and thymocytes (data not shown). After elimination of the critical cross-reactivity of ALS-EP with the hormone-secreting islet cells by additional absorptions with

Table 1. Pancreas islet cell reactivity in fluorescence (FU) and cytotoxicity (CT)

Target cells Cell type and % of visible cells			Test sera and % of reactive cells					
			α RT1B		SAL		SAM	
			FU	CT	FU	CT	FU	CT
 β-cells (?)	hormone-secreting cells	~70–80	–	–	–	–	–	–
 α-cells (?)		~10–20	–	–	–	–	–	–
	macrophage	~ 3– 5	30**	–	–	–	100***	100
	dendritic cell	~ 1	–	–	–	–	100****	100
	lymphocyte	~ 5	50**	–	100***	100	–	–

Fluorescence was graded by its intensity: ****, very strong; ***, strong; **, medium; –, no detectable fluorescence

adult or fetal liver cells, approximately 5% of the lymphocytes could be detected in the islet-cell suspension (see Table 1). Furthermore, SAL proved to be 100% cytotoxic when used in combination with rat complement.

Reactivity with SAM

Exhaustive absorptions with erythrocytes, thymocytes, and lymphocytes yield an antiserum (AMS-ETL) which specifically reacts with control macrophages and dendritic reticulum cells of the spleen (data not shown). After elimination of the critical cross-reactivity of AMS-ETL with the hormone-secreting islet cells by additional absorptions with adult or fetal liver cells, 3%–5% of the macrophages and approximately 1% of cells showing a characteristic dendritic structure (both of which reacted strongly with SAM) could be detected in the suspension (see Table 1). Similar to SAL, SAM was 100% cytotoxic when used in combination with rat complement.

Characterization of an Islet Dendritic Cell

In fresh single-cell suspensions I-DCs are only observed when islet dispersion is performed with dispase of very low activity. Their serological pattern is Ia^-, SAL^-, SAM^{++++} and they are easily lysable in cytotoxicity tests. They are of typical dendritic appearance as can be seen from single cells in immuno-

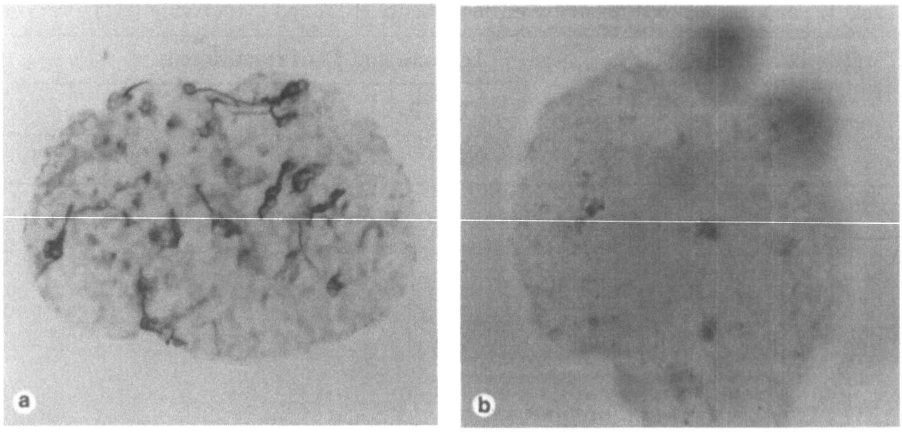

Fig. 1 a, b. Immunoperoxidase test with freshly isolated pancreatic rat islets. **a** SAM-positive dendritic cells within the islet. **b** No staining with normal rabbit serum as first antibody. ×250

fluorescence as well as phase contrast (data not shown) and from total islets in immunoperoxidase tests (see Fig. 1). Morphologically they are often irregularly shaped cells frequently with very long plasmatic extensions. They develop only after 1–2 h following preparation of cell suspension. I-DCs never stain in background fluorescence or in immunoperoxidase control tests using normal rabbit serum as first antibody.

In Vivo Data Following SAL and SAM Incubation of Islet Grafts: A Study in Progress

A preliminary finding in a major histocompatibility complex (MHC) fully and biologically strongly histoincompatible rat strain combination ($RT1^c \rightarrow RT1^l$ in which grafts are always rejected between 5 and 7 days) suggests that fresh islets pretreated with SAL plus SAM and complement are accepted in the long run (> 50 days) without any immunosuppression of the graft recipient, provided the antiserum treatment is performed with sufficiently large amounts of antibodies over a sufficiently long period.

Discussion

Immunofluorescence tests and phase contrast microscopy with dispase-dispersed rat islet cells show that clean, handpicked islets contain a distinct number of lymphocytes and macrophages, each of which contain Ia^+ cells. The existence of such cells within the islet of Langerhans has previously been suggested (Zitron et al. 1981); however, precise data have not been published so far.

An unexpected result obtained from our studies is the identification of an islet dendritic cell which appears to be Ia^-, SAL^-, and SAM^{++++}. This is in contrast to the observation of Hart and Fabre (1981), who identified Ia^+ interstitial dendritic cells in rat islet frozen tissue sections. (The serologically and morphologically detailed characterization of isolated cells will be of greater value for further evaluation of the I-DC as an immunogenic cell than histological studies.) The Ia negativity of a cell with obvious dendritic morphology is a striking phenomenon, since it is generally believed that this cell serves for antigen presentation requiring the expression of class II antigens. This observation may be due to the methodological procedure having caused a damaging effect on the Ia determinant since the cells were studied within only few hours after enzymatic dispersion. In vitro culture prior to immunofluorescence may help to restore antigen expression (study in progress).

On the other hand, lymphocytes as well as macrophages clearly contain Ia^+ cells following the serum preparation procedure, which suggests that I-DCs are at least much more labile Ia-antigenic determinant expressors than those cells. This may be seen in context with increasing evidence from the recent literature that Ia expression on certain cell surfaces may be variable, and dependent on immunological (de Waal et al. 1983) or nonimmunological stimuli (Klareskog et al. 1980). From this labile expression of Ia it could be inferred that the approach of islet incubation in SAM (plus SAL), i.e., with cell-group specific antibodies, may well prove to be superior for allogeneic islet grafting.

The I-DCs clearly differ from islet fibroblasts, which are also Ia^- but are numerous, have a typical spindle-like morphology, and react only weakly with SAM. Pretreatment of mouse islet allografts with Ia antibodies has been shown to be effective in an MHC-histoincompatible model (Faustman et al. 1981) which, however, appears to have been relatively weak in biological terms (as inferred from control graft survival data). In contrast, our rat islet allograft model appears to be much stronger (consistent control graft rejection time, 5–7 days). It is hoped that the preliminary findings obtained in this model and using SAL and SAM for islet incubation can be extended and further analyzed (with regard to the relative roles of SAM and SAL), provided that sufficient penetration of antibodies and complement into the islet can methodologically be assured.

Acknowledgment. The author wishes to thank Mrs. T.T.H. Ly, Mrs. D. Hoffmann, and Mrs. M.v. Pein-Böhm for skillful technical assistance.

References

Bowen KM, Lafferty KJ (1980) Reversal of diabetes by allogeneic islet transplantation without immunosuppression. Aust J Exp Biol Med Sci 58:441–447

De Waal RMW, Bogman MJJ, Maass CN, Cornelissen LMH, Tax WJM, Koene RAP (1983) Variable expression of Ia antigens on the vascular endothelium of mouse skin allografts. Nature 303:426–429

Faustman D, Hauptfeld V, Lacy P, Davie J (1981) Prolongation of murine islet allograft survival by pretreatment of islets with antibody directed to Ia determinants. Proc Natl Acad Sci USA 78:5156–5159

Hart DJN, Fabre JW (1981) Demonstration and characterization of Ia-positive dendritic cells in the interstitial connective tissues of rat heart and other tissues, but not brain. J Exp Med 153:347–361

Klareskog L, Forsum U, Peterson PA (1980) Hormonal regulation of the expression of Ia antigens on mammary and epithelium. Eur J Immunol 10:958–963

Lacy P, Kostianovsky M (1967) Method for the isolation of intact islets of Langerhans from the rat pancreas. Diabetes 16:35–39
Müller-Ruchholtz W, Wottge HU, Müller-Hermelink HK (1976) Bone marrow transplantation in rats across strong histocompatibility barriers by selective elimination of lymphoid cells in donor marrow. Transplant Proc 8:537–541
Zitron IM, Ono J, Lacy P, Davie JM (1981) The cellular stimuli for the rejection of established islet allografts. Diabetes 30:242–246

Heterotopic Auxiliary Liver Transplantation with Respect to Its Clinical Application

O. Zelder

The idea of permanent or temporary organ substitution through a heterotopic auxiliary transplant is often proposed in spite of not very encouraging results in transplantation surgery. These are exemplified by the latest reports on such substitution with the heart.

Except for a very few successes in this field auxiliary liver transplantation have been doomed to failure (Fortner et al. 1979; Houssin et al. 1980). The failures have been caused by problems such as: lack of space for two livers; competition of two livers from a functional viewpoint, defined as the so-called inter-liver competition; and technical difficulties with the anastomosis, resulting in outflow block and thus supply of blood from the portal vein (Hess et al. 1972; van der Heyde et al.1966). When it then became possible to carry out liver transplantations in rats in large series through microsurgical operative techniques both heterotopic and auxiliary, we were also able to accomplish this successfully (Hess et al. 1972; Lee and Edgington 1966; Zelder 1974, 1984).

Of interest to us was first the technical capability of carrying out such an operation and second the behavior of key enzymes of the various metabolic pathways of the two competing livers in one organism. We had already carried out experiments in this direction on the normal and cirrhotic rat liver with and without various portacaval anastomoses together with the research group of Bode and Bode in Marburg (Bode et al. 1975; Zelder 1974, 1984).

With the histomorphological investigations that were carried out solely by Jerusalem from Nijmegen, local oxygen pressure measurements on the liver surface that we carried out in common with Thermann and Jostarndt in Marburg all served as valuable supplements or confirmations of our collected data (Hess et al. 1972; Jostarndt et al. 1978; Thermann et al. 1975; Zelder 1974, 1984). In all, about 500 heterotopic auxiliary liver transplantations were carried out, mainly by the methods of Hess (Hess et al. 1972; Zelder 1974, 1984) (Fig. 1 g).

Poor initial results were due to lack of training, lack of expertise, and technical mistakes. No extended survival time was found in the roughly 60 transplantations according to Lee's method (Lee and Edgington 1966). Our observations showed that outflow blocks in the grafts were responsible for the early deaths. We believe that these blocks were brought on by suprahepatic vena cava cuffs on the grafts that were too long. For this reason, the technique was modified. By using the tech-

Microsurgical Models in Rats
for Transplantation Research
Eds. A. Thiede et al.

nique of Hess et al. (1972) with a short subhepatic cava cuff placed almost at the liver level, and with an outflow anastomosis as high as possible subhepatically, these difficulties seemed to be resolved. But the excellent results of Hess et al. (1972), who worked with rats from closed breeding colonies under specific pathogen free (SPF) conditions, remained unreproducible in Marburg. Of course in a few cases outflow blocks were still seen due to technical mistakes. A part from acute and chronic rejection reactions, however, infections and microcirculatory defects in the grafts and the remnant livers were the main cause for graft and animal losses. Differences between SPF and normal laboratory conditions must be a key factor, particularly since the operational technique was performed in an identical manner. About 30% of the normal animals survived a transplantation for 50 days. The longest survival period in rats with healthy livers was 395 days, in cirrhotic animals 329 days. In later experiments the animals were killed earlier for various examinations. Enzyme activities were measured in the grafts and remnant livers within the first 7–14 days during the so-called adaptation period.

It would take up too much space to mention the multitude of individual results or to discuss them individually. Surprisingly enough, we discovered parallel changes of enzymes in both graft and host livers. Enzymes of the aerobic metabolic pathway decreased, while glycolytic enzymes increased. This reaction was explained by morphological changes: beginning as soon as the third postoperative day and increasing to an acute cellular rejection by the 7th day and after or with focal necrosis. Enzymes involved in energy-producing metabolism had several things in common with other situations, such as the condition after portal vein ligation, and portacaval anastomosis in normal rats or in rats with liver cirrhosis (Bode et al. 1975; Zelder 1974, 1984).

The changes in enzyme activity must be interpreted as an adaptation of the individual metabolic pathways to an increased anaerobic energy production. The decrease in arginase is not a result of a reduced nutritional protein-uptake course of pair-feeding, but explanable as induced by a diminished substrate supply through a reduction in general blood circulation. This may be clarified by oxygen-pressure measurements. Immediately after the operations, the portally supplied graft is aligned with that of a normally supplied rat liver. This must be connected with the increased blood circulation of the donor parenchyma, which has been resected ba 30%. The figures for oxygen determination, which decrease later and restabilize after 40–60 weeks, are apparently linked to the special circulatory conditions and the microarchitectural changes of the hepaton during the functional adaptation of compensational hypertrophy (Jostarndt et al. 1978; Thermann et al. 1975; Zelder 1974, 1984).

Over the past years several experimental surveys have been conducted on dogs and pigs with toxic or ischemic liver damage or heterotopic auxiliary liver transplantation (either permanent of temporary liver replacement) (von der Heyde et al. 1966; Kuster and Woods 1972; Zelder 1974, 1984). Therefore we used a rat model of liver damage by peroral treatment of thioacetamide (Fig. 1). Depending on dosage and time span of application, a cirrhosis of the small to large nodular type was produced (Fig. 1 a, b), which showed histologically hyperplastic or hyperplastic-hypertrophic (Fig. 1 e), adenoma-like (Fig. 1 f), and often even hepatoma-like cells (Fig. 1 d) and signs of portal hypertension (Fig. 1 c). In about 30%

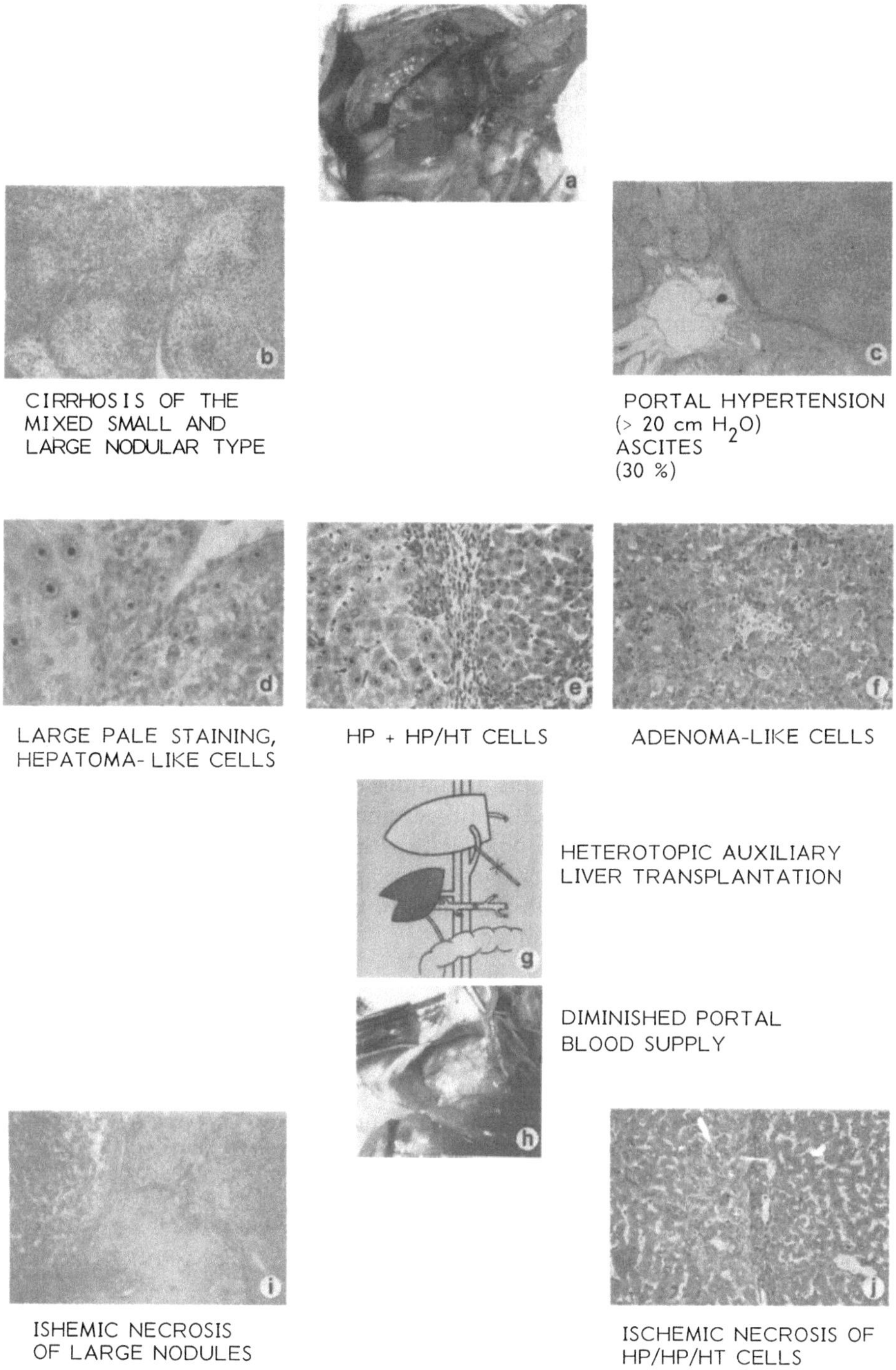

Fig. 1 a–j. Rat model of liver chirrhosis and heterotopic auxiliary liver transplantation: *a* surface of the host liver before transplantation; *b–f* histological findings in the host liver before transplantation; *g* scheme of transplantation procedure by the technique of Hess; *h* surface of the host liver 3 days after transplantation; *i, j* histological findings in the host liver after transplantation ascites was found. Lymphatic vessels, especially those around the hepatic portal, were clogged (Fig. 1 c). The portal hypertension was double-checked through hemodynamic measurements together with Thermann and Jostarndt. It was remarkable that the local oxygen pressure of the liver tissue was significantly diminished in cirrhotic livers (Jostarndt et al. 1978; Thermann et al. 1975; Zelder 1974, 1984). Of the cirrhotic animals 70% survived for about 14 days, and one even for 329 days. A part from acute rejection of the graft, total necrosis of the cirrhotic host liver was seen most often. Neither the application of a larger graft nor an additional treatment with neomycin could improve these results. Further leading surveys with three randomized cirrhotic groups showed even worse results. Even the heterotopic transplantation of 70% of a normal liver after total hepatectomy of the recipient presented no alternative.

Severe morphological changes of the cirrhotic host liver up to total collapse or complete necrosis were characteristic (Fig. 1 h–j). They must be due to the low and strictly arterial blood supply. We consider this to lead to an infiltration by toxic decomposition products with corresponding circulatory reactions, which must be considered or at least discussed as a cause of death in these animals of generally worse condition. All grafts were in good condition. Signs of right-sided heart failure other than outflow block were only discovered in the vessel system of the graft. The space problem played a major part in these transplantations. In one series, five to ten grafts could be implanted either not at all or only with extreme difficulty. They belong to the group of technically faulty transplantations (Zelder 1984).

Although we are aware of all the limitations of trying to transfer results from animal experiments to the clinic, it is still our opinion that further evidence is necessary to show that it is hardly possible experimentally and particularly in the clinic to use heterotopic auxiliary liver transplants with much success as temporary organ substitutions for badly damaged livers. Because of this experiments in this direction have been discontinued.

References

Bode JC, Zelder O, Bode C, Paidlick A, Jerusalem CR (1975) Changes in enzyme activities in the regenerating heterotopic auxiliary liver graft in the rat. In: Lesch R, Reutter W (eds) Liver regeneration after experimental injury. Stratton, New York, pp 195–199

Fortner JG, Yeh SDJ, Kim DK, et al. (1979) The case for and technique of heterotopic liver grafting. Transplant Proc 11:269–273

Hess F, Jerusalem C, von der Heyde MN (1972) Advantages of auxiliary homotransplantations in rats. Arch Surg 104:75–80

Houssin D, Berthelot P, Franco D, Bismuth H (1980) Heterotopic liver transplantation in end-stage HBSAg-positive cirrhosis. Lancet I 990–993

Jostarndt L, Thermann M, Jerusalem CR, Zelder O (1978) Hemodynamic parameters and blood gas analyses in the normal and the cirrhotic rat. Res Exp Med (Berl) 173:187–191

Kuster G, Woods JE (1972) Auxiliary liver transplantation in the dog as a temporary support in acute fulminating hepatic necrosis. Ann Surg 176:732–735

Lee S, Edgington TS (1966) Liver-transplantation in the rat. Surg Forum 17:220–222

Thermann M, Zelder O, Hess F, Jostarndt L, Jerusalem CR, Hamelmann H (1975) Studies on the oxygen supply of heterotopic auxiliary liver grafts in rats. Eur Surg Res 7:156–163
Von der Heyde MN, Schalm L, Vink M (1966) The role of functional competition in auxiliary transplantation. Transplantation 5:78–80
Zelder O (1974) Zur Chirurgie im Bereich des Pfortadersystems und zur heterotopen auxiliären Lebertransplantation. Görich and Weiershäuser, Marburg.
Zelder O (1984) Heterotopic auxiliary liver transplantation in the rat. In: Olszewski W (ed) Handbook of microsurgery. CRC, Boca Raton, vol 2, pp 377–395

Accelerated But Not Hyperacute Rejection of Liver Allografts in Hypersensitized Rats *

D. Houssin, J. Gugenheim, M.D. Brunaud, M. Gigou, M. Charra, S. Crougneau, and *H. Bismuth*

The privileged status of liver allografts compared with other organ allografts is illustrated by the spontaneous tolerance of liver allografts in pigs (Calne 1973) and in several combinations of inbred rats (Houssin et al. 1979; Zimmermann et al. 1979). Another remarkable peculiarity of liver transplantation is the absence of hyperacute rejection of liver allografts in man in case of an ABO-incompatible transplantation or when preformed lymphocytotoxic antibodies directed toward the donor are present in the recipient (Starzl et al. 1982). The reason for the relative resistance of the liver to cytotoxic antibodies is still unclear. In order to study further the effect of cytotoxic antibodies on the liver, the fate of donor-specific liver allografts was compared with that of donor-specific heart allografts in hypersensitized rats.

Animals and Methods

Animals

Male inbred LEW ($RT1^l$) or BN ($RT1^n$) rats, body weight 250 g, were purchased from the CNSEAL (Orléans La Source, France).

Grafting Procedures

Skin grafting was carried out by removing full-thickness ventral abdominal skin from LEW donor rats under ether anesthesia. The 2.5-cm circular skin grafts were fastened with wound clips on the dorsum of BN recipient rats. Heterotopic intra-abdominal heart transplantation was performed according to the technique described in rats by Ono and Linsey (1969). Liver transplantation was performed orthotopically using cuff techniques for portal and infrahepatic vena cava anastomoses (Kamada 1983). No arterial anastomosis was performed. A choledochocholedochal end-to-end anastomosis was used for biliary reconstruction employing the same technique as that used for ureteral reconstruction after kidney transplantation in rats (Thoenes et al. 1974).

* Study by the Groupe de Recherche de Chirurgie Hépatique, INSERM U 17, Hôpital Paul Brousse, Villejuif, France

Microsurgical Models in Rats
for Transplantation Research
Eds. A. Thiede et al.

Detection of Lymphocytotoxic Antibodies

Complement-dependent antibody assay was done using the two-step incubation method as previously described (Charpentier et al. 1979). LEW spleen cells (10^5/well) were incubated with various dilutions of serum from hypersensitized BN rats for 20 min at 22 °C. Guinea pig complement previously adsorbed on rat red blood cells and rat spleen cells was further added at one-fifth final dilution. After 60 min incubation at 37 °C, dead cells were counted using the trypan blue exclusion test. Sera tested for specific cytotoxic antibodies were taken immediately before and 30 min after heart or liver allografting in hypersensitized BN rats.

Histological Examination

After liver transplantation, a biopsy of the graft was made 30 min after revascularization. At the time of rejection, heart and lifer grafts were removed for histological, examination which was performed using hematoxylin and Masson's stains. The following features were recorded and graded: inflammatory infiltration with mononuclear cells, infiltration with polynuclear leukocytes, foci of ischemic necrosis, hemorrhage, and edema.

Experimental Protocol

BN rats were grafted with a LEW skin graft on three successive occasions at 10-day intervals. Between 15 and 20 days after the third skin graft, these hypersensitized BN rats were grafted either with a LEW heart allograft (group 1) or with a LEW liver allograft (group 2). In the control groups, unsensitized BN rats were grafted with a LEW heart allografts (group 3) or with a LEW liver allograft (group 4). Criteria chosen for rejection were the cessation of beatings after heart allograft and the death of the recipient after liver allograft. At the time of rejection, heart recipients were killed and liver recipients were autopsied in order to detect a technical complication. At that time, the graft was removed for histology.

Results

The survival time of LEW liver allografts in previously hypersensitized BN rats was significantly shorter than the survival time of LEW liver allografts in unsensitized BN rats (Table 1). However, the survival time of LEW liver allografts in previously hypersensitized BN rats was significantly longer than the survival time of LEW heart allografts in hypersensitized BN rats (Table 1). The level of specific cytotoxic antibodies was markedly elevated in hypersensitized BN rats. It remained unmodified after LEW heart allografting (Table 2). On the contrary, the level of cytotoxic antibodies was markedly reduced following LEW liver allografting (Table 3). Histological examination of a biopsy of LEW liver allografts 30 min after revascularization revealed a nearly normal parenchyma in 11 animals out of 12. In one animal, multiple foci of ischemic necrosis with hemorrhages were present.

Table 1. Survival times of LEW heart and liver allografts in previously sensitized or in unsensitized BN rats

Experimental groups	Graft survival time (h)
Group 1 (LEW heart in presensitized BN rats)	1, 1, 2, 2, 3, 3, 4, 5, 6, 7, 7, 12, 12, 96[a]
Group 2 (LEW liver in presensitized BN rats)	1, 7, 8, 9, 24, 30, 36, 48, 72, 72, 96, 144, 216[a, b]
Group 3 (LEW heart in nonsensitized BN rats)	144, 144, 168, 168, 168, 192, 192, 192, 192, 216
Group 4 (LEW liver in nonsensitized BN rats)	30, 48, 120, 144, 144, 264, 288, 368, 384, tolerant[b]

[a] $P<0.01$ between group 1 and group 2
[b] $P<0.01$ between group 2 and group 4

Table 2. Lymphocytotoxic antibodies before and after grafting a LEW heart in BN rats previously sensitized with three LEW skin grafts (group 1)

Heart allografted rats (rat number)	Lymphocytotoxic antibody level[a]	
	Before	After
52	80	69
56	71	62
57	60	49
61	62	87
69	95	90
70	95	95
74	62	67
77	85	48
83	51	45
103	72	71

[a] Results expressed as a percentage of dead cells at one-fourth dilution

At the time of rejection in hypersensitized BN rats, histological examination of LEW heart allografts revealed an inflammation with mononuclear cells (9/12) [+(4/9), + +(2/9), + + +(3/9) (slight, medium, and strong, respectively)], foci of ischemic necrosis (5/12) [+(1/5), + +(2/5), + + +(2/5)], edema (8/12) [+(2/8), + +(5/8), + + +(1/8)], hemorrhages (3/12) [+ +3/3)], infiltration with polynuclear leukocytes (5/12) [+(1/5), + +(4/5)], and a nearly normal tissue in one animal. Histological examination of LEW liver allografts at the time of rejection in hypersensitized BN rats revealed an inflammation with mononuclear cells (4/12) [+ + +4/4)], foci of ischemic necrosis (7/12) [+ +(1/7), + + +(6/7)], edema (1/12) [+ + + 1/1)], hemorrhages (2/12) [+ 1/2), + +(1/2)], infiltration with polynuclear leukocytes (2/12) [+ +(2/2)], and a nearly normal parenchyma (4/12).

Table 3. Lymphocytotoxic antibodies before and after grafting a LEW liver in BN rats previously sensitized with three LEW skin grafts (group 2)

Liver allografted rats (rat number)	Lymphocytotoxic antibody level[a]	
	Before	After
25	80	10
38	82	< 5
46	78	< 5
50	74	< 5
51	90	24
62	57	2
86	82	34
89	59	15
96	82	24
97	75	6
104	61	< 5

[a] Results expressed as a percentage of dead cells at one-fourth dilution

Discussion

This study shows the absence of hyperacute rejection of liver allografts in a previously described (Guttmann 1974) rat model of allogeneic hypersensitization using repeated skin allografts. This study confirms the remarkable resistance of the liver to hyperacute rejection which has been previously underlined in man in case of an ABO incompatibility or when preformed cytotoxic antibodies directed toward the donor were present in the recipient at the time of transplantation (Starzl et al. 1982). The mechanism of the resistance of the liver to cytotoxic antibodies is still unclear. In our study, the dramatic decrease of the level of specific cytotoxic antibodies early after revascularization of liver allografts in hypersensitized recipients contrasts with the maintenance of a high level of cytotoxic antibodies after heart allografting in these animals. This decrease following liver allografting could be the consequence of a massive and nontoxic absorption of the antibodies by the liver allograft. However, preliminary studies in immunofluorescence (unpublished data) do not indicate a massive deposition of immunoglobulins in the liver graft. Another explanation for the rapid decrease of cytotoxic antibodies following LEW liver allografting in hypersensitized BN rats could be the formation of antigen-antibody immune complexes caused by a massive release of soluble major histocompatibility complex (MHC) antigens by the graft. As a matter of fact, it was recently shown using molecular biology techniques that murine hepatocytes (Cosman et al. 1982), and perhaps also human hepatocytes, could have the property to secrete products of the MHC in a soluble form.

Whatever the still unknown exact mechanism of the resistance of the liver to cytotoxic antibodies, this property provides an interesting prospect for liver transplantation in man and more generally underlines a possibly new aspect of the immunological role of the liver.

References

Calne RY (1973) Allografting in the pig. In: Calne RY (ed) Immunological aspects of transplantation surgery. Medical and Technical Publishing, Lancaster, pp 296–316
Charpentier B, Houssin D, Lang P, Fries D (1979) In Vitro and in vivo correlates of anti-idiotypic immunization in inbred strains of rats. Transplant Proc 11:1613
Cosman D, Kress M, Khoury G, Jay G (1982) Tissue specific expression of an unusual H-2 (class I) related gene. Proc Natl Acad Sci USA 79:4947–4951
Guttmann RD (1974) Genetics of acute rejection of rat cardiac allografts and a model of hyperacute rejection. Transplantation 17:383–389
Houssin D, Gigou M, Franco D, Szekely AM, Bismuth H (1979) Spontaneous long-term survival of liver allografts in inbred rats. Transplant Proc 11:567–570
Kamada N (1983) Technique in the rat. In: Calne RY (ed) Liver transplantation. Grune and Stratton, New York, pp 25–34
Ono K, Lindsey E (1969) Improved technique of heart transplantation in rats. J Thorac Cardiovasc Surg 57:225–228
Starzl TE, Iwatsuki S, Van Thiel DH, Carlton Gartner J, Zitelli BJ, Malatack JJ, Shade RR, et al. (1982) Evolution of liver transplantation. Hepatology 2:614–636
Thoenes GH, Urban G, Doering I (1974) Kidney transplantation between congenic versus standard inbred strains of rats. The significance of H-1 and non H-1 gene differences. Immunogenetics 3:329–335
Zimmerman FA, Butcher GW, Davies HS, Brons G, Kamada N, Turel O (1979) Techniques for orthotopic liver transplantation in the rat and some studies of the immunological responses to fully allogeneic liver grafts. Transplant Proc 11:571–577

Orthotopic Versus Auxiliary Liver Transplantation

G.H. Müller, A. Wunderlich, U.T. Hopt, and *H. Bockhorn*

Of the two types of liver transplantation, orthotopic in situ replacement and heterotopic auxiliary insertion, the latter has the technical advantage. In the course of experimental research on rats, both types of transplantation have been carried out in different models.

In our experiment, we examined the immunological response to a second vascularized allograft following a primary auxiliary liver transplantation, in order to compare these results with the findings of Kamada's experiments with orthotopic liver grafts (Kamada et al. 1981). Kamada describes a reversal of transplantation immunity by liver grafting. Immunized rats not only failed to respond after liver grafting, but they also showed donor-specific tolerance and subsequent skin grafts could survive indefinitely.

For this study, the combination of the DA strain as donor and the PVG strain as recipient was used, a combination well known for spontaneous tolerance to liver grafts; kidney or heart grafts in the same combination are rejected. Liver grafts were performed using the recipient's renal vessels for end-to-end anastomoses (Müller 1983): the renal artery to the portal vein and the renal vein to the inferior vena cava. The bile duct was anastomosed to the ureter (Fig. 1). Thirty days later, a heterotopic cardiac allograft was performed, again using a DA donor. Hearts were dissected out, leaving only the pulmonary artery and the aorta for end-to-side anastomoses with the vena cava and the aorta. The day of rejection was taken as being the day the graft stopped beating. Diagnosis of rejection was confirmed by histological examination.

The following results were observed: by day 30, all the liver grafts were seen to be severely damaged, under histological examination. No hepatic parenchyma remained. DA livers were not accepted as described for orthotopic grafts, and auxiliary liver grafts from DA donors were acutely rejected. DA and PVG isografts show that this is not an effect produced by the surgical technique or by the arterialization of the graft. Graft decay is caused by the host's immune response. Isografts in this model show a certain amount of atrophy, but a microstructure which is intact.

The results of the second set response to heart allografts are not surprising. DA heart allografts were rejected more rapidly. After 5–7 days, no beat could be felt. Control allografts were rejected between day 7 and 9. The difference between

Microsurgical Models in Rats
for Transplantation Research
Eds. A. Thiede et al.

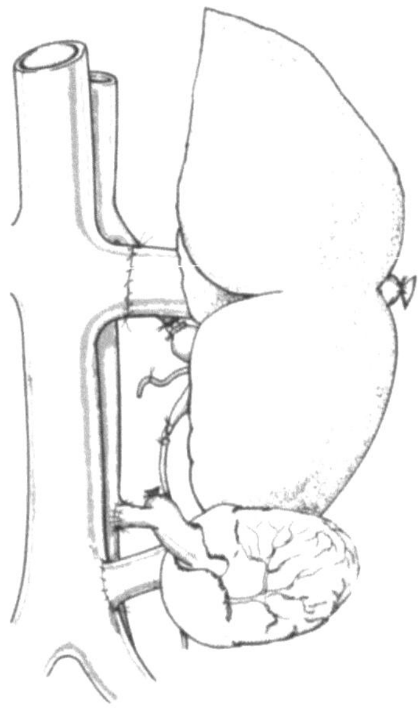

Fig. 1. Heart transplant after primary auxiliary liver graft. Heterotopic position for the liver using the renal vessel and heterotopic implantation end-to-side for cardiac grafts

these groups, established using the Wilcoxon rank sum test, is statistically significant.

Orthotopic liver grafts overcome sensitization and induce tolerance, whereas auxiliary grafts not only fail to become accepted, but they also induce a second set response; they immunize the host. The difference between these grafts is considered here: auxiliary grafts are not life supporting, in contrast to orthotopic grafts, which are. Auxiliary grafts were rejected by a host with its own liver functioning and its immune system intact, in contrast to a recipient deprived of its own liver and whose immune system had been partially removed (the Kupffer's cell system).

We showed earlier that even these auxiliary grafts have a positive, prolonging effect on allograft survival, when the transplantation of the auxiliary liver is carried out at the same time as a heart or kidney allograft (Müller and Morris 1982). However, as a result of this study, we had to admit that these grafts do not produce the tolerance described by Kamada in orthotopic grafts (Kamada et al. 1981).

Our results are similar to those published by Gugenheim and Houssin; by presenting another auxiliary liver model, they were able to demonstrate the role of the host's own liver (Gugenheim et al. 1981). They come to the same conclusion. The host's liver seems to be the key to the peculiar immunological role of the liver grafts. If the liver is removed, tolerance can be induced. If not, sensitization occurs. The working mechanism of this has yet to be investigated. It may well be due to a transfer of immune competent cells within the liver graft and/or to a loss of soluble antigens and to the removal of immune competent cells resulting from the host's hepatectomy.

References

Gugenheim J, Houssin D, Tamisier D, Franco D, Martin E, Lang P, Bismuth H (1981) Spontaneous long-term survival of liver allografts in inbred rats. Transplantation 32:445–450

Kamada N, Davies H, Roser B (1981) Reversal of transplantation immunity by liver grafting. Nature 292:840–842

Müller GH (1983) A simple technique for heterotopic auxiliary liver transplantation in the rat. Transplantation 36:221–222

Müller GH, Morris PJ (1982) Auxiliary liver allografts produce minimal prolongation of kidney and heart allografts in the rat. Langenbecks Arch Chir 357/3:168

Fine-Needle Aspiration Cytology of Orthotopic Liver Allografts in Rats

F. A. Zimmermann, T. Schmid, Ursula Schmid, J. M. Gokel, T. Siegel, and *C. Hammer*

Introduction

There are now various reliable models of orthotopic liver transplantation in the rat available (Engemann et al. 1982; Houssin et al. 1980; Kamada and Calne 1983; Lee et al. 1975; Limmer et al. 1978; Miyata et al. 1980) and many data of the somewhat peculiar behavior of rat allografts have been reported (Engemann et al. 1983; Houssin et al. 1980; Kamada et al. 1981; Zimmermann et al. 1979). We have modified the original techniques of Lee et al. (1975) for orthotopic liver transplantation in the rat. In particular, a new method of bile duct reconstruction has been developed which has eliminated complications previously experienced with bile duct anastomosis (Zimmermann et al. 1981). This technique is now used by most other workers. Our transplantation technique made possible the first long-term survival of recipients given fully allogeneic liver grafts (Zimmermann et al. 1979) and has formed the basis for immunological studies by ourselves and others. The graft survival is strictly dependent upon strain combination of donor and recipient and varies from acute rejection through delayed rejection to long-term survival (Zimmermann et al. 1984). This contrasts with the uniformly rapid rejection of vascularized heart and kidney grafts in these combinations. The manifestation of recipient-strain-dependent levels of immune responsiveness to donor transplantation antigens is an unique property of liver grafts among transplantable vascularized organs. Our technique for liver grafting in the rat is not a significant variable affecting immunological and histological events following transplantation and provides an excellent model for studying the biology of graft rejection. Thus, we felt that an analysis of graft infiltrating cells could be a helpful approach in the understanding of the cellular immune response in situ. Cell analysis of allografts by fine-needle aspiration cytology was inaugurated by the Helsinki transplant group (Häyry and von Willebrand 1981; von Willebrand 1980) for monitoring renal allograft rejection in humans. We have tried to apply this approach to liver grafts in rats. Early results have already been reported elsewhere (Zimmermann et al. 1983). This paper deals with further results of fine-needle aspiration cytology in rat liver allografts.

Microsurgical Models in Rats
for Transplantation Research
Eds. A. Thiede et al.

Material and Methods

Orthotopic liver grafts were performed in the isograft combination BN to BN ($RT1^{n}$) as controls, in the allograft combination DA ($RT1^{av1}$) to BN as an example of an acutely rejecting combination, in LWE ($RT1^{l}$) to BN as an example of delayed rejection, and in BN to DA as an example of a spontaneously tolerant combination. Furthermore, allografted animals in the DA/BN combination received cyclosporin A (CyA) (10 mg/kg per day) for 4 weeks. In a further group of this combination (DA/BN) CyA was withdrawn for a short interval (day 4–12) and given again for another month. All animals were obtained from the Zentralinstitut für Versuchstierzucht der DFG, Hannover, Federal Republic of Germany.

Fine-Needle Aspiration Cytology, Histology, and Blood Specimens

In overlapping intervals the livers were punctured with a thin pediatric spinal needle and the cells aspirated. Specimens of peripheral blood were taken simultaneously with the biopsies. Both samples were drawn into Hank's tissue culture medium containing albumin and heparin. Both specimens were processed identically: the cells were cytocentrifuged onto slides (Cytocentrifuge, Shandon Scientific Corp., London, UK) and stained with May-Grünwald-Giemsa stain (Häyry and von Willebrand 1981). The parenchymal and the infiltrative cells were examined independently by two investigators and the data compared. The counts of the peripheral blood cells and the white-cell content of the aspirate were compared and the increment calculated. The increment method has been described by Häyry and von Willebrand (1981) and was shown to describe truly the inflammatory events in situ. For histological examinations small specimens from the livers were taken after laparotomy.

Results

The frequent fine-needle biopsies as well as the normal biopsies following laparotomy were well tolerated by the animals. Isografts and cyclosporin-A-treated animals survived for over 200 days; however, untreated grafts of the combination DA/BN were rejected within 15 days and in the LEW/BN combination allografts were rejected in a delayed fashion (mean survival time, 48 days) as reported previously (Zimmermann et al. 1984). More than 300 punctures were performed on 38 recipients of liver grafts and only one animal which was punctured daily died from multiple liver abscesses. In 6% we were not able to gain sufficient material from the punctures for cytological analysis.

Fine Needle Aspiration Cytology Analysis of the Inflammatory Events

All major inflammatory components of rejection were recorded. A fine-needle biopsy was proved to be sufficient if hepatocytes were present in the sample.

Figure 1 A describes the cellular pattern of isografts. There is only a slight and occasional increment of infiltrating cells. The distribution of aspirated cells cor-

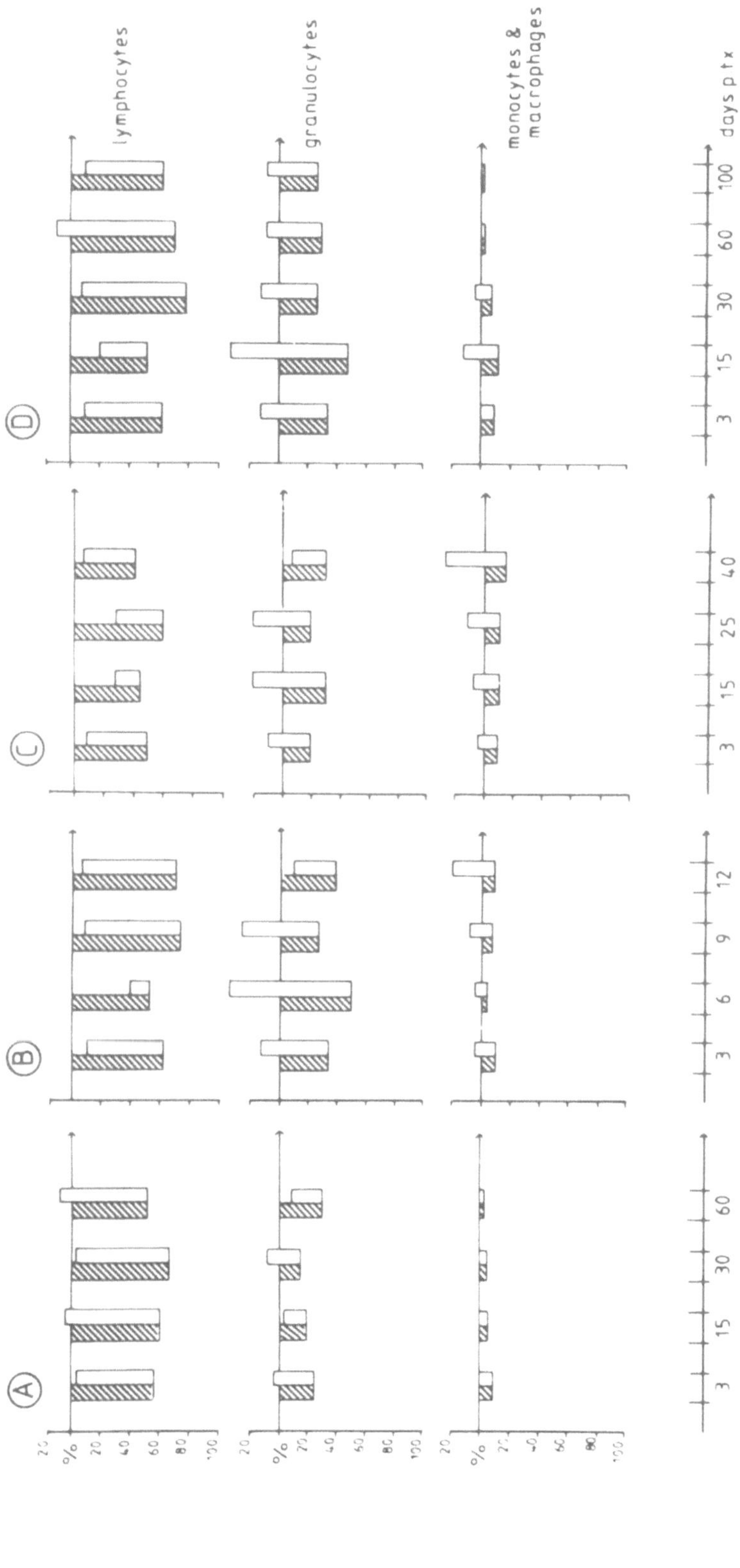

Fig. 1A–D. The *hatched columns* describe the peripheral white blood counts. The *plain* columns indicate the number of cells found in fine-needle biopsy. The rise above the zero line represents the increment and shows the ratio of infiltrating cells within the liver. **A** Isografts (BN/BN). There is only a slight and occasional increment. **B** Acutely rejecting combination (DA/BN) (see "Results"). **C** Chronic rejecting combination (LEW/BN) (see "Results"). **D** Spontaneously tolerant combination (BN/DA) (see "Results")

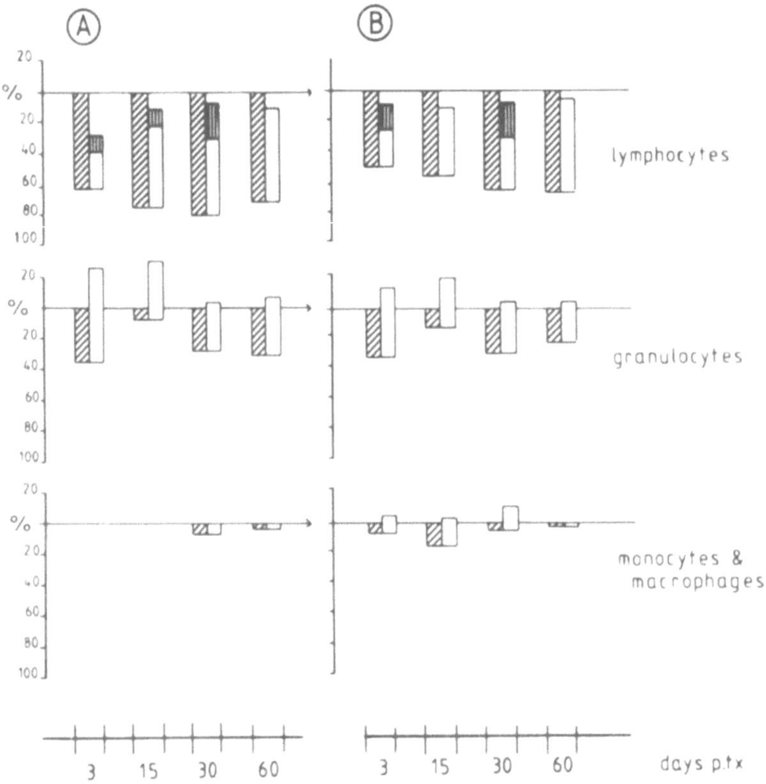

Fig. 2 A, B. Experiments with CyA. In **A** CyA is given for 4 weeks (10 mg/kg); in **B** CyA is withdrawn between days 5 and 12. In both **A** and **B** the *vertical hatched areas* represent the amount of vacuolized lymphocytes seen exclusively under CyA treatment

responds almost exactly with the normal peripheral white blood count. Figure 1 B describes the inflammatory component in acutely rejecting allografts (DA/BN). Up to the 9th day after transplantation granulocytes predominate in the liver, whereas monocytes, macrophages, granulated cells and blast cells increase toward the end of the rejection, presenting a pluriform picture. Figure 1 C describes the cytological changes in allografts which suffered delayed rejection (LEW/BN). In contrast to the preceding group granulocytes predominate in the liver up to the 25th day post-transplantation. Later on effector cells accumulate in the graft toward the end of the rejection. Figure 1 D describes the situation in the spontaneously tolerant combination BN/DA. There is only a slight rise of monocytes and macrophages from the 2nd to the 4th week after transplantation, but later on cellular distribution of aspirated cells returns to normal. Figure 2 A shows experiments in which cyclosporin A is given for 4 weeks. Distinct inflammatory changes were seen in the first period. During cyclosporin A treatment a major portion of vacuolized lymphocytes is seen which is totally restricted to the cellular component of the graft itself. Such cells are never seen in the peripheral cell population either in the controls of nontransplanted liver or in isografted animals re-

Table 1. Histological alteration in sequential biopsy specimens

Group	Days post transplantation	Findings: Parenchymal changes – Cytoplasmatic	Findings: Parenchymal changes – Portal structure	Findings: Inflammation
I	4			
	8		(+)	+
	24	(+)	(+)	(+)
	60	(+)	(+)	
II	4		+	+
	8	++	+++	+++
	12	+++	+++++	+++++
III	4	(+)	(+)	(+)
	8	+	++	++
	30	+++	++++	++++
IV	4		(+)	(+)
	8	+	+	++
	24	(+)	(+)	+
	60		(+)	(+)
V	4			(+)
	8	+	+	+(+)
	30	(+)	(+)	+
	60	(+)	(+)	
VI	4			
	8	+	+	+
	20	+(+)	++	++(+)
	30	+	+	+(+)
	60		(+)	+

Group I, isografts (DA/DA); group II, acutely rejecting combination (DA/BN); group III, delayed rejecting combination (LEW/BN); group IV, spontaneously tolerant combination (BN/DA); group V, DA/BN with cyclosporin A treatment; group VI, DA/BN with intermittently withdrawn cyclosporin A treatment
Symbols: (+), occasionally; + distinctive; ++, spread; +++, widespread; ++++, extensive

ceiving the same dose of cyclosporin A for an equivalent period. Figure 2 B shows the results of experiments in which cyclosporin A was withdrawn for 7 days. Animals which were not put on cyclosporin A again died after 37 days from rejection. If cyclosporin A was withdrawn intermittently, the condition of the animals deteriorated, liver function tests showed hepatic damage, and the animals lost weight, indicating a rejection episode. After the onset of cyclosporin A therapy they recovered. During the withdrawal of cyclosporin A the vacuolized lymphocytes disappeared, monocytes and macrophages increased, but the rejection was prevented after cyclosporin A was given again. The cellular picture returned to normal and corresponded well with the distribution of the different cell types seen in isografts, continuously treated animals, or even normal peripheral white blood counts, indicating a lack of infiltrating cells.

Correlations to Biopsy Histology

The findings in histological sections are summarized in Table 1. In the rejecting combinations edema and mononuclear-cell infiltration of the portal tracts and the central parts of the lobules are seen. Cells invade the walls of the vessels. The cells appear to be a mixture of macrophages, large and small lymphoid cells, and plasma cells. Therefore cytological and histological signs of rejection correspond well. In cyclosporin-A-treated animals vacuolized lymphoid cells are present.

Discussion

The cytological profile of inflammation in liver allograft rejection in the rat is very similar to the profiles reported earlier for renal transplant rejection in rats (von Willebrand 1980) and more recently in pig liver grafts (Häyry, personal communication). The appearance of vacuolized lymphocytes was typical in our experiments in cyclosporin-A-treated grafts. Some authors have already reported vacuolized tubular cells in kidney grafts and have explained them as an expression of cytotoxic effects. We wonder whether this phenomenon may be due to the toxic effects of cyclosporin A on lymphoid cells and this vacuolization was a precursor of cell death. We are sure it has nothing to do with the solvent of cyclosporin A because such cells were never found in cyclosporin-A-treated controls. Aspiration cytology supports histological examination of liver grafts and makes possible representative results on cellular rejection following grafting. The method is obviously of value in this small animal model. Because all types of immune response following liver grafting can be encountered, the model provides an excellent tool for studying the biology of graft rejection and tolerance. A more advanced method of the differentiation of infiltrating cells using selected monoclonal antibodies or the *Staphylococcus aureus* assay is being established in our laboratory. Using these assays, we hope we shall be able to learn more about the patterns of cellular rejection of liver allografts.

References

Engemann R, Ulrichs K, Thiede A (1982) Value of a physiological liver transplant model in rats. Transplantation 33:566

Engemann R, Ulrichs K, Thiede A, Müller-Ruchholtz W, Hamelmann H (1983) Induction of liver graft tolerance in a primarily nontolerant rat strain combination with temporary treatment of CyA. Transplant Proc 15:2986

Häyry P, von Willebrand E (1981) Practical guideline for fine needle aspiration biopsy of human renal allografts. Ann Clin Res 13:288

Houssin D, Gigon M, Franco D, Bismuth H (1980) Specific transplantation tolerance induced by spontaneously tolerated liver allografts in inbred strain of rats. Transplantation 29:418

Kamada N, Calne RY (1983) A surgical experience with five hundred and thirty liver transplants in the rat. Surgery 93:64

Kamada N, Davies HS, Roser B (1981) Reversal of transplantation immunity by liver grafting. Nature 292:840

Lee S, Charters AC III, Orloff MH (1975) A simplified technique for orthotopic liver transplantation in the rat. Am J Surg 130:38

Limmer J, Zimmermann FA, Dunn DC, Calne RY (1978) Orthotopic rat liver transplantation using a cuff technique for portal vein anastomosis (Abstr). Eur Surg Res [Suppl 1] 10:21

Miyata M, Fischer JH, Fuhs M, Isselhard W (1980) A simple method for orthotopic liver transplantation in the rat. Transplantation 30:335

Von Willebrand E (1980) Fine needle aspiration cytology of human renal transplants. Clin Immunol Immunopathol 17:309

Zimmermann FA, Butcher GW, Davies HS, Brons G, Kamada N, Turel O (1979) Techniques for orthotopic liver transplantation in the rat and some studies of the immunological response to fully allogeneic liver grafts. Transplant Proc 11:571

Zimmermann FA, Obermüller K, Gokel JM, Dorn-Kling S (1981) Die Gallengangsrekonstruktion bei der Ratte durch Choledocho-Choledochostomie über einen verlorenen Drain. Z Exp Chir 14:241

Zimmermann FA, Schmid T, Siegel T, Vossenkuhl U, Gokel JM, Hammer C, Feifel G (1983) Monitoring of rejection of rat liver allografts using fine needle biopsies and aspiration cytology. Eur Surg Res [Suppl 1] 15:44

Zimmermann FA, Davies HS, Knoll PP, Gokel JM, Schmid T (1984) Orthotopic liver allografts in the rat. The influence of strain combination on the fate of the graft. Transplantation 37:406

Immunosuppressive Effects of Liver Grafts in the Rat

N. Kamada and *T. Shinomiya*

DA major histocompatibility complex [(MHC) haplotype, RT1[a]] hearts or skin grafts into PVG (RT1[c]) recipients have always been rejected between 7 and 9 days after transplantation. On the other hand, DA livers orthotopically grafted into PVG animals have survived indefinitely (more than 100 days), even though no immunosuppression has been used (Kamada et al. 1981).

In the experimental group, DA heart was implanted into PVG (in the neck using the cuff method) (Heron 1971), followed by orthotopic (DA) liver grafting (Kamada and Calne 1983) performed in the same recipient. All (six cases) of these heart and liver grafts were not rejected and survived indefinitely as shown in Table 1, group 1.

In another experiment, DA heart grafts were heterotopically implanted into PVG recipients. Five or six days later, DA liver grafts were orthotopically implanted into these PVG recipients. It was expected that such DA heart grafts would be rejected because, at this stage, these grafts showed progressive clinical rejection signs (weak beating and swollen) and histological signs (massive cellular infiltrations and tissue necroses). Unexpectedly, 6 heart grafts out of 13 cases were protected from rejection and survived indefinitely (Table 1, groups 2, 3).

In another seven cases (Table 1, groups 2, 3), all liver grafts were rejected; however, heart grafts showed strong beating just before the recipient's death (due to

Table 1. Survival of heart grafts using no immunosuppression

Number of group	Combination of heart grafting	Days of (DA) liver transplantation after heart grafting	Survival (days) of heart grafting
1	DA into PVG	0	>100, >100, >100, >100, >100, >100
2	DA into PVG	5	>100, >100, >100, 19[a], 20[a], 21[a], 25[a]
3	DA into PVG	6	>100, >100, >100, 18[a], 20[a], 23[a]
4	WAG into PVG	0	10, 10, 11, 11, 20, 28
5	WAG into PVG	6	8, 9, 9, 10, 11, 24

[a] No histological evidence of rejection in heart – host died following rejection of liver graft

Microsurgical Models in Rats
for Transplantation Research
Eds. A. Thiede et al.

Table 2. Survival of heart grafts using cyclosporin A

Starting days of CyA injection after heart grafting	Survival (days) of heart grafts
0	>100, >100, >100, >100, >100, >100
5	7, 8, 8, 8, 9, 9, 10
6	7, 7, 8, 9, 9, 9

liver graft rejection). In all 13 heart grafts, it was observed that the signs of rejection were promptly eliminated clinically within a few days, and that histology at biopsy or postmortem examination revealed almost normal heart grafts.

In the same way, third party (WAG, RT1^{u}) heart grafts were all rejected, but showed significantly prolonged survival (Table 1, groups 4, 5) compared with the controls; in the combination WAG to PVG, all WAG heart grafts were rejected within 9 days after transplantation.

In the next experiment, cyclosporin A (CyA) was used to compare the immunosuppressive effects. DA heart had been implanted into PVG, then injection of CyA (30 mg/kg per day, i.m.) was started 5 or 6 days after heart grafting. In these animals, CyA failed to stop the progressive immune response, and all (11 cases) DA heart grafts were rejected with normal first-set fashion as shown in Table 2.

It has been previously reported (Kamada et al. 1981) that DA skin grafts (nonvascularized organ) transplanted to PVG at the same time as DA liver grafting were all rejected. From this result, it has been summarized that it takes more than 5 days (Kamada et al. 1981) after liver transplantation to induce systemic unresponsiveness to donor-type antigens. Nevertheless, the results described here using vascularized organ (heart) as an indicator clearly suggest that liver grafts immediately induce systemic unresponsiveness and even act as a powerful immunosuppression to protect grafts (hearts) which have well-established signs of rejection. It is interesting to utilize vascularized skin grafts (Muller et al. 1983) as an indicator in this system.

Acknowledgments. We thank Miss A. Jones for secretarial assistance.

References

Heron I (1971) A technique for accessory cervical heart transplantation in rabbits and rats. Acta Pathol Microbiol Scand 79:366–372

Kamada N, Calne RY (1983) A surgical experience with five hundred thirty liver transplants in the rat. Surgery 93:64–69

Kamada N, Davies HS, Roser B (1981) Reversal of transplantation immunity by liver grafting. Nature 292:840–842

Muller GH, Hettich R, Hopt UT, Bockhorn H (1983) Vascularised skin graft in the rat – a new microsurgical and immunological model. Eur Surg Res [Suppl] 15:59–60

Graft Tolerance After Orthotopic Liver Transplantation in a Primarily Nontolerant Strain Combination Following Temporary Cyclosporin A Treatment *

R. Engemann, K. Ulrichs, A. Thiede, W. Müller-Ruchholtz, and *H. Hamelmann*

In liver transplantation, especially in some inbred rat strain combinations, the phenomenon of spontaneous long-term survival of animals with fully major histocompatibility complex (MHC) allogeneic liver grafts has been described (Engemann et al. 1983; Houssin et al. 1979; Kamada et al. 1980; Zimmermann et al. 1979). To study further possibilities for manipulation of the liver allograft rejection, a donor/recipient combination was found in which all animals died of graft rejection. Manipulation of rejection was performed with cyclosporin A (CyA), a nonsteroidal fungal metabolite which was first used for an immunosuppressive regimen by Calne and White (1977) and has proved to be a powerful immunosuppressive agent in various species (Morris 1981). Though the drug has been used in clinical liver transplantation with promising results (Starzl et al. 1981), analytical data on CyA in rat liver transplantation are not yet available. Therefore the aim of the present study was to investigate survival, liver morphology, and cell-mediated lymphocytotoxicity in the DA ($RT1^{av1}$) – LEW ($RT1^{l}$) combination after CyA therapy.

Material and Methods

Male LEW rats ($RT1^{l}$), weighing 250–300 g, were used as recipients. Male BDV ($RT1^{d}$), F344 ($RT1^{lv1}$), BN ($RT1^{n}$), or DA ($RT1^{av1}$) were used as liver donors and DA as skin and F344 as third party skin donors respectively. Using microsurgical techniques, orthotopic rat liver transplantation (ORLT) was performed with reanastomosis of the hepatic artery, as described in detail by Engemann et al. (in this volume). For immunosuppression CyA (Sandoz AG, Basel) was suspended in olive oil at 50 mg/ml and given orally in a concentration of 15 mg/kg body weight/day from the day of operation to the 13th postoperative day only. As in in vivo studies survival times were evaluated and in long-term survivors tolerance was tested by grafting donor-specific and third party skin at day 100 postoperatively. For in vitro monitoring of graft recipients cell-mediated reactivity was tested in the microcytotoxicity assay (MCA) (Takasugi and Klein 1970) accord-

* This work was supported by the Deutsche Forschungsgemeinschaft, SFB 111, Projekt B 9

Microsurgical Models in Rats
for Transplantation Research
Eds. A. Thiede et al.

ing to the procedure of Bröcker et al. (1977): Reactivity of the recipients' lymphocytes of the various lymphatic compartments (blood, spleen, peripheral lymph nodes, mesenteric lymph nodes, Peyer's patches) was tested against donor and third party strain fibroblasts in Terasaki microtest plates for 48 h. The result was read microscopically and indicated as an integral of cell-mediated lysis. In vitro reactivity of the allogeneic graft recipients was always compared with background reactivity of the syngeneic control animals. Liver histology was performed with biopsies usually taken on days 14, 20, and 40, on the day of death, and on the day of in vitro testing. Stainings were done with hematoxylin and eosin.

To study the pattern of liver rejection followed by the death of the animals, various strain combinations were investigated: BDV–LEW (*n*, 5), F344–LEW (*n*, 5), BN–LEW (*n*, 12), DA–LEW (*n*, 10), and LEW–LEW (*n*, 17) as syngeneic control group. Of all groups tested only the DA–LEW combination was selected for experiments, since none of the graft recipients survived beyond day 14 while in all other combinations long-term survivors could be observed. According to these pilot studies the following experimental groups were chosen: group I, DA–LEW (without CyA treatment) with a total number of 17 animals, and group II, with a total number of 32 CyA-treated liver-grafted animals to date.

Results

Survival and Tolerance

In the syngeneic LEW–LEW control group 80% of the animals survived for more than 100 days (*n*, 17). In group I without CyA treatment none of the LEW rats bearing a DA liver allograft survived beyond day 14 (medium survival time, 10.5 ± 2.3) in a survival study (*n*, 10). In contrast to this finding, in the CyA-treated group II long-term survivors beyond day 100 could be obtained (survival study: see Table 1). Long-term survivors, who received donor-specific (DA) and third party (F344) skin grafts on the 100th day postoperatively, showed specific unresponsiveness, since the DA skin was accepted while F344 skin was rejected in a normal way (*n*, 5).

Table 1. DA-LEW combination with CyA treatment: In the seven animals retained for the long-term surviving group, the day and cause of death as well as liver histology on the day of death are indicated

Animal No.	3	4	1	2	6	5	7
Day of death	14	38	40	53	149	150	153
Cause of death	LI	LI	LI	LI	Test	Test	Test
Liver histology on the day of death	[a]	[b]	[c]	[b]	[b, c]	[a]	[b, c]

LI, lung infection; Test, cell-mediated in vitro reactivity

[a] No periportal mononuclear-cell infiltrates, no fibrosis

[b] Slight periportal mononuclear-cell infiltrates

[c] Slight fibrosis, restricted to the periportal field

Cell-Mediated In Vitro Reactivity

From previous observations (Leapman et al. 1981) we expected that cell-mediated in vitro reactivity of the recipient's lymphocytes against the allograft would be suppressed in CyA-treated rats compared with nonimmunosuppressed rats. Four liver graft recipients in each experimental group were tested for cell-mediated in vitro reactivity at day 10; additionally three tolerant rats of the CyA-treated group were tested on days 149, 150, and 153. Analysis yielded the following findings: (1) Group I versus group II at day 10: In all five lymphatic compartments investigated, reactivity of the recipient's lymphocytes against donor antigen as well as against third party antigen was remarkably stronger in nonimmunosuppressed individuals of group I than in CyA-treated individuals of group II. (2) Group II at day 10 versus day 150: (2a) Reactivity in the lymphatic transport compartment (blood) was clearly stronger at day 10 than at day 150, when it was almost zero. (2b) In contrast, reactivity in the lymphatic tissue compartment spleen was clearly lower at day 10 than at day 150. (2c) Reactivity in the three other lymphatic tissue compartments, i.e., peripheral and mesenteric lymph nodes and Peyer's patches, was similar at both test times. (2d) Specificity of cell-mediated in vitro reactivity: The long-term survivors were alloantigen specific since reactivity against a third party strain (BN) was not elevated above the background reactivity of the syngeneic control animals; however, it was alloantigen nonspecific in the early postoperative phase, since third party reactivity (BN) was elevated above background reactivity.

Liver Macroscopy and Histology

(1) LEW–LEW. At all stages of their postoperative life (days 20, 50, 150, and 270) the syngeneic graft recipients showed a normal appearance of the liver (*n*, 17). The liver lobes had sharp edges and possessed few connective tissue adhesions. Histologically in this syngeneic control group evidence of liver alterations could not be found. (2) DA–LEW without CyA treatment. In contrast to the above, the DA livers of the nonimmunosuppressed, nonsurviving LEW rats looked spotted and were swollen with round-edged lobes. Histology of these livers revealed severe periportal as well as intrahepatic mononuclear-cell infiltrates, which finally led to the destruction of the organ (Fig. 1). (3) DA–LEW with CyA treatment. The livers of the CyA-treated LEW rats bearing a DA graft, including some long-term survivors, were only slightly swollen and possessed round-edged liver lobes. Four out of the seven rats listed in Table 1 died of lung infections (histologically documented) on days 14, 38, 40, and 53. No death occurred due to graft rejection. Investigated at the time of death, animal No. 1 showed slight fibrosis restricted to the periportal field. Animals No. 2 and 4 had slight periportal mononuclear-cell infiltrates and animal No. 3 possessed a totally unaltered liver morphology. In the three long-term survivors periportal mononuclear-cell infiltrates could not be detected at day 14; only a few were present at day 20, and at day 40 round-cell infiltrates appeared to be restricted to the periportal area. Histological examination of these rats at days 140, 150, and 153 revealed only very slight liver fibrosis and a few mononuclear-cell infiltrates (see Table 1, animals No. 6 and 7), similar to those found at earlier stages. Absolutely no liver alteration could be detected in

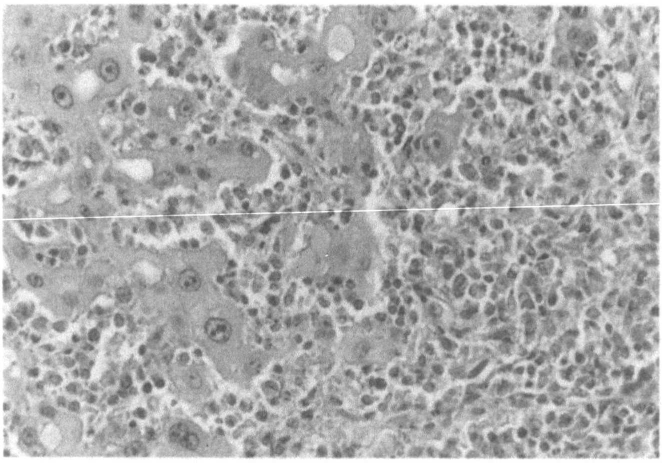

Fig. 1. Histology of an allogeneic untreated DA–LEW liver transplant at day 9 postoperatively: severe allograft rejection as documented by periportal and intrahepatic mononuclear cell infiltrates. Hematoxylin and eosin, × 100

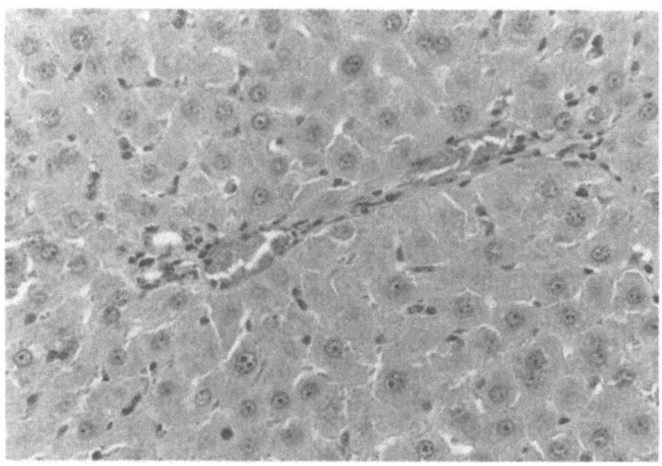

Fig. 2. Histology of an allogeneic CyA-treated DA–LEW liver transplant at day 150 postoperatively: compared with the syngeneic control normal structure of the periportal field and liver parenchyma. Hematoxylin and eosin, × 100

rat No. 5 (Fig. 2): the DA graft, having functioned in a CyA-treated LEW recipient for 150 days, had the appearance of a normal liver.

Discussion

Experimental and human liver transplantation still yield heterogeneous and partly conflicting results. At present some parameters causing this heterogeneity are becoming more transparent and may have to be carefully considered if liver transplantation is to be consistently successful in the future. In liver transplantation in several fully MHC-allogeneic rat strain combinations (e.g., DA–PVG or DA–LEW), long-term graft survival occurs, as well as rapid, genetically determined graft rejection. So far very little is known about the nature of this genetic determination in the experimental rat system and even less about its equivalent in humans. In strain combinations with genetically determined graft rejection (rejection in spite of an optimized technique) it is obvious that the transplantation barrier can be overcome by treating the recipient with an immunosuppressive drug such as CyA. So far promising results have been achieved with this drug in other transplantation models, including rats bearing kidney (Homan et al. 1980) or heart (Nagao et al. 1982) allografts and also in human liver transplantation (Starzl et al. 1981). In the clinical situation, a systematic analysis of these parameters is extremely difficult to achieve. Therefore, it seemed desirable to investigate them further in our experimental model.

In the survival study of group II four out of seven rats died from lung infection rather than from graft rejection. Therefore, it cannot be excluded that the dosage, based on the standard regimen for successful graft survival in the rat, was too high and that a less immunosuppressive dosage would improve the results. We now use only 10 mg/kg body weight and have less infection morbidity.

Histological data of organ-grafted animals with CyA treatment are rarely documented in the literature. In our experiments all rats who did not die from lung infection lived in very good health until the day of testing. Nonoccurrence or alternatively the occurrence and precise side of the (almost negligible) morphological expression of immunological reaction to the target organ liver could only be documented adequately by histological examination. In the CyA-treated liver recipients in vitro cell-mediated reactivity was clearly reduced compared with the nontreated animals. The nearly complete lack of reactivity in the blood, combined with reactivity in the four tissue compartments at day 150, may represent a "locking-in phenomenon" of graft-damaging cytotoxic effector lymphocytes. This is supported by the observation that the decrease in blood reactivity is accompanied by an increase in splenic reactivity. Two similar "locking-in phenomena" (nonreactivity in the blood versus positive reactivity in the splenic compartment) have been observed in fully allogeneic, cyclophosphamide and cross-transfused rats, specifically tolerant to skin allografts (Schienle et al. 1975) and in F_1 rats which had survived a graft versus host reaction crisis after receiving a small-intestine graft (Ulrichs et al. 1982). Further analysis of this most interesting finding is in progress.

Specific unresponsiveness in organ-transplanted rats, induced by a short course of CyA, is already known for heart (Nagao et al. 1982) and kidney (Homan et al. 1980) and has been confirmed with regard to another organ, namely liver, according to the results presented in this paper. One major aspect of unresponsiveness, its kinetics, has become a matter of growing interest (Morris 1980). Kinetic studies have recently been carried out in CyA-treated and heart-allografted rats (Nagao et al. 1982) with the result that there appears to be a stable nonspecific suppression in the early postoperative stage, which further develops into a weak specific and finally into a strong long-lasting donor-specific suppression. From the data presented here it may be concluded that a similar time course of unresponsiveness also develops in liver-grafted, CyA-treated rats. The strong nonspecific cell-mediated in vitro reactivity of group I may be correlated with the lethal rejection crisis of these animals. The relatively lower nonspecific in vitro reactivity in group II at the early stage may be explained by the fact that CyA is still administered at day 10 postoperatively. However, skin grafts, donor type and third party (F344) transplanted on the 20th postoperative day, were rejected in a slightly delayed fashion (data not shown). When these animals were tested at about day 40 postoperatively four-fifths showed nonspecific cell-mediated in vitro reactivity and one-fifth specific cell-mediated in vitro reactivity. A similar phenomenon was observed in mixed leukocyte culture experiments (unpublished data) where the spleen cells of these liver graft recipients were tested for suppressor-cell activity: in four-fifths there was indication of nonspecific suppressor cells and in one-fifth specifically acting suppressor cells could be found. This has also been reported after rat heart transplantation by others (Bordes-Aznar et al. 1983; Hutchinson et al. 1981). Though our long-term survivors have not yet been tested for suppressor-cell activity so far, there are other parameters (skin grafts and cell-mediated in vitro reactivity) which indicate the occurrence of specific unresponsiveness in the late postoperative stage. This is in good concordance with our previous data (Engemann et al. 1983), showing that in the BN–LEW combination, in which long-term survival occurs without immunosuppressive regimen, specific unresponsiveness develops after an early nonspecific phase. So it may be possible that this time course of unresponsiveness is a general mechanism in rat liver transplantation.

Acknowledgments. The excellent technical assistance of Frauke Blunck and Linda Hoffmann is gratefully acknowledged.

References

Bordes-Aznar J, Lear PA, Strom TB, Tilney NL, Kupiec-Weglinski JW (1983) Kinetics of cyclosporin-A-induced unresponsiveness to cardiac allografts in rats. Transplant Proc 15:500–503

Bröcker EB, Kuhlencordt KM, Müller-Ruchholtz W (1977) Microcytotoxicity test in allograft immunity. Int Arch Allergy Appl Immunol 53:234–241

Calne RY, White DJG (1977) Cyclosporin A – a powerful immunosuppressant in dogs with renal allografts. IRCS Med Sci 5:595

Engemann R, Ulrichs K, Thiede A, Müller-Ruchholtz W, Hamelmann H (1983) A mechanism of tolerance in arterialized rat liver transplantation. Transplant Proc 15:729–733

Homan WP, Fabre JW, Williams KA, Millard PR, Morris PJ (1980) Studies on the immunosuppressive properties of cyclosporin A in rats receiving renal allografts. Transplantation 29:361–366

Houssin D, Gigou M, Franco D, Szekely AM, Bismuth H (1979) Spontaneous long-term survival of liver allografts in inbred rats. Transplant Proc 11:567–570

Hutchinson IF, Shadur CA, Duarte A, Baldwin WM, Strom TB, Tilney NL (1981) Mechanisms of cardiac allograft prolongation by cyclosporin A. Transplant Proc 13:412–416

Kamada N, Brons G, Davis HS (1980) Fully allogenic liver grafting in rats induces a state of systemic nonreactivity to donor transplantation antigens. Transplantation 29:429–431

Leapman SB, Filo RS, Smith EJ, Smith PG (1981) Differential effects on cyclosporin A on lymphocyte subpopulations. Transplant Proc 13:405–409
Morris PJ (1981) Cyclosporin A. Transplantátion 32:349–354
Nagao R, White DJG, Calne RY (1982) Kinetics of unresponsiveness induced by a short course of cyclosporin A. Transplantation 33:31–35
Schienle W, Bröcker EB, Müller-Ruchholtz W (1975) Zelluläre und humorale Reaktivität bei funktioneller Immuntoleranz nach Alkylans-Behandlung. Z Immunitätsforsch 150:235
Starzl TE, Klintmalm GBG, Porter KA, Iwatsuki S, Schröter GPJ (1981) Liver transplantation with the use of cyclosporin A and prednisone. N Engl J Med 305:266–269
Takasugi M, Klein E (1970) A microassay for cell-mediated immunity. Transplantation 9:219–227
Ulrichs K, Deltz E, Thiede A, Müller-Ruchholtz W (1982) In: Nieuwenhuis P (ed) In vivo immunology. Histophysiology of the lymphoid system. Plenum, New York, p 591
Zimmermann F, Butcher GW, Davies HS, Brons G, Kamada N, Türel O (1979) Techniques for orthotopic liver transplantation in the rat and some studies of the immunologic responses to fully allogenic liver grafts. Transplant Proc 11:571–577

Liver Transplantation and Cyclosporin A Therapy: Induction of Tolerance in a Strong Rejection Combination

G. H. Müller, W. Schareck, A. Wunderlich, and *J. Bonner*

Cyclosporin A (CyA) is a powerful immunosuppressive drug which has been used for preventing rejection of allografts in rats. After only a short course of CyA treatment some strains tolerated allografts without subsequent reaction.

In a strong rat model, DA to Lewis, heart and kidney grafts were rejected acutely. Using the same strain combination, the same thing happened to liver grafts. Auxiliary liver grafts were totally destroyed by day 14 (Müller 1983). No functioning hepatic parenchyma remained. However, a short course of 10 mg/kg body weight of CyA can induce a tolerance to liver grafts in this combination. In order to study the specificity in vivo, different organs were transplanted following the liver transplantation (Fig. 1). Kidney grafts were performed using the orthotopic position in the recipient. End-to-end anastomoses were used for the renal artery, renal vein, and the ureter. A nephrectomy was performed on day 7; the day of rejection was taken as being the day the animal dies of ureamia. Heart allografts were performed using Ono's method (Ono and Lindsay 1969); grafts were dissected out and used for end-to-side anastomoses to the aorta or vena cava, or for end-to-end anastomoses to the renal vessels. The day of rejection was taken as being the day a beat could no longer be felt. Auxiliary liver grafts were transplanted using the left renal vessels as described above (Müller 1983). Thirty days after the liver transplantation, a second graft was performed: either a kidney or a heart was transplanted. DA or BN specimens were used as donors (Fig. 2). One

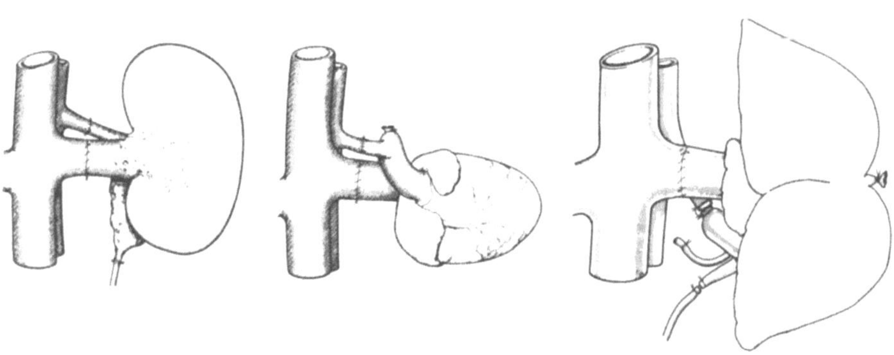

Fig. 1. Kidney, heart, and liver grafts: site of implantation using the renal vascular supply

Microsurgical Models in Rats
for Transplantation Research
Eds. A. Thiede et al.

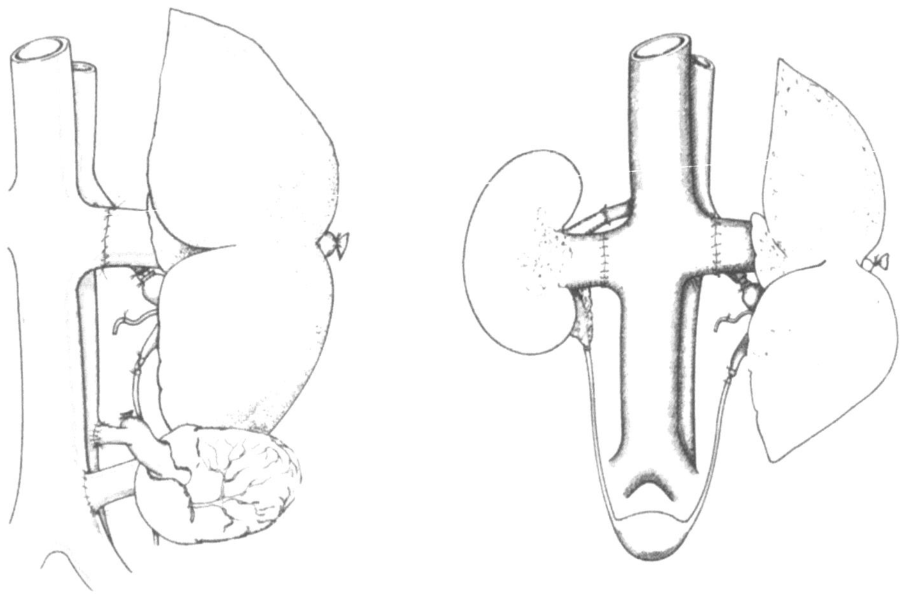

Fig. 2. Secondary heart and kidney grafts 30 days after primary liver transplantation. The orthotopic site for renal grafts end-to-end was used – for cardiac grafts, end-to-side anastomoses below the liver were performed

hundred days after the liver graft, a skin graft was performed, again using a DA rat as the donor.

The following results were observed: all single-control grafts, heart allografts, kidney allografts, liver allografts, and skin allografts were acutely rejected. After a short course of CyA, liver grafts were tolerated by the Lewis recipients. Histological sections taken at this time showed cellular infiltration periportally but a microstructure which was intact. A second graft – a heart allograft or a kidney allograft – transplanted on day 30 after the primary liver transplantation showed no signs of rejection. All second DA grafts survived more than 70 days. Third party grafts (PVG grafts) were rejected in an acute fashion. Surprisingly, DA skin grafts performed on day 100 after the liver transplantation were rejected: when compared with the control graft the skin grafts survived slightly longer, for 12–16 days.

The donor-specific tolerance of the host's immune response is demonstrated by another experiment: 30 days after the DA liver grafts a BN heart allograft was rejected acutely. Nevertheless, a second heart allograft, in this case a DA heart allograft, implanted in the same Lewis recipient on day 50, was again tolerated.

We conclude, therefore, that in a strong rejection combination in the rat model, CyA is a powerful immunosuppressive drug which prevents rejection of vascularized allografts. Following liver grafts, CyA induces a tolerance which is donor specific and stable. In contrast to orthotopic grafts with CyA therapy, there is no tolerance for skin grafts carried out 100 days after auxiliary liver grafts. Taking into account recent experiments with kidney grafts and CyA, and with heart grafts and CyA treatment, which did not produce a tolerance to another organ, we must conclude that liver grafting and CyA produces an additional effect which leads to specific unresponsiveness.

References

Müller GH (1983) A simple technique for heterotopic auxiliary liver transplantation in the rat. Transplantation 36:221–222

Ono K, Lindsay ES (1969) Improved technique of heart transplantation in rats. J Thorac Cardiovasc Surg 57:225–229

Study of the Cellular Immunity of Tolerance Induced by Liver Grafting in the Rat

N. Kamada

In the early days of experimental liver transplantation, the cellular immunity of liver recipients was not carefully investigated and has remained unclear. This paper describes investigations into the cellular alloreactivity of PVG [major histocompatibility complex (MHC) haplotype $RT1^c$] rats bearing DA ($RT1^a$) liver grafts. In this combination, DA liver grafts are not rejected even though no immunosuppression is used (Kamada et al. 1981).

This set of experiments has used quantitative adoptive transfer assays and 5×10^6 thoracic duct lymphocytes (TDLs) from normal or liver-grafted recipient rats. Sublethally irradiated (800 rad) PVG recipient rats grafted with DA (specific) or AO (third party $RT1^u$) skin grafts received: (1) TDLs from liver (DA)-grafted PVG rats, (2) normal PVG TDLs, or (3) no cellular inoculum.

Thoracic duct lymphocytes from liver-grafted rats rejected third party skin grafts in normal first set fashion (Fig. 1 G), but DA skin grafts were rejected with the same tempo (Fig. 1 C) as those on recipients receiving no cellular inoculum (Fig. 1 D).

These results suggest that the recirculating pool of alloreactive cells in PVG rats bearing DA liver grafts is specifically deleted of those clones recognizing DA alloantigens. These adoptive transfer experiments give no evidence for the existence of suppressor cells in liver-grafted rats, since there are no cases of prolonged or permanent survival of DA skin grafts (Fig. 1 C), as may be seen, for example, in similar adoptive transfer of lymphocytes from neonatally tolerant rats. The reasons for donor-specific deletion of alloreactive cells from the circulating lymphocyte pool of the liver graft recipient remain unclear. It is possible that such cells are destroyed progressively during their interaction with the grafted liver: the mechanism could involve interactions between donor Kupffer cells and recipient lymphocytes. This possibility is open to further investigations.

In the same strain combination, however, lymph node cells or TDLs were still capable of mounting strong graft versus host (GVH) responses against MHC antigens of the graft type (Fig. 2). Thus, in PVG rats orthotopically transplanted with DA livers, the clones of cells required for graft rejection were specifically deleted from the circulation, but the proliferative response (GVH reaction) to DA alloantigens was still retained. A further explanation could be that different MHC antigens and thus different T-cell populations are involved in graft rejection and

Microsurgical Models in Rats
for Transplantation Research
Eds. A. Thiede et al.

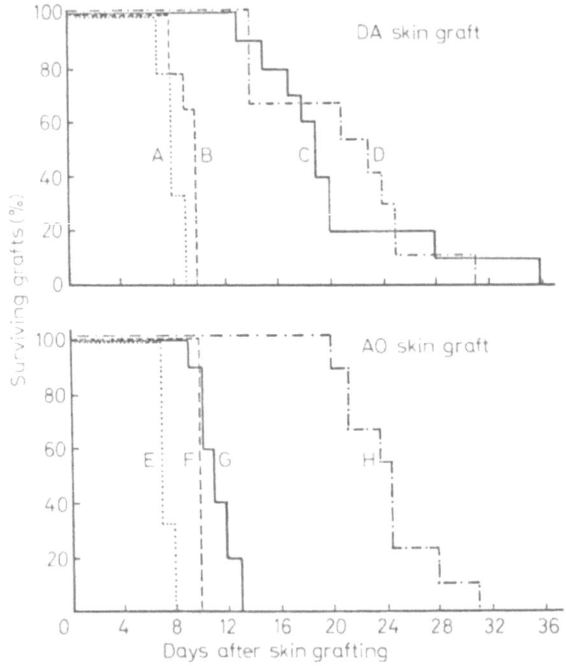

Fig. 1. Survival of skin grafts. PVG recipients of DA livers, which had survived more than 30 days after transplantation, were used. These PVG recipients of DA liver grafts were cannulated to obtain thoracic duct lymphocytes (TDLs). *A*, survival of DA skin grafts (n, 9) placed on normal PVG rats with no further treatment (8 ± 0.7 days); *B*, survival of DA skin grafts (n, 9) on irradiated (800 rad) PVGs treated with 5×10^7 TDLs from normal PVGs (9.4 ± 0.8 days); *C*, survival of DA skin grafts (n, 10) on irradiated PVGs injected i.v. with 5×10^7 TDLs from liver-grafted rats (20.5 ± 6.3 days); *D*, survival of DA skin grafts (n, 9) on irradiated PVGs with no further treatment (21.2 ± 5.7 days); *E*, survival of AO skin grafts (n, 9) on normal PVGs with no further treatment (7.3 ± 0.5 days); *F*, survival of AO skin grafts (n, 5) on irradiated PVGs injected i.v. with 5×10^7 TDLs from normal PVGs (10 days); *G*, survival of AO skin grafts (n, 10) on irradiated PVGs injected i.v. with 5×10^7 TDLs from PVG rats with DA liver grafts (10.8 ± 1.3 days); *H*, survival of AO skin grafts (n, 9) on irradiated PVGs with no further treatment (23.9 ± 3.5 days). Results as mean survival time $\pm$SD)

in GVH reaction. Rejection probably involves mainly recognition of RT1.*A* (class I) antigens by cytotoxic T cells, while GVH reactivity mainly involves proliferative responses of other T-cell subpopulations to RT1.*B* (class II) antigens (Bach et al. 1977). These results suggest that there is a form of split tolerance (Sprent et al. 1975). Thus, deletion of one subpopulation, reactive to RT1.*A* antigens, may lead to the systemic and donor-specific unresponsiveness.

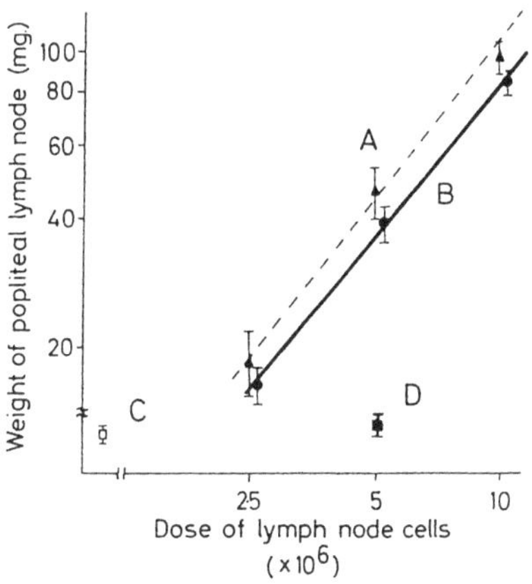

Fig. 2. Graft versus host reactivity of LN cells from normal PVG rats and from long-surviving PVG recipients of DA liver grafts. The reactive potency of LN cells was assayed by injection s.c. of one preparation into the right footpad and the other into the left footpad, using groups of six recipient (DA × PVG)F_1-hybrid rats. Log lymph node weight is plotted against log cell dose. The regression lines are shown in *A* and *B*. *Vertical bars* represent the SDs. ▲, mean popliteal lymph node response to LN cells from normal PVG rats. ●, response to LN cells from long-surviving PVG recipients of DA liver grafts. Control experiments (injection of syngeneic (DA × PVG)F_1 LN cells; injection of medium DAB/FCS only) were carried out. Each experiment used groups of six recipient rats. □, control – popliteal lymph node weight after injection of DAB/FCS (0.1 ml) into the left footpad (*C*, 4.8 ± 0.6 mg); ■, control – popliteal lymph node weight after injection of 5×10^6 (DA × PVG)F_1 cells into the right footpad (*D*, 7.5 ± 1.9 mg)

Acknowledgments. We thank Miss A. Jones for secretarial assistance.

References

Bach FH, Grillot-Courralin C, Kuperman OJ, Sollinger HW, Hayes C, Sondel PM, Alter BJ, Bach ML (1977) Antigenetic requirements for triggering of cytotoxic T lymphocytes. Immunol Rev 35:76–96

Kamada N, Davies HS, Roser B (1981) Reversal of transplantation immunity by liver grafting. Nature 292:840–842

Sprent T, von Boehmer H, Nabholz M (1975) Association of immunity and tolerance to host H-2 determinants in irradiated F1 hybrid mice reconstituted with bone marrow cells from one parental strain. J Exp Med 142:321–331

Mechanisms of Tolerance in Rat Liver Transplantation *

R. Engemann, K. Ulrichs, A. Thiede, W. Müller-Ruchholtz, and *H. Hamelmann*

When the technique for orthotopic rat liver transplantation became available to several microsurgical research groups in the second half of the last decade, it turned out that when inbred strain combinations were used, even with a fully allogeneic donor recipient combination, the animals with grafted livers survived long-term without immunosuppression (Engemann et al. 1982; Houssin et al. 1979; Kamada and Calne 1979; Lie et al. 1980; Limmer et al. 1980; Zimmermann et al. 1979). Moreover, it was more difficult to find a combination in which strong rejection occurred (Engemann et al. 1983; Zimmermann et al. 1982). Even though it was well established that tolerance can be induced in rats more easily than in humans, this situation provokes at least the following questions: How does this in vivo finding of long-term graft acceptance correlate with the reactivity in various conventional in vitro assays [mixed leukocyte culture (MLC), microcytotoxicity assay (MCA)]? Does truly specific unresponsiveness develop, and what are the kinetics of unresponsiveness? These questions were investigated at an early point in the recipients' lives and at two later intervals by skin grafting, MLC, and MCA.

Materials and Methods

The transplantation technique is described in detail by Engemann in this volume (p. 69). For reasons discussed there we use the arterialized model, with which, with increasing skill of the surgeons involved, we now have a long-term survival rate in the syngeneic control group of more than 90%. As allogeneic combination we used a BN ($RT1^n$)→LEW ($RT1^l$) combination. Full-thickness grafts of donor and third party origin (F344 : RT1vl, AS : $RT1^l$) were transplanted at different times postoperatively (14–30, 120–150, >250 days). In the BN→LEW combination we now have a long-term survival rate of 90% for more than 100 days.

* This work was supported by the Deutsche Forschungsgemeinschaft, SFB 111/B9

Microsurgical Models in Rats
for Transplantation Research
Eds. A. Thiede et al.

Results

MLC Reactivity

While the cellular immune response of splenic lymphocytes of liver graft recipients to donor antigen BN and the third party antigen DA is nonspecifically suppressed at 14–30 days postoperatively, by 120–150 days postoperatively it has been completely restored to normal. At 250 days, however, the following alteration in reactivity is observed. The response to third party antigen is normal, but the response to the donor antigen is altered, i.e., it is specifically raised in four out of six animals and specifically suppressed in two out of six animals. Thus, specifically altered cellular reactivity appears to be a late event in the recipient's life.

Skin Graft Survival

To analyze the cellular immune response of liver graft recipients in vivo, skin grafts of donor strain origin and from two third party strains of varying degrees of histocompatibility were performed. Donor-type skin grafts (BN) survive indefinitely when grafted as early as day 14. The survival of F344 skin grafts [major histocompatibility complex (MHC) variant subregion to LEW] is significantly prolonged at the early time interval (medium survival time (MST) 18 ± 0.8 days at 14–30 days] but decreases constantly with progressive survival of the liver allograft and is normal (MST, 12.4 ± 0.8 days) when tested at >250 days. A similar phenomenon is observed with skin from a second third party strain, AS, which differs from LEW only in non-MHC antigens. Significantly delayed rejection occurs when it is grafted at 14–30 days (MST, 20.4 ± 0.8 days), but rejection is normal when the graft is performed at >250 days. Thus, specific nonrejection of donor-type grafts is time dependent and is found only at the late time interval.

Microcytotoxicity assay (MCA)

Lymphocytes from the blood, speen, and peripheral and mesenteric lymph nodes of liver recipients were tested for in vitro effector-cell reactivity in microcytotoxicity assays using adherent embryonic fibroblasts of donor (BN) and third party strain origin (DA) as target cells (Fig. 1). To ensure that the observed reactivity is not due to the operation itself, but to immune reactivity, syngeneic control animals were included in the test. As can be seen in Fig. 1, in syngeneic recipients MCA reactivity was never elevated above background. When the experimental data are analyzed at 14–30 days, cytotoxic effector cells are found in the various compartments investigated. They act nonspecifically, since both donor strain and third party target cells are destroyed. At both later periods, cell mediated cytotoxicity (CMC) reactivity is found predominantly in the spleen and is a highly specific process. Thus, specificity of CMC also appears to be a late event. Furthermore, reactivity becomes cell compartment restricted.

Suppressor-Cell Reactivity

Mitomycin-C-treated splenic lymphocytes from liver graft recipients of varying ages were added to either a related (LEW anti-BN) or a nonrelated (LEW anti-

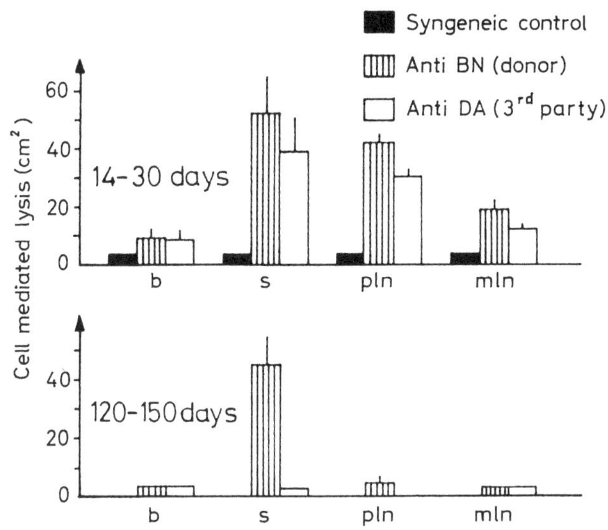

Fig. 1. MCA reactivity of liver graft recipients at various time intervals. *b*, blood; *s*, spleen; *pln*, peripheral lymph nodes; *mln*, mesenteric lymph nodes

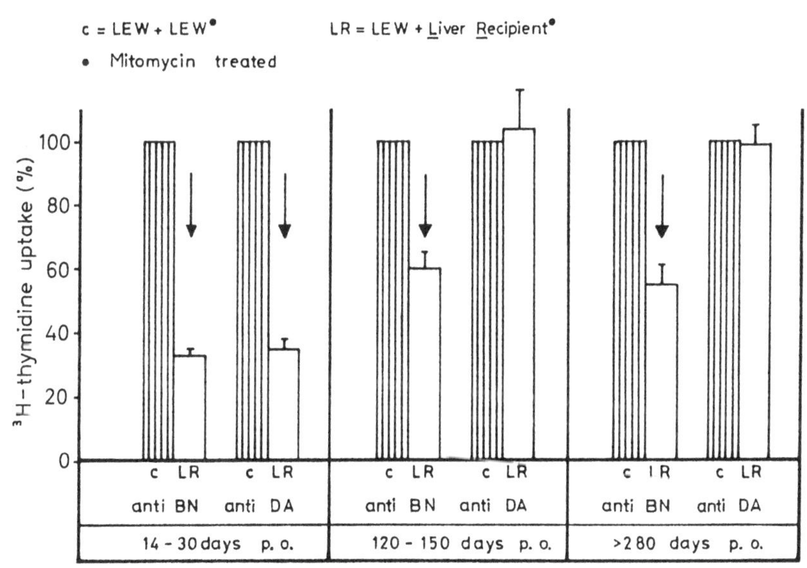

Fig. 2. Reactivity of splenic lymphocytes from liver graft recipients in an indicator MLC taken from graft recipients at various time intervals

DA) MLC indicator system (Fig. 2). The cellular immune response of the LEW indicator lymphocytes is significantly suppressed by the cells from the liver recipients at every time interval after liver grafting when they are stimulated with donor antigen. A different pattern is found, however, when the MHC-different third party antigen DA is used. While cells from recently grafted recipients (14–30 days postoperatively) significantly suppressed the immune response of the indicator cells, lymphocytes from older graft recipients (120–150 days) do not act similarly. Thus two types of suppressor cells were observed, cells that act nonspecifically at the early time interval and cells that act specifically at later time intervals.

Discussion

In both clinical and experimental transplantation the induction and maintenance of specific unresponsiveness remains a challenging subject. In experimental organ transplantation specific unresponsiveness can be achieved by application of powerful immunosuppressive regimens, e.g., total lymphoid irradiation of cyclosporin A (CyA). In rat liver transplantation we are confronted with the fact that in certain donor-recipient combinations (BN to LEW, DA to PVG) rejection does not occur even without immunosuppression, i.e., the liver graft is able to overcome the ongoing rejection, and the grafted organ is accepted in the long run. The impact of such genetic determination is underlined by the fact that in other fully allogeneic strain combinations, e.g., DA→LEW, the recipients die by day 15 postoperatively (Engemann et al. 1983). The problem of the biological relevance of in vitro assays, e.g., MLC and MCA, is again demonstrated by the findings described above. Specific unresponsiveness to the donor antigen BN at the later time interval, as expressed by the specific acceptance of donor-type skin grafts, is seemingly contradicted by the reactivity in vitro: in both the MLC and the MCA (Fig. 1) splenic lymphocytes from liver recipients respond very strongly to the BN antigen. Keeping in mind the obvious contrast between these findings and the acceptance of both liver and skin grafts, we should interpret these results as evidence that reactivity is present, and not absent, which would indicate the deletion/abortion type of unresponsiveness. The specific unresponsiveness found in rats with liver grafts [strain combination DA ($RT1^a$) to PVG ($RT1^c$)] has been claimed by Kamada et al. (1981) to be deletion-type unresponsiveness. He used adoptively transferred thoracic duct lymphocytes (TDLs). He could not find cytotoxic lymphocytes in the blood using the ^{51}Cr-release assay. Since TDLs belong to the recirculating pool, this could explain the discrepancy.

The suppressive reactivity of splenic lymphocytes from liver recipients in the indicator MLC may provide evidence of the existence of two types of suppressor cells, one population acting nonspecifically in the first stage, and a second acting specifically in the second stage. Evidence of suppressively acting cells was also found by Houssin et al. (1981).

A time-dependent transition from nonspecific (early) stage I to specific (late) stage II unresponsiveness similar to that described here for liver transplantation without initial immunosuppression has been repeatedly observed with initial immunosuppression, e.g., in another fully allogeneic liver transplantation model (DA→LEW) with initial suppression with CyA (Engemann et al. 1983), in mice after total lymphoid irradiation and bone marrow transplantation (Strober et al. 1981), and in CyA-treated rats with allogeneic heart transplants (Nagao et al. 1982). With regard to the mechanism(s) of specific unresponsiveness it should be

Table 1. Mechanisms under discussion for the development and/or maintenance of spontaneous tolerance after rat liver transplantation (alphabetical order)

Author	Mechanism(s)	Year
Engemann et al.	Suppressor cells	1982
Houssin et al.	Suppressor cells, antibodies	1981
Kamada et al.	Clonal deletion, release of RT1.*A* antigen	1981
Lie et al.	Early stage: nonspecific serum factors; late stage: enhancing antibodies	1983
Zimmermann et al.	Formation of immune complexes in antigen excess, release of RT1.*A* antigen	1979

considered that MCA reactivity lacks in the transport compartment, e.g., blood, despite positive MCA reactivity in the tissue compartment, e.g., spleen. This may mean that in long-term liver recipients we have something similar to the "locking-in phenomenon" of cytotoxic effector cells (nonreactivity in the transport compartment versus reactivity in the tissue compartment) recently found in small-intestine recipients that had successfully survived a graft versus host crisis (Deltz et al. 1983).

Other mechanisms besides deletion or suppression types of unresponsiveness may, however, play an important role in rat liver transplantation (for an overview see Table 1). For the DA→PVG combination the presence of RT1 $-A^a$ antigen and/or immune complexes in antigen excess in the serum of PVG rats bearing DA liver grafts has also been suggested (Zimmermann et al. 1979), and they may act as early as day 1 after transplantation (Kamada et al. 1981). When serum from such rats with liver grafts was added to CML from normal PVG lymph node lymphocytes after in vitro stimulation with DA antigen, it showed *no* specific inhibition (Kamada et al. 1981). Finally, nonspecifically acting serum factors in the early stage after liver transplantation may also play a role (Lie et al. 1983). We were able to detect donor-specific antibodies in some LEW rats that survived for a long time with a BN liver, using an immunofluorescence assay. However, the evaluation of the question whether or not these antibodies are essential for the development of spontaneous tolerance is not yet finished. Thus it is not clear so far whether the above-mentioned mechanisms are related to the different donor-recipient combinations that are used or whether they are all more or less responsible for the phenomenon of spontaneous tolerance after rat liver transplantation.

References

Deltz E, Ulrichs K, Schack T, Friedrichs B, Thiede A (1983) Verteilungsmuster und Kinetik cytotoxischer T-Lymphozyten bei florider and supprimierter GVHR nach Dünndarmtransplantation. Langenbecks Arch Chir [Suppl Chir Forum] 151–155

Engemann R, Ulrichs K, Thiede A, Müller-Ruchholtz W, Hamelmann H (1982) Value of physiological liver transplant model in rats. Transplantation 33:566–568

Engemann R, Ulrichs K, Deltz E, Thiede A, Müller-Ruchholtz W, Hamelmann H (1983) Induction of liver graft tolerance in a primarily non tolerant rat strain combination with temporary treatment of cyclosporine. Eur Surg Res 15 [Suppl 1]:2986–2991

Houssin D, Gigou M, Franco D, Szekely AM, Bismuth H (1979) Spontaneous long-term survival of liver allografted rats. Transplant Proc 11:567–570

Houssin D, Charpentier B, Lang P, Tamisier D, Gugenheim J, Gigou M, Bismuth H (1981) In vivo and in vitro correlates of the specific transplantation tolerance induced by spontaneously tolerated liver allografts in inbred strains of rats. Transplant Proc 13:619–622

Kamada N, Calne RY (1979) Orthotopic liver transplantation in the rat. Transplantation 28:47–50

Kamada N, Davies HS, Roser B (1981) Reversal of transplantation immunity by liver grafting. Nature 292:840–842

Lie TS, Rasche A, Gundermann KJ, Kim YZ, Bernhardt-Huth D (1980) Is the humoral transfer factor responsible for the prolonged survival of rat liver allografts? Eur Surg Res 12 [Suppl 1]:23–24

Lie TS, Galkowska H, Jaeger K, Niehaus KJ (1983) Immunmechanismen nach Lebertransplantation. Langenbecks Arch Chir 360:17–27

Limmer J, Herbertson BM, Calne RY (1980) Orthotopic rat liver transplantation using different combinations of four inbred strains. Eur Surg Res 12:343–348

Nagao T, White DJG, Calne RY (1982) Kinetics of unresponsiveness induced by a short course of cyclosporin A. Transplantation 33:31–35

Strober S, King DP, Gottlieb MS, Hoppe RT, Kaplan HS (1981) Cellular mechanism of tolerance after total lymphoid irradiation (TLI). Transplant Proc 13:556–561

Zimmermann F, Butcher GW, Davies HS, Brons G, Kamada N, Türel O (1979) Techniques for orthotopic liver transplantation in the rat and some studies of the immunologic responses to fully allogeneic liver grafts. Transplant Proc 11:571–577

Zimmermann F, Knoll PP, Davies HS, Gokel JM, Schmid T (1982) The fate of orthotopic liver allografts in different rat strain combinations. Transplant Proc 15:1272–1275

Decreased Graft Versus Host Reaction After Portal Venous Drainage of Spleen Grafts in Inbred Strains of Rats

J. Gugenheim, D. Houssin, E. Martin, and *H. Bismuth*

It has been suggested that the liver may have the ability to reduce the strength of the graft versus host (GVH) reaction (Eloy et al. 1976). Other studies have shown that portal venous drainage of various organ allografts resulted in delayed rejection compared with caval drainage (Boeckx et al. 1975; Fukuda et al. 1969; Sakai 1970) and that liver allografting was liable to induce a state of specific unresponsiveness in several combinations of inbred rats (Gugenheim et al. 1981; Kamada et al. 1980; Lie et al. 1980). All these facts favor the hypothesis that the liver may have some immunosuppressive properties. In order to analyze further the protective capacities of the liver against a GVH reaction, we developed a technique of spleen transplantation with portal venous drainage in the rat and compared the strength of the GVH reaction observed when the spleen was drained through the portal vein rather than through the vena cava.

Animals and Methods

Animals

Inbred adult male Lewis (LEW, $RT1^{l}$) and (Lewis × Brown Norway) F_1-hybrid (LEW × BN) F_1 ($RT1^{l/n}$) rats were purchased from the CNRS (Orléans, France).

Experimental Groups

Graft versus host reaction was created by grafting a LEW spleen into (LEW × BN) F_1-hybrid rats. Two experimental groups were made. Animals of group A (17 rats) received a spleen graft with portal venous drainage. Animals of group B (17 rats) received a spleen graft with caval venous drainage. These two groups were subdivided as follows: In groups A1 and B1 (nine rats for each group), the spleen graft was a nonauxiliary whole spleen graft; in groups A2 and B2 (eight rats for each group), the spleen graft was a half-spleen and the spleen of the recipient was kept in place. It is known that, in the latter situation (auxiliary half-spleen graft), the strength of the GVH reaction is probably less important, because of the decrease in the number of grafted lymphoid cells (Okunewick et al. 1980) and because the recipient's own spleen is retained (Bitter-Suermann 1974).

Microsurgical Models in Rats
for Transplantation Research
Eds. A. Thiede et al.

Surgical Procedures

Orthotopic and heterotopic spleen grafting were performed on animals anesthetized with ether, with the help of an operating microscope. Spleen transplantation with portal venous drainage was performed according to an original technique. The donor splenic artery and vein were dissected free of tributaries up to the aorta and portal vein respectively. The graft was cooled and washed in vivo via the aorta with 5 ml of a 4 °C Hartmann's solution. When the spleen graft was implanted orthotopically, an aortic cuff bearing the celiac axis of the donor was anastomosed end-to-side to the abdominal aorta of the recipient just below the left renal vein and the venous drainage of the graft was ensured by a porto-portal end-to-side anastomosis. Heterotopic spleen transplantation was performed according to the technique of Lee and Orloff (1969). The spleen graft was implanted heterotopically below the recipient's right kidney. An end-to-side aortoaortic anastomosis was performed and the portal vein of the donor was anastomosed end-to-side to the recipient's vena cava. Anchorage of the spleen graft to the abdominal wall, in order to prevent vessel twisting, completed the procedure. In groups A2 and B2, the spleen graft was reduced to one-half by ligating the upper pedicles and by resecting the upper pole after ligation of the spleen at its middle part. In groups A1 and B1, the spleen graft was rendered nonauxiliary by a splenectomy of the recipient's own spleen. The immunosuppressive regimen and blood transfusion were not administered during or after the operation.

Assessment of GVH Reaction

The GVH reaction was assessed by daily examination of recipient's for general appearance (development of skin erythema and progressive emaciation) and for weight changes. The criterion chosen for evaluating the strength of the GVH reaction was survival time after transplantation. For every animal, an autopsy was performed in order to detail technical complications, and the animals in which autopsy revealed a technical complication (thrombosis or hemorrhage at the anastomotic sites) were not included in the study. A skin biopsy was done and the following organs were removed for histological examination: liver, small bowel, grafted spleen, and the recipient's own spleen in groups A2 and B2. Hematoxylin and eosin stain, Masson's trichrome stain, and Perl's method were used.

Statistical Analysis

Student's test was used for comparisons of the mean survival time (MST $\pm$ SEM) in the different groups.

Results

The clinical course of the spleen-grafted animals was as follows: no outward clinical symptom was observed until day 6–9 after the operation, when the skin became erythematous and swollen. At that time, some hair loss was observed; ulcerations appeared on the snout, nostrils, and paws; and the animals started losing

Table 1. Survival time of (LEW × BN) F_1 rats transplanted with a LEW spleen graft according to the venous drainage of the graft (MST + SEM in days)

Groups of animals	Nonauxiliary whole-spleen graft	Auxiliary half-spleen graft
Group A: portal drainage	17.3 + 2.1	18.5 + 1.1[a]
Group B: caval drainage	16.0 + 0.3	15.7 + 0.5[a]

[a] Difference statistically significant ($P<0.05$)

weight. Finally, they became increasingly cachectic and died with signs of severe dyspnea. Autopsy revealed atrophy of the liver, lymph nodes, and the recipient's own spleen in groups A2 and B2. Histological examination of the skin revealed an epidermolysis with complete dissolution of the basal layer of epidermal cells. A decrease of lymphocytic cells was observed in the liver and in the small bowel of the recipient. Examination of the spleen grafts revealed viable splenic tissue with atrophy of the malpighian corpuscles and an increase of large pyroninophylic mononuclear cells.

Survival times of the spleen-grafted rats with caval venous drainage were compared with those of rats with portal venous drainage (Table 1). The difference between the mean survival time of rats of group A1 (nonauxiliary whole spleen graft with portal venous drainage) and that of group B1 (caval venous drainage) was not significant: 17.3 ± 2.1 days versus 16.0 ± 0.3 days. In group A2 (auxiliary half-spleen graft with portal venous drainage), the mean survival time was 18.9 days (SEM ± 11 days). In group B2 (caval venous drainage), the mean survival time was 15.7 days (SEM ± 0.5 days). This difference between groups A2 and B2 was statistically significant ($P<0.05$).

Discussion

The main result of this study is that in the LEW into (LEW × BN) F_1 combination of inbred rats, portal venous drainage of an auxiliary half-spleen graft was able to reduce the strength of the GVH reaction: the mean survival time of animals with an auxiliary half-spleen graft with portal venous drainage was significantly ($P<0.05$) longer than the survival time of rats in which the graft had caval venous drainage. On the contrary, when the graft was a nonauxiliary whole-spleen graft, no difference in the survival time could be demonstrated between rats with portal and caval venous drainage. It is possible that in the latter situation, the high intensity of the GVH reaction may not have permitted the demonstration of the effect of the portal drainage.

In the other situation, when the graft was an auxiliary half-spleen graft, the strength of the GVH reaction is probably less important due to the decrease of the number of grafted lymphoid cells (Okunewick et al. 1980) and the retention of the recipient's own spleen (Bitter-Suermann 1974). Other studies have suggested the ability of the liver to reduce the strength of the GVH reaction. Eloy et al. (1976) have observed that intrahepatic implantation of 4×10^7 lymphoid cells induced a mild GVH disease without mortality whereas intrarenal implantation of the same number of lmyphoid cells led to a mortality of 75%. In the LEW into (LEW × BN) F_1 combination of inbred rats, Kort et al. (1973) showed that rats could survive with a small-bowel graft with portal venous drainage. On the contrary, in the same combination, Monchlik and Russel (1971) have observed a constantly lethal GVH reaction after small-bowel transplantation with caval venous drainage. Confrontation of these two results suggests that portal venous drainage may reduce the strength of the GVH reaction induced by small-bowel grafting. Our study is in line with that hypothesis, but suggests that the liver is liable to exert its protective role only when the strength of the GVH reaction is not too important.

References

Bitter-Suermann H (1974) Different ways of using graft-versus-host reactivities to achieve acceptance of rat-spleen allografts in the same donor-recipient pairing. Transplantation 18:515–519

Boeckx W, Sobis H, Lacquet A, Gruwez J, Vandeputte M (1975) Prolongation of allogeneic heart graft survival in the rat after implantation on portal vein. Transplantation 19:145–159

Eloy R, Vuitton D, Vaultier JP, Pousse A, Grenier JF (1976) Decreased graft-versus-host reaction after intrahepatic lymphoid tissue implantation. Cell Immunol 21:236–242

Fukuda A, Hanaoka T, Solowey AC, Rapaport FT (1969) Inhibition of second set renal allograft responses by portal vein drainage. Transplant Proc 1:602–604

Gugenheim J, Houssin D, Tamisier D, Franco D, Martin E, Lang P, Bismuth H (1981) Spontaneous long-term survival of liver allografts in inbred rats. Influence of the hepatectomy of the recipient's own liver. Transplantation 32:445–450

Kamada N, Brons G, Davies HS (1980) Fully allogeneic liver grafting in rats induces a state of systemic non-reactivity to donor transplantation antigens. Transplantation 29:429–431

Kort WJ, Westbroeck DL, MacDicker I, Lameder T (1973) Orthotopic non auxiliary total small bowel transplantation in the rat. Eur Surg Res 5:81–89

Lee S, Orloff MJ (1969) A technique for splenic transplantation in the rat. Surgery 3:436–439

Lie TS, Rasche A, Gundermann KJ, Kim YS, Bernhardt-Huth D (1980) Is the humoral transfer factor responsible for the prolonged survival of rat liver allograft? Eur Surg Res 12:23–24

Monchlik GJ, Russel PS (1971) Transplantation of small bowel in the rat. Technical and immunological considerations. Surgery 70:693–702

Okunewick JP, Mieredith RF, Brozovich BJ, Seeman PR, Magliere K (1980) Exponential relationship between spleen cell concentrations and fatal graft-versus-host response after transplantation of allogeneic spleen marrow cells mixtures. Transplantation 29:507–510

Sakai A (1970) Role of the liver in kidney allograft rejection in the rat. Transplantation 9:333–334

Do Transplants with Venous Drainage into the Portal System Undergo Delayed Rejection?

G.H. Müller, A. Wunderlich, W. Schareck, U.T. Hopt, and *H. Bockhorn*

Introduction of antigens into the portal vein or the organ grafting with drainage into the portal vein has shown varying results. The different results observed may be produced by the immunological models, experimental methods, or surgical techniques used for these studies in the past.

In our study, heterotopic heart allografts were performed using the well-defined rejection model, the DA to Lewis model. The immune response of the host was screened by a first set and a second set rejection. Rejection was taken as being the day a beat could no longer be felt.

The graft was removed from the donor, with the aorta, common carotid artery, and right pulmonary artery left intact. Using Ono's technique (Ono and Lindsay

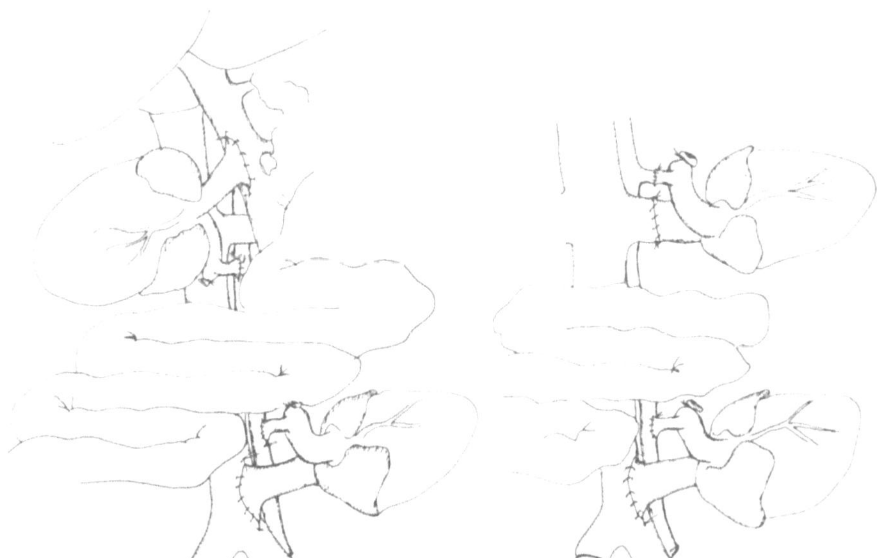

Fig. 1. Experimental protocol for first and second set rejection of heart allografts – *on the left* heart with drainage to the portal vein and *on the right* graft with systemic drainage to the inferior vena cava

Microsurgical Models in Rats
for Transplantation Research
Eds. A. Thiede et al.

1969), the DA heart was implanted into the Lewis recipient systemically using the renal vessels for end-to-end anastomoses. Periods of ischemia usually lasted less than 25 min. Fourteen days later, a second heart was transplanted below the first heart allograft end-to-side to the aorta and vena cava (Fig. 1). In a second group, the DA heart was implanted end-to-side to the aorta and end-to-side to the portal vein. In this group, a second heart allograft was performed 14 days later systemically end-to-side to the aorta and vena cava.

DA hearts were rejected after 7–8 days when transplanted systemically. A second set rejection showed DA hearts surviving only 5 days. The difference is statistically significant (Wilcoxon $P<0.05$). Heart allografts in this strong model with portal drainage were rejected uniformly on day 7. The second set rejection in these animals occurs on day 5 or 6. The difference is again significant (Wilcoxon $P<0.05$). In contrast to this, there is no significant difference between the groups with portally versus systemically transplanted heart allografts.

We conclude, therefore, that portal drainage in a strong rejection model does not alter the immune response of the host. There is no induction of tolerance by antigens which are drained into the portal system. This type of drainage does not avoid sensitization of the recipient. A second set rejection in portally drained recipients and in systemically transplanted recipients is similar. The conflicting results concerning portal drainage mentioned above probably arise from the experimental protocols used. Side effects of blood transfusion, cell transfer, ischemia time, and use of genetically poorly defined animals have not been considered in the past. Thus, in the model we presented, a peculiar immunological function may well be associated with the liver, but not with portal vein drainage.

Reference

Ono K, Lindsay ES (1969) Improved technique of heart transplantation in rats. J Thorac Cardiovasc Surg 57:225–229

Transplantation of Chimeric Organs: Implications for the Passenger-Cell Concept *

G. Filler, J. Klempnauer, B. Steiniger, and *K. Wonigeit*

Introduction

The major contribution of marrow-derived Ia-positive dendritic cells (DCs) in initiating alloimmune reaction both in vivo (Elkins and Guttmann 1968) and in vitro (Steinman and Witmer 1978) forms the basis of the passenger concept of allograft rejection. Elimination of dendritic cells is thought to be a useful approach in reducing graft immunogenicity (Lafferty 1980). Pretreatment of grafts with cytotoxic antibodies directed against Ia-positive cells of islets in vitro has yielded promising results (Faustman et al. 1981) but is not feasible in vascularized organ transplantation. Many attempts to deplete these cells using donor pretreatment by cytotoxic agents (Guttmann et al. 1969) failed possibly because the grafts were harvested prior to complete depletion of DCs (Hart and Fabre 1981 a). The interval between the cytoreductive pretreatment and graft harvesting is limited by hemopoietic failure and may be overcome by allogenic bone marrow reconstitution (Tutschka and Santos 1975) enabling grafts to be removed later on. Furthermore, chimeric animals provide a model to investigate the role of dendritic cells in graft rejection (Esses and Halloran 1983; Woodward et al. 1982). The relative contribution of donor DCs to the rat heart and pancreas allografts was studied by transplantation of chimeric organs. Allograft chimeras were created to have the majority of DCs syngenic with prospective allograft recipients. Gradual disappearance of recipient type and repopulation with donor-type Ia-positive cells taking several weeks could be demonstrated. Transplantation of chimeric organs was performed across incompatibility of either DCs or parenchymatous cells. Disparity of DCs alone did not provoke rejection of heart and pancreas grafts. Despite a significant reduction of allogenic DCs no prolongation of allograft survival was noted.

Materials and Methods

Rats. Male LEW.1U ($RT1^u$) rats weighing 250 g received a supralethal dose of busulfan given i.v. at a dose of 35 mg/kg body weight (Santos and Tutschka 1974 a). In the rat busulfan is myelotoxic but not immunosuppressive (Santos and

* This work was supported by the Stiftung Volkswagenwerk, AZ 1/38 510

Microsurgical Models in Rats
for Transplantation Research
Eds. A. Thiede et al.

Tutschka 1974b). Bone marrow reconstitution was performed 24 h later with 64×10^6 viable nucleated bone marrow cells collected from the femurs and tibias of male, age-matched LEW ($RT1^l$) rats. Immunosuppression consisted of cyclosporin A at a dose of 15 mg/kg given i.m. from day 0 to 13. Cyclosporin A was the most appropriate immunosuppressive agent because it had been reported to prevent graft versus host (GVH) reactions without affecting immune reconstitution (Tutschka et al. 1979). Graft function was monitored by daily clinical observation combined with white blood cell counts and blood smears.

Chimerism. Animals were killed on days 14, 28, and 42 following marrow reconstitution. Chimerism was demonstrated serologically using conventional cytotoxic alloantibodies (LEW anti LEW.1U and LEW.1U anti LEW) raised in our laboratory. For histological examination unfixed fresh-frozen sections were incubated with monoclonal mouse anti rat antibodies against RT1.*B* (Ia-like or class II) and stained by immunoperoxidase labeling using F 17-23-2 (Hart and Fabre 1981 b) and MRC OX 4 (McMaster and Wiliams 1979). In inbred rats F 17-23-2 reacts with the *l* haplotype but not with the *u* haplotype (Hart and Fabre 1981 b); MRC OX 4 reacts with both haplotypes (McMaster and Wiliams 1979).

Transplantation of Chimeric Organs. On day 42 following allogenic marrow reconstitution chimeric animals served as donors of heart and pancreas grafts. Heterotopic auxiliary heart transplantation was performed with unmodified LEW.1U and LEW rats as graft recipients. Rejection was defined as cessation of palpable heart beats and confirmed by conventional histology (Abbot et al. 1964). Prospective LEW and LEW.1U pancreas recipients were rendered diabetic by a single i.v. injection of 55 mg/kg streptozotocin. Diabetes was confirmed by repetitive blood glucose levels above 18 mmol/liter. Duct-ligated whole organ transplantation was performed as described by Lee et al. (1971). Reccurrence of hyperglycemia above 14 mmol/liter was taken as rejection. Following an observation period of 100 days transplants were removed in isografts. Prompt recurrence of hyperglycemia proved that normoglycemia was entirely due to the presence of the graft.

Results

Production and Characterization of Chimeras

In order to establish long-term chimeras the following control experiments were performed: Rats receiving busulfan only (n, 6) died from severe hypoplastic anemia and granulocytopenia within 13 days. Another six animals were killed for histology on day 8 following injection of busulfan and revealed no depletion of Ia-positive cells at all. Isogenic bone marrow reconstitution (n, 6) protected the animals from aplastic anemia. All animals returned to normal blood cell counts within 3 weeks and remained healthy within the observation period of 100 days but one which died from pneumonia on day 78. Untreated bone marrow allografts developed signs of severe acute GVH reactions such as weight loss, diarrhea, pancytopenia, and clinical signs of gross infections. All animals (n, 6) died within 13–16 days.

Table 1. Serologically determined chimerism in bone marrow and lymph nodes

Group	Percentage of specific lysis[a] (mean ± SD)			
	Bone marrow		Lymph nodes	
	Anti donor	Anti recipient	Anti donor	Anti recipient
LEW.1U controls (*n*, 6)	12.0±7.3	84.0±9.8	13.7±4.2	93.2± 8.5
LEW controls (*n*, 6)	65.4±4.9	10.8±3.0	91.4±7.2	9.6± 0.8
Day 14 (*n*, 6)	31.6±3.2	42.4±7.7	19.9±3.4	82.4± 6.2
Day 28 (*n*, 6)	41.6±1.9	10.7±2.3	35.5±4.5	68.6±10.1
Day 42 (*n*, 12)	47.5±4.4	15.5±2.5	55.0±7.9	41.0±11.6

[a] Lysis of individual animals was calculated by the mean of specific lysis of alloantisera dilutions 2^{-2} and 2^{-4}

Long-term chimeras were created by addition of cyclosporin A following allogenic marrow transfusion. They were protected from acute GVH reactions as long as the drug was administered. Following cessation, however, the animals developed clinical signs of lethal wasting disease including weight loss, allopecia, and lymph node enlargement and died after a mean of 44.5±2.4 days following marrow reconstitution.

Serological investigations (summarized in Table 1) in cyclosporin-A-treated allograft recipients killed on days 14, 28, and 42 after marrow reconstitution revealed a marked increase of donor-type marrow cells while recipient-type cells were at background levels from day 28 onward. The majority of the lymph node cells tested was of donor type by day 42.

Immunohistological examination confirmed our serological findings. On day 14 only very few donor-type Ia-positive cells could be detected on unfixed fresh frozen sections, mainly around small arteries. By day 28 their number increased considerably in the pancreas and celiac lymph nodes and by day 42 even exceeded normal levels. When stained with MRC OX 4, which detects both donor- and recipient-type Ia-positive cells, the overall density of DCs was indistinguishable from the picture obtained with F 17-23-2 staining.

Transplantation of Chimeric Organs

On day 42 following marrow reconstitution chimeric organs were transplanted to both marrow donor and recipient strain and rejection times were compared with unmodified control iso- and allografts. The results are shown in Fig. 1. Disparity of DCs alone did not provoke rejection of heart and pancreas allografts. Despite a significant reduction of allogenic DCs no prolongation of allograft survival was noted.

Discussion

The relative contribution of DCs to allograft rejection could be investigated in a fully major histocompatibility complex (MHC)-incompatible strain combination

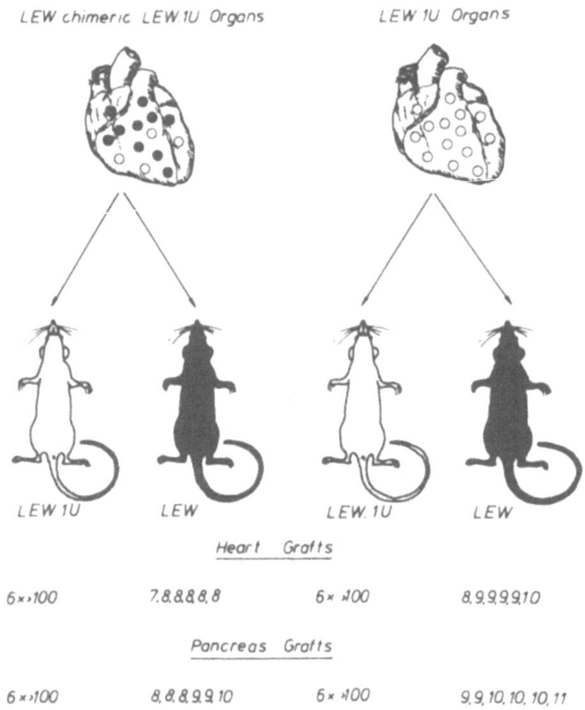

Fig. 1. Rejection times (days) of heart and pancreas grafts from chimeric donors transplanted to unmodified rats of both marrow recipient and donor strains compared with unmodified control iso- and allografts

by using cyclosporin A to establish long-term chimerism. Use of the antibody F 17-23-2 enabled us to visualize the expression of donor-type Ia-positive cells in chimeric organs. We were able to demonstrate that by day 42 the vast majority of DCs were of the donor type. The exact ratio of recipient-type Ia-positive cells could not be detected in this model because there is no appropriate antibody available. However, in comparable experiments, where LEW ($RT1^l$) rats received DA ($RT1^a$) bone marrow, recipient-type cells could still be demonstrated by MRC OX 3. Therefore it is likely that a certain number of recipient-type Ia-positive cells were still present in the grafts.

There are several possible explanations for the incomplete chimerism: First, a very slow turnover of recipient-type Ia-positive cells has to be considered. No exact data are available on the half-life of dendritic cells yet. In skin radiation chimeras it has been estimated that about 3 months are necessary for a complete replacement of donor-type DCs (Steinmuller 1981). It may well be that by altering both timing and dosage of cyclosporin A treatment the development of lethal wasting disease may be postponed or even abrogated. Second, it remains debatable whether donor pretreatment with cytoreductive agents is able to produce complete chimerism since only cells in the G_1- and S-phase are affected and surviving precursor cells may initiate a certain degree of repopulation (Wheeler

1967). This problem may be overcome by protocols involving repetitive cytoreductive treatment and marrow reconstitution. Third, our own preliminary studies indicate that expression of Ia antigens on cells previously negative for Ia determinants may be induced by GVH reactions. This problem may be solved by using F_1 hybrids as donors which on the other hand complicate the use of monoclonal xenoantibodies. Furthermore F_1 hybrids do not represent a fully allogenic model.

Although the chimerism in the model used was not complete the overwhelming majority of Ia-positive cells of the grafts harvested on day 42 were of the donor type. When transplanted to the marrow recipient strain incompatible at DCs alone neither heart nor pancreas grafts were rejected. Steinmuller and others (Steinmuller 1981; Woodward et al. 1982) also failed to provoke rejection of skin isografts taken from SD- and Ia-incompatible chimeric donors. It is likely that allogenic Ia-positive cells may contribute significantly to immunization of the host but are not capable of inducing rejection of heart and pancreas grafts since the appropriate target of the alloimmune reaction is not present on the parenchymatous cells of the graft.

The failure to demonstrate prolonged survival of DC-depleted allografts may be explained by three different mechanisms: First, the persisting recipient-type Ia-positive cells, although reduced in number, still provided a sufficient stimulus to cause rejection. The remarkable potential of these cells to stimulate mixed lymphocyte reaction has been demonstrated by Steinman and Witmer (1978). Second, graft recipient-type DCs (possibly those derived from the grafted marrow) have already processed and presented alloantigen and activated graft recipient T-helper cells. This would correspond to route II of the dual route of sensitization which is active in the absence of allogenic DCs described by Lechler and Batchelor (1982). Third, it may be that DCs do not contribute significantly to the immunogenicity of rat heart and pancreas, possibly as in the situation of the skin (Steinmuller 1981). In this respect it is noteworthy that in the case of RT1.*B* compatibility, disparity exclusively at RT1.*A* (class I) was capable of provoking acute rejection in rat pancreas transplantation (Klempnauer et al. 1983). In rat heart transplantation experiments have been reported suggesting that donor pretreatment may reduce heart immunogenicity (Nowygrod et al. 1979; Oluwole et al. 1980). It has been pointed out, however, that there is no correlation between allograft survival and the degree of passenger-cell depletion (Oluwole et al. 1980).

To conclude, it may well be that in appropriate strain combinations depletion of dendritic cells is capable of reducing graft immunogenicity. Organ-specific aspects must be considered. In the strain combination we used donor pretreatment and reconstitution with marrow matched with the prospective recipient along with subsequent depletion of DCs was insufficient to prolong heart and pancreas allograft survival.

Acknowledgment. We appreciate the excellent technical assistance of Margot Giesecke and Petra Peschel.

References

Abbot CP, Lindsay ES, Creech OJ (1964) A technique for heart transplantation in the rat. Arch Surg 89:645–652
Elkins WL, Guttmann RD (1968) Pathogenesis of a local graft vs. host reaction: immunogenicity of circulating host leukocytes. Science 159:1260–1261
Esses SI, Halloran PF (1983) Donor marrow derived cells as immunogens and targets for the immune response to bone and skin allografts. Transplantation 35:169–174
Faustman D, Hauptfeld V, Lacy P, Davie J (1981) Prolongation of murine islet allograft survival by pretreatment of islets with antibody directed to Ia determinants. Proc Natl Acad Sci USA 78:5156–5159
Guttmann RD, Lindquist RR, Ockner SA (1969) Renal transplantation in the inbred rat. IX. Hemopoietic origin of an immunogenic stimulus of rejection. Transplantation 8:472
Hart DNJ, Fabre JW (1981 a) Demonstration and characterization of Ia positive dendritic cells in the interstitial connective tissues, but not brain. J Exp Med 154:347–361
Hart DNJ, Fabre JW (1981 b) Major histocompatibility antigens in rat kidney, ureter and bladder. Transplantation 31:318–325
Klempnauer J, Wonigeit K, Steiniger B, Günther E, Pichlmayr R (1983) Pancreas whole organ transplantation in the rat: differential effect of individual MHC regions. Transplant Proc 15:1308–1310
Lafferty KJ (1980) Immunogenicity of foreign tissue. Transplantation 29:179–182
Lechler RI, Batchelor JR (1982) Restoration of immunogenicity to passenger cell depleted kidney allografts by the addition of donor strain dendritic cells. J Exp Med 155:31–41
Lee S, Chandler JG, Lakey NT, Rosen H, Orloff NJ (1971) Relief of diabetes mellitus by heterotopic transplantation of whole pancreas in inbred rats. Surg Forum 22:75–77
McMaster R, Wiliams AF (1979) Identification of Ia glycoproteins in rat thymus and purification from rat spleen. Eur J Immunol 9:426–433
Nowygrod R, Hardy MA, Todd GJ, Oluwole S, Reemtsma K (1979) Donor pretreatment in heart allografts. Transplant Proc 9:1462–1464
Oluwole S, Hardy MA, Wang T, Satake K, Todd G, Reemtsma K, Nowygrod R (1980) Donor pretreatment: rat allograft survival and measurement of passenger leukocyte depletion with indium-111. Transplantation 30:31–33
Santos GW, Tutschka PJ (1974a) Marrow transplantation in the busulphan treated rat: Preclinical model of aplastic anemia. JNCI 53:1781–1785
Santos GW, Tutschka PJ (1974b) Effect of busulphan on antibody production and skin allograft survival in the rat. JNCI 53:1775–1780
Steinman RM, Witmer MD (1978) Lymphoid dendritic cells are potent stimulators of the primary mixed lymphocyte reaction in mice. Proc Natl Acad Sci USA 75:5132–5136
Steinmuller D (1981) Passenger leukocytes and the immunogenicity of skin allografts: a critical reevaluation. Transplant Proc 13:1094–1098
Tutschka PJ, Santos GW (1975) Bone marrow transplantation in the busulphan treated rat. Transplantation 20:101–106
Tutschka PJ, Beschorner WE, Allison A, Burns WH, Santos GW (1979) Use of cyclosporin A in allogenic bone marrow transplantation in the rat. Nature 280:148–153
Wheeler GP (1967) Some biochemical effects of alkylating agents. Fed Proc 26:885–892
Woodward JG, Shigekawa BL, Frelinger JA (1982) Bone marrow derived cells are responsible for stimulating I region incompatible skin graft rejection. Transplantation 33:254–259

Orthotopic Rat Renal Transplantation Using the Cuff Method

N. Kamada

The technique for rat renal transplantation was originally described by Fisher and Lee in 1965 using a right heterotopic transplant with end-to-side anastomoses of the renal artery and vein. Since then, various techniques of this model have been well established using suture anastomoses of renal vessels. In 1976, Dunn reported the technique using a cuff for anastomosis of the renal artery in the rabbit.

This paper describes the techniques for rat renal transplantation using cuffs to anastomose end-to-end the renal artery and vein.

The cuff for renal vein anastomosis consisted of a 2-mm length of cuff body and 2-mm cuff extension (polyethylene tube, 5 FG, outside diameter 1.65 mm, Portex Ltd, United Kingdom) and was applied to the donor side (Fig. 1); the cuff for renal artery anastomosis consisted of a 1.5-mm length of cuff body and a 2-mm cuff extension (polyethylene tube, 3 FG, outside diameter 1.02 mm, Portex Ltd, United Kingdom) and was applied to the recipient side (Fig. 2).

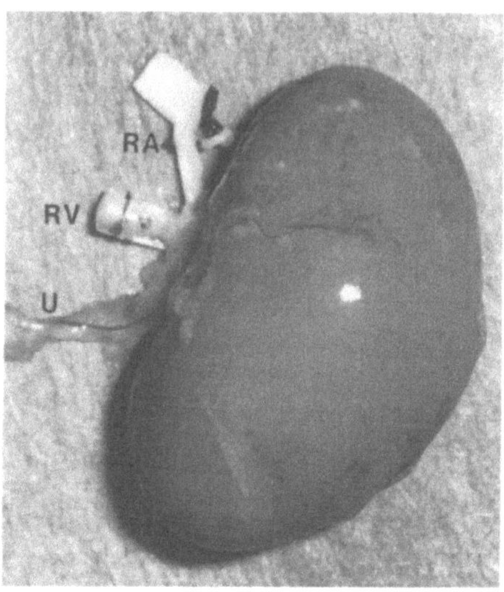

Fig. 1. Final stage of donor operation. Suprarenalis arteries and veins intern spermatic arteries and veins were ligated and divided. The donor kidney was perfused through the renal artery with cold saline solution including heparin. The cuff preparation of the renal vein was performed in an iced saline bath (Kamada and Calne 1983). *RA*, renal artery with patch of aorta; *RV*, renal vein with the cuff; *U*, ureter

Microsurgical Models in Rats
for Transplantation Research
Eds. A. Thiede et al.

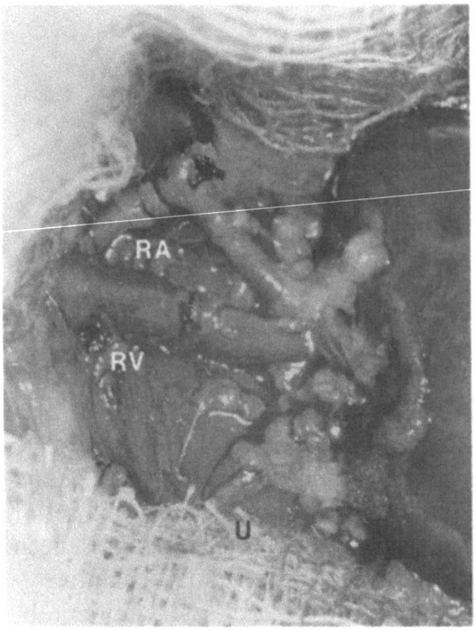

Fig. 2. Completed anastomosis. *RA*, renal artery; *RV*, renal vein; *U*, ureter

Both the cuffs described above are the same as the ones I use in heterotopic rat heart transplantation into the recipient's neck.

The methods for reconstruction of the urinary tract tried during the initial experimental term were: (a) using a polyethylene tube, (b) implanting the ureter into the bladder, and (c) anastomosing end-to-end using a 10-0 suture (four or five independent stitches).

A total of 29 syngeneic (DA or PVG) renal graftings were performed using the above methods. Urinary anastomosis using (a) resulted in a high incidence of hydronephrosis due to stenosis of the tract, and no recipients (a total of ten cases) survived beyond 100 days. Anastomosis using (b) induced better results than with (a) but showed evidence of hydronephrosis (two out of ten cases), necrosis of tract (two cases), and infection of the bladder (one case). Anastomosis using (c) resulted in a higher success rate, and seven out of nine recipients survived for more than 100 days two dying by leakage or stenosis of the tract due to technical errors. From these results in recent renal allografting, only suture anastomosis (c) for urinary tract reconstruction has been used.

The cuff method for the anastomoses of both renal vessels resulted in less operative time compared with conventional suture anastomosis methods. No bleeding after declamping both vessels and no stenosis or leakage, even in long-term surviving recipients, was observed. It was especially useful for both liver and kidney operations, or for the heart and kidney, in the same recipient.

Acknowledgments. I thank Miss A. Jones for secretarial assistance and Miss N. Townley for taking the photographs.

References

Dunn DC (1976) Orthotopic renal transplantation in the rabbit. Transplantation 22:427–433

Fisher B, Lee S (1965) Microvascular surgical techniques in research, with special reference to renal transplantation in the rat. Surgery 58:904–914

Kamada N, Calne RY (1983) A surgical experience with five hundred thirty liver transplants in the rat. Surgery 93:64–69

Effect of Cyclosporin on the Function of the Proximal Tubule and Loop of Henle in the Rat Kidney

H.-U. Gutsche, M. Lüneburg, K. H. Gnutzmann, E. Deltz, and *W. Niedermayer*

Renal toxicity is one of the major side effects of immunosuppression with cyclosporin. Apart from an almost regularly observed decrease in glomerular filtration rate (GFR), toxic alterations prevail within the proximal tubule and Henle's loop. Although detailed descriptions of morphological defects have been published, including vacuolization of the proximal tubular cells, microcalcifications, formation of giant mitochondria together with pathologic vascular and interstitial alterations (Marbet et al. 1980; Shulman et al. 1981; Mihatsch et al. 1981, 1983), little is known about functional characteristics of the nephron segments involved (Thiel et al. 1981, 1982).

It was the aim of the present study to investigate the electrolyte transport of the end-proximal tubule and of the loop of Henle by analyzing the electrolyte concentration of early distal tubular fluid obtained from surface nephrons of rat kidneys during free flow and pulse stop flow micropuncture experiments (Gutsche 1980; Gutsche et al. 1980, 1983). After 8–10 days of cyclosporin administration early distal ion content of tubular fluid was significantly raised above normal during free-flow sampling and during artificial loop perfusion. These results are believed to reflect a transport impairment within the proximal tubule and/or the loop of Henle.

Methods

Lewis rats of between 231 and 264 g body weight were pretreated with either 15 or 75 mg cyclosporin (Sandimmun, Sandoz, Basel, Switzerland) per kilogram body weight by gavage for 8–10 days. The respective quantity of the drug was dissolved in 0.5 ml olive oil. Control animals received the same amount of drug-free olive oil (*Deutsches Arzneibuch*). The preparation of the animals for micropuncture following Inactin anesthesia has been detailed previously (Gutsche 1980). The following two series of micropuncture experiments were performed:

Conductivity Measurements Under Free-Flow Conditions. Early distal tubules were identified by brief injections of lissamine-green-stained Ringer's solution into randomly selected maid-proximal tubules. A conductivity microprobe was

Microsurgical Models in Rats
for Transplantation Research
Eds. A. Thiede et al.

then inserted into the first accessible distal loop and tubular fluid was continuously aspirated at a rate of 5 nl/min or less when necessary. After a steady conductivity reading was established, tubular fluid from the various loop segments was propelled toward the distal measuring site by application of a perfusion burst of up to 90 nl/min into the mid-proximal tubule. The resulting transient conductivity fluctuations of the aspirated fluid related to the different ion composition of fluid from the respective loop segments were recorded continuously.

Conductivity Measurements During Pulse Stop Flow Microperfusion. Early distal nephron segments were identified as described above; however, proximal fluid flow was interrupted by intratubular injection of solid paraffin wax (Gutsche et al. 1975b, 1980) upstream from the mid-proximal perfusion pipette. During continuous microperfusion of 10–50 nl/min Ringer's solution, early distal tubular fluid conductivity was measured. Then flow rate was kept constant for 60 s at 10 or 20 nl/min respectively. After sudden onset of a 90-nl/min perfusion burst, tubular fluid ion concentrations which had established along the loop during these flow rates were measured as the respective conductivity fluctuations.

Results

Rats treated with the lesser dosis of 15 mg/kg body weight displayed a conspicuously altered kidney surface at a linear magnification of ×100 under the stereomicroscope. The tubular cells were swollen and less transparent, a patchy inhomogeneity of the tissue blood supply was apparent, and many tubules were collapsed. Lissamine green transit times were considerably prolonged and most distal surface tubules did not clear until up to 5 min after treatment. Clearing normally occurs within 45–60 s. Nevertheless, type I and II conductivity measurements could be performed in several surface nephrons of these rats. Free-flow conductivity was 8.24 mS/cm, equivalent to a sodium chloride concentration of 62.3 mEq/liter in controls, and 13.08 mS/cm, corresponding to a 98.9-m*M* NaCl concentration in 15-mg/kg body weight cyclosporin-treated rats. The perfusion burst method showed a NaCl concentration of 54.6 m*M* in the diluting segment of cyclosporin-pretreated rats and of 29.2 m*M* in controls. The results of artificial loop perfusion experiments in rats treated with 15 mg cyclosporin/kg body weight are displayed in Fig. 1. The flow-dependent distal concentrations are considerably higher in 15-mg/kg-treated rats than in controls. The data of the pulse stop flow experiments are represented by open symbols in Fig. 1. These data are connected with broken lines. The inhibition of dilution within the thick ascending limb after prolonged cyclosporin treatment at a dose of 15 mg/kg per day is clearly demonstrated.

The kidney surface of rats treated with 75 mg/kg cyclosporin showed more severe lesions under the stereomicroscope; the surface tubule cells were swollen and looked opaque, the luminal diameters were inhomogeneous in size, and most tubules were collapsed. Only a few partially collapsed tubules were accessible for micropuncture. No distal flow could be observed and experiments were restricted to pulse flow micropuncture measurements. Almost no NaCl reabsorption occur-

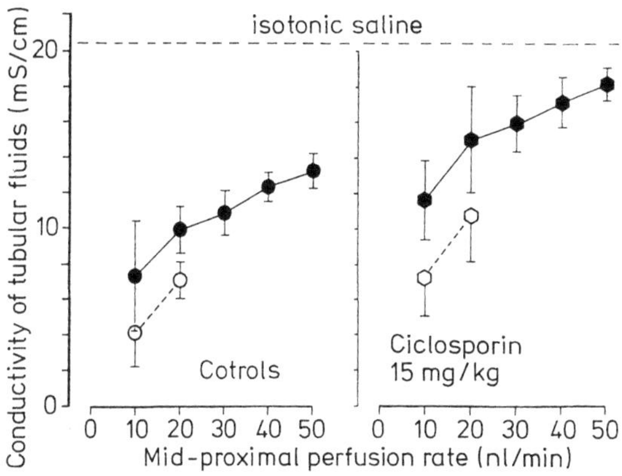

Fig. 1. *Distal conductivity measurements. On the right* are data from intratubular conductivity measurements in three cyclosporin-treated rats (15 mg/kg body weight for 8–10 days) compared with five control animals (*left*). Mid-proximal nephrons were perfused at constant flow rates of 10–50 nl/min and the resulting conductivity values of early distal tubular fluid were recorded continuously (*closed symbols*). Cyclosporin treatment (*hexagons* represent mean values ± SDs obtained from seven tubules) resulted in a significant increase in early distal conductivity ($P < 0.01$, Wilcoxon test) and hence in total ion content compared with controls (*circles, left side*, mean values + SDs of nine tubules). The respective conductivity values of fluid at the macula densa (i.e., from the diluting segment of the ascending limb of Henle's loop) during flow rates of 10 and 20 nl/min are obtained with the pulse stop flow method (*open symbols*). Significant differences in the ion concentration in this particular nephron segment can be inferred. The increase in ion concentration at the macula densa as well as in the early distal tubule is believed to reflect a transport impairment within the proximal tubule and/or the loop of Henle

red along the mid-proximal tubules and the loop of Henle. Due to the inhomogeneity of the surface nephrons and to the arbitary selection of semi-collapsed tubules for micropuncture, the absence of reabsorption may not be representative for all nephrons of the kidneys of high-dose cyclosporin-treated rats. These data are therefore not further detailed here.

Discussion

A considerable number of investigators have observed that cyclosporin treatment leads to a rise in creatinine concentration due to glomerular filtration rate (GFR) decrease. This was reported not only from kidney transplant patients but also from patients with liver, pancreas, or bone marrow transplantation and even from patients treated with cyclosporin for polyarthritis (Calne et al. 1979; Hows et al. 1982; Klintmalm et al. 1981; Marbet et al. 1980; Shulman et al. 1981; Sweny et al. 1981, Thiel et al. 1983). Comparison of kidney transplant patients under conventional immunosuppression with those receiving cyclosporin shows that serum creatinine concentrations are lower and GFR values generally higher in the conventional treatment group (Thiel et al. 1983). Apparently an intrinsic GFR-decreasing effect is operative in cyclosporin therapy even when morphologic signs of toxicity are absent. Thiel (1983) called this phenomenon the "basic nephrotoxicity" of cyclosporin and the question arose whether a dose-dependent transport inhibition in a nephron segment proximal to the macula densa, i.e., the beginning of the distal tubule, might be the primary cause. The vacuolization of the cells together with giant mitochondria formation would suggest at least some degree of functional defect of the proximal salt and water reabsorption. The present experiments disclosed transport defects by direct assessment of an increased early distal NaCl concentration under free-flow conditions. Of importance in respect of the frequently observed GFR decrease in cyclosporin-treated animals as well as in man we also found a significant increase in ion concentration at the macula densa region (see Fig. 1). The macula densa at the beginning of the distal tubule segment operates as the sensor of the tubuloglomerular feedback mechanism. This feedback loop operates by adapting the glomerular filtration rate of each nephron to the individual transport capacity of the proximal tubule and of the loop of Henle (Thurau and Schnermann 1965; Wright and Briggs 1979). Thus the transport mechanism of each nephron cannot be overcharged by inadequate filtration rates, which could rapidly lead to a dangerous salt and water loss. A compensating GFR decrease results from proximal transport impairment after ischemia or mercury poisoning (Mason et al. 1978) and during administration of acetazolamide, a diuretic compound effective on proximal sodium transport (Persson and Wright 1974). The physiological stimulus of the macula densa feedback to reduce the single-nephron GFR is an increase of the early distal NaCl concentration above 50 m*M* (Müller-Suur et al. 1973, Schnermann et al. 1976; Müller-Suur and Gutsche 1979).

In order to examine the prerequisites of a feedback-induced GFR decrease during cyclosporin treatment, transport efficiency and diluting ability of the nephron segment proximal to the macula densa were analyzed. Our results proved that NaCl transport was impaired in one or more nephron segments upstream from the macula densa after rats were kept under an oral cyclosporin treatment of more than 15 mg/kg body weight for 8–10 days. Under free-flow conditions as well as at a physiological flow rate of between 10 and 20 nl/min during artificial loop perfusion the NaCl concentration at the macula densa was considerably above 50 m*M*. These findings infer that feedback-mediated GFR decrease should become operative during cyclosporin treatment since the basic requirements are met for a continuous stimulation of the tubuloglomerular feedback mechanism. A dose increase to 75 mg/kg led to an almost complete transport defect in surface nephrons, and no luminal fluid movement could be detected in the cortical nephrons observed. Although the probability of feedback stimulation during cyclosporin treatment is known, no direct evidence yet exists whether or not feedback stimulation is the primary mechanism causing the reversible GFR decrease at slightly elevated blood levels of cyclosporin. Feedback inhibitors like saralasin, captopril, or Ca-antagonists that suppress feedback operation to a large extent (Gutsche et al. 1975a; Stowe et al. 1979; Ploth et al. 1978; Tucker and Blantz 1978) might be able to reduce the "basic nephrotoxicity" of cyclosporin if feedback operation is involved. Further studies are being performed in this laboratory to examine this hypothesis.

References

Calne RY, Rolles K, White DJG, et al. (1979) Cyclosporin A initially as the only immunosuppressant in 34 recipients of cadaveric organs: 32 kidneys, 2 pancreas, and 2 livers. Lancet 2:1033–1036

Gutsche H-U (1980) Micro-stop flow experiments: a new method for in vivo analysis of the function of the diluting segment. Contrib Nephrol 19:231–239

Gutsche H-U, Müller-Suur R, Schurek HJ (1975a) Ca-antagonist prevents feedback induced SNGFR decrease in rat kidney. Kidney Int 8:477

Gutsche H-U, Müller-Suur R, Hegel U, Hierholzer K, Lüderitz S (1975b) A new method for intratubular blockade in micropuncture experiments. Pflügers Arch 354:197–202

Gutsche H-U, Müller-Suur R, Hegel U, Hierholzer K (1980) Electrical conductivity of tubular fluid of the rat nephron. Pflügers Arch 383:113–121

Gutsche H-U, Peterson LN, Levine DZ (1983) In vivo evidence of impaired transport by the thick ascending limb in potassium-depleted rats. J Clin Invest 73:908–916

Hows JM, Palmer S, Gordon Smith EC (1982) Use of cyclosporin A in allogenic bone marrow transplantation for severe aplastic anemia. Transplantation 33:382–386

Klintmalm GBG, Iwatsuki S, Starzl TE (1981) Nephrotoxicity of cyclosporin A in liver and kidney transplant patients. Lancet 1:470–471

Marbet UA, Graf U, Mihatsch MJ, Gratwohl A, Müller W, Thiel G (1980) Renale Nebenwirkungen der Therapie mit Cyclosporin A bei chronischer Polyarthritis und nach Knochenmarktransplantation. Schweiz Med Wochenschr 110:2017–2020

Mason J, Takabatake T, Olbricht C, Thurau K (1978) The early phase of acute renal failure. III. Tubuloglomerular feedback. Pflügers Arch 373:69–76

Mihatsch MJ, Olivieri W, Marbet U, Thiel G, Harder F, Zollinger HU (1981) Giant mitochondria in renal tubular cells and cyclosporin A. Lancet 1:1162–1163

Mihatsch MJ, Thiel G, Loertscher R, Spichtin HP, Oberholzer M, Brunner FP, Harder F, et al. (1983) Morphological findings in kidney transplants after treatment with cyclosporin A. In: Kahan BD (ed) Cyclosporine. Biological activity and clinical applications. Grune and Stratton, New York

Müller-Suur R, Gutsche H-U (1978) Effect of intratubular substitution of Na and Cl ions on the operation of the tubuloglomerular feedback. Acta Physiol Scand 103:353–362

Müller-Suur R, Gutsche H-U, Hegel U (1973) Triggering factors of the tubulo-glomerular feedback. Pflügers Arch 343:R45

Persson AEG, Wright FS (1974) Reduction of glomerular filtration rate by intrarenal feedback during acetazolamide diuresis. Fed Proc 33:347

Ploth DW, Rudulph J, Lagrange R, Navar LG (1978) Nephron function after converting enzyme inhibition in the rat. Proceedings of the 7th International Congress on Nephrology, Montreal, p F-9

Schnermann J, Ploth DW, Hermle M (1976) Activation of tubulo-glomerular feedback by chloride transport. Pflügers Arch 362:229–240

Shulman H, Striker D, Deeg HJ, Kennedey M, Storb R, Thomas ED (1981) Nephrotoxicity of cyclosporin A after allogenic marrow transplantation. Glomerular thrombosis and tubular injury. N Engl J Med 305:1392–1395

Stowe N, Schnermann J, Hermle M (1979) Feedback regulation of nephron filtration rate during pharmacologic interference with the renin-angiotensin and adrenergic system in rats. Kidney Int 15:473–486

Sweny P, Hopper J, Gross M, Vargese Z (1981) Nephrotoxicity of cyclosporin A. Lancet 1:663

Thiel G (1983) Communication. Symposium über Cyclosporin A-Therapie bei Nierentransplantation, Med. Hochschule Hannover

Thiel G, De Rougemont D, Torhorst J, Brunner FP (1981) Renal toxicity of cyclosporin A in the rat. Kidney Int 20:141

Thiel G, Torhorst J, Brunner FP (1982) Renal toxicity of cyclosporin A in the rat. Proc Clin Dial Transplant Forum 10:62

Thiel G, Harder F, Lörtscher R, Brünisholz M, Landman J, Brunner F, Follath F, et al. (1983) Cyclosporin A used alone or in combination with low-dose steroids in cadaveric renal transplantation. Klin Wochenschr 61:991–1000
Thurau K, Schnermann J (1965) Die Natriumkonzentration an den Macula densa Zellen als regulierender Faktor für das Glomerulumfiltrat. Klin Wochenschr 43:410–413
Tucker BJ, Blantz RC (1978) Inhibition of tubulo-glomerular feedback induced reduction of nephron filtration rate with saralasin. Kidney Int 14:784
Wright FS, Briggs JP (1979) Feedback control of glomerular blood flow, pressure, and filtration rate. Physiol Rev 59:958–1006

Cyclosporin A Blood Levels in Rats After Different Modes of Administration

G. Leimenstoll, H. Jessen, R. Engemann, and *W. Niedermayer*

Cyclosporin A (CyA), a new immunosuppressive drug, is widely used to prevent rejection after experimental and human transplantation (Beveridge 1982; Squifflet et al. 1982). CyA inhibits the production of interleukin 2, a product of the T-helper cell. Interleukin 2 is a signal for the proliferation of the cytotoxic effector-cell (Fidelus et al. 1982). In vitro, this inhibition is dose dependent and an effect is seen at a concentration of as low as 10 ng/ml (Fidelus et al. 1982).

The correct dosage therefore may determine the success of a transplantation experiment. A high dose may lead to toxic effects. A low dose does not suppress the immune system sufficiently. Therefore we have systematically tested the pharmacokinetic properties of CyA after different modes of administration, in order to find out the best method to achieve desirable and reproducible blood levels of CyA in rats.

Methods

Eight groups of Wistar rats received CyA as single doses of 10 and 15 mg/kg body weight intravenously (i.v.), intramuscularly (i.m.), intraperitoneally (i.p.), and orally (p.o.), each group being given one dose/route combination. Blood levels of CyA were measured after 2, 4, 6, 8, and 24 h. Another three groups of five rats were treated with CyA i.m., i.p., and p.o. respectively every 24 h for 14 days. Blood was sampled before administration of the next dose at days 1, 5, 10, and 14 p.o. Using anesthesia with ether, blood was collected from the tail vein (500 μl blood in a heparinized tube). Blood levels of CyA were measured by radioimmunoassay (Sandoz, Basel) using sheep antibody (Abisch E. 1982, ciylosporin Ria Kit, batch 1282, to monitor plasma, serum and blood concentrations. Document, Sandoz Ltd, Basel, unpublished).

Orally, CyA was given by stomach tube (750 mg CyA was dissolved in 22.5 ml olive oil). For the intravenous and intramuscular routes we used the CyA solution provided by Sandoz Ltd (Basel). The desired amount was solved in physiological saline solution. To administer CyA intraperitoneally 400 mg of the drug was dissolved in 90 ml Intralipid and 10 ml of 96% alcohol. The pharmacokinetic parameters were calculated using a one-compartment model. For calculation of the

Microsurgical Models in Rats
for Transplantation Research
Eds. A. Thiede et al.

total areas under the blood concentration versus time curves the trapezoidal rule was used (Gibaldi and Perrier 1975).

Results

With a single dose of CyA in the four different modes of application (Fig. 1), the peak blood-levels were reached after 3 h after p.o. and i.p. administration and after 7 h when administered i.m. The same results were achieved after administering 15 mg CyA/kg body weight (results not shown). The blood levels after oral and intraperitoneal administration were lower than those after intravenous and intramuscular application.

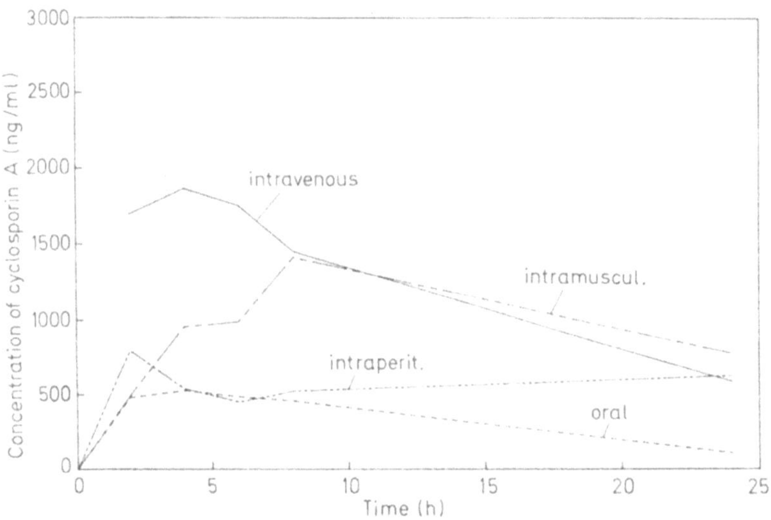

Fig. 1. Blood concentration versus time curves of CyA after intravenous, intramuscular, intraperitoneal, and oral administration (mean values of five Wistarrats, 10 mg/kg body weight (CyA)

Table 1. Pharmacokinetic parameters, calculated after a single dose of CyA, according to a one-compartment model (mean values of five Wistar rats). K_a, absorption rate constant; K_{el}, elimination-rate constant; V', volume of distribution per kilogram body weight; AUC_T, total area under the blood concentration versus time curve

	10 mg/kg CyA				15 mg/kg CyA			
	$K_a\,[h^-]$	$K_{el}\,[h^-]$	$V'\left[\frac{1}{kg}\right]$	$AUC_T\left[\frac{mg \cdot h}{1}\right]$	$K_a\,[h^-]$	$K_{el}\,[h^-]$	$V'\left[\frac{1}{kg}\right]$	$AUC_T\left[\frac{mg \cdot h}{1}\right]$
i.v.	–	0.04	3.97	51.23	–	0.03	3.86	72.2
i.m.	0.53	0.03	–	49.73	0.32	0.03	–	65.00
i.p.	–	–	–	(30.06)	–	–	–	(10.06)
p.o.	–	–	–	11.80	–	–	–	27.93

In Table 1 the pharmacokinetic parameters are shown. The elimination-rate constant, calculated from the blood levels after intravenous and intramuscular injection, was 0.03/h according to an elimination half-life of 23 h. The areas under the blood concentration versus time curves, which are proportional to the amount of drug reaching the central circulation, are nearly the same after intravenous and intramuscular injection but are much higher than after the other two modes of administration. Compared with the 100% effectivity of the intravenous injection only 20%–30% of the oral dose reaches the central circulation. After intraperitoneal injection it was not possible to calculate the areas exactly, because the blood levels of CyA differed widely from one animal to the other. The absorption-rate constant (k_a) after intramuscular injection was about tenfold higher than the elimination-rate constant.

When CyA was administered daily for 14 days, the following results were achieved:

1. When given at a dose of 10 mg/kg body weight intramuscularly, the drug accumulated. The blood level after 1 day was 800 ng/ml, after 10 days 1500 ng/ml, and after 14 days over 3000 ng/ml. When the same dose was administered orally, a steady state was achieved after 5 days, with blood levels ranging between 600 and 800 ng/ml. Nearly the same blood levels, between 400 and 900 ng/ml, resulted after intraperitoneal injection.
2. When given at a dose of 15 mg/kg body weight intramuscularly, blood levels reached a concentration of 2500 ng/ml after 5 days and a concentration of over 3000 ng/ml after 10 days. When the same dose was administered intraperitoneally, blood levels were 1500 ng/ml after 10 days and over 3000 ng/ml after 15 days. Given orally, a steady state was reached after a few days, with blood levels ranging between 1400 and 1800 ng/ml.

When these blood levels were calculated using the equation $C_{ss} = F\, x_o / V'\, K_{el} T$ (where C_{ss} is average blood concentration, F is fraction of the dose reaching the central circulation, x_o is amount of the drug administered, V' is volume of distribution per kilogram body weight, k_{el} is elimination-rate constant, and T is dosing interval) (Gibaldi and Perrier 1975) they agreed well with those measured experimentally.

Discussion

The phenomenon of first-pass effect has been observed after administering CyA orally to rats as in humans (Beveridge et al. 1981). This is evident when comparing the areas of the blood-concentration versus time curves. Therefore about 70% of the oral dose is metabolized in the liver or in the tissues of the intestine. Unfortunately an exact calculation of the areas after intraperitoneal administration was not possible. Had this been the case, a comparison of these areas with those after oral administration would have made possible the calculation of the rate of absorption in the gastrointestinal tract.

From our results, the intramuscular route is the most favorable method of administration. The interindividual differences in the blood levels are low, and the amount of CyA reaching the systemic circulation is independent of absorption and of the first-pass effect. Severe disorders of the liver, however, may influence the pharmacokinetic parameters.

There is still one more problem to be solved: What is the desirable blood level? According to Riffel et al. (1983) 14 mg/kg orally per day is highly immunosuppressive, but may cause severe side effects. According to the experience of Professor Thiede's laboratory (personal communication), an oral dose of 10 mg/kg may be sufficient, resulting in trough blood-levels ranging from 500 to 1000 ng/ml, a value which is probably desirable also in humans. To achieve such a concentration intramuscularly, an initial loading dose of 6 mg/kg is needed, followed by a maintenance dose of 3 mg/kg. Our next experiment will have to prove this calculation.

References

Beveridge T (1982) Steroidfreie Behandlung nierentransplantierter Patienten mit Cyclosporin A. Klin Wochenschr 60:1137–1142

Beveridge T, Gratwohl A, Michot F, Niederberger W, Neusch E, Nussbaumer K, Schaut P, Speck B (1981) Cyclosporin A: pharmacokinetics after a single dose in man and serum levels after multiple dosing in recipients of allogen bone marrow grafts. Curr Ther Res 30/1:5–18

Fidelus RK, Ferguson RM, Widmer MB, Wee S, Bade FH, Orosz C (1982) Effect of cyclosporin A on murine and human T-helper-cell clones. Transplantation 34/5:308–311

Gibaldi M, Perrier D (1975) Pharmacokinetics. Deccer, New York

Ryffel B (1982) Experimental toxicological studies with cyclosporin A. In: White DJG (ed) Cyclosporin A. Elsevier Biomedical, Amsterdam, p 45

Squifflet JP, Sutherland DER, Rynasiewiez JJ, Field J, Heil J, Najarian JS (1982) Combined immunosuppressive therapy with cyclosporin A and azathioprin. A synergistic effect in three of four experimental models. Transplantation 34/6:315–318

Orthotopic Transplantation of the Fallopian Tube, Ovary, and Segment of the Uterus in Inbred Rats

P. H. Scheidel, M. Korell, T. Strowitzki, U. Noss, R. A. Wiedemann, and *H. K. Hepp*

Introduction

Gynecology is not really a latecomer in the field of experimental and clinical transplantation. The first attempt to transplant the human fallopian tube was undertaken in 1946 (Ritala 1946). Ovarian transplantation was reported as early as in the last century (Wood 1981). There have been reports on the experimental transplantation of the uterus, e.g., in the ewe (Baird et al. 1976), but no reports on transplantations in humans.

Microsurgical transplantations of the fallopian tubes have been recently reported in five cases, but pregnancy did not occur (Paterson 1981). The future of tubal transplantation is questionable because progress in extracorporeal fertilization and embryo transfer has now provided a successful method of treatment for patients with badly damaged or absent tubes (Trounson and Wood 1981). The transplantation of the ovary, however, may offer an optimal treatment for patients after bilateral oophorectomy, irradiation of the lower pelvis for treatment of Hodgin's disease, or other pelvic neoplasms (Cantor 1983). The uterus is still the genital organ which cannot be replaced by hormones or in vitro fertilization, especially if the transfer of a fertilized egg into a foreign mother (substitute mother) is not acceptable due to legal and ethical reservations.

Clinical transplantations of the genital organs are still shadowed with doubt. The two major reasons are:

1. Genital organs are not life-sustaining (but "life-producing").
2. At this time there is no information available on the immunological problems after gynecological transplantations although rejections have been described after transplantation of the fallopian tube (Wood et al. 1978).

Although successful microsurgical transplantations of the ovary have been reported in primates (Scott et al. 1981) and of the fallopian tube in rabbits (Winston and McClure Brown 1974), pigs (Cohen 1974), and other animals, the problem of uterine transplantation is still unsolved. We therefore wanted to establish an immunologically well-defined model for the study of all the possible problems involved in and resulting from transplantation of the genital organs. The only model in experimental transplantation which provides these advantages and

Microsurgical Models in Rats
for Transplantation Research
Eds. A. Thiede et al.

which is also reasonable in terms of cost is the orthotopic transplantation in inbred rats (Thiede et al. 1978).

Organ Preservation

As there are no data available from the literature on the preservation of the genital organs, we studied the method of initial perfusion and hypothermic storage at 4 °C as recommended for kidney preservation by Collins et al. (1969) and others. In 115 rats (weight 300 g and anesthesia by 1.6% chloral hydrate) different solutions were used for initial perfusion via the lower aorta (Table 1).

Table 1. Osmolarity of different solutions used for initial perfusion and storage at 4 °C

Ringer's solution + heparin (4000 U/liter)	270 mOsm/liter
Tissue culture medium	~300 mOsm/liter
Euro-Collins' solution	355 mOsm/liter
Tubal perfusate medium (TPM II)	358 mOsm/liter

Table 2. Morphological changes to be observed in the mucosa of the ampulla (fallopian tube in the rat)

	Histological grading
Grade I	No morphological changes
Grade II	Intracellular edema, intact epithelium
Grade III	Detachment of epithelium, slight vacuolation
Grade IV	Severe detachment of epithelium, partial necrosis (epithelium)
Grade V	No more intact epithelium (no muscle necrosis)

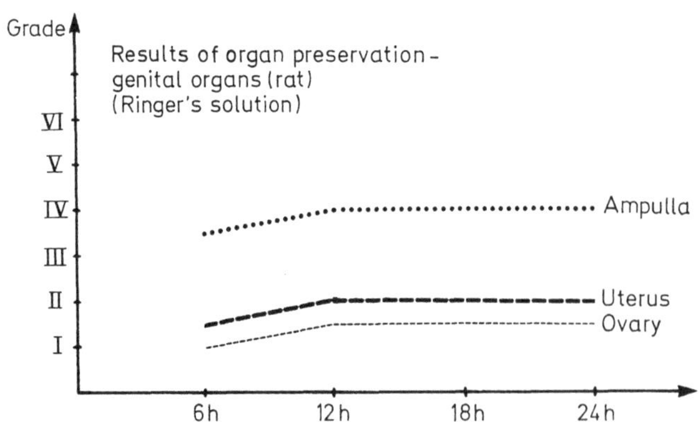

Fig. 1. Microscopic alterations of the different genital organs after perfusion and hypothermic storage (according to the histological grading in Table 2)

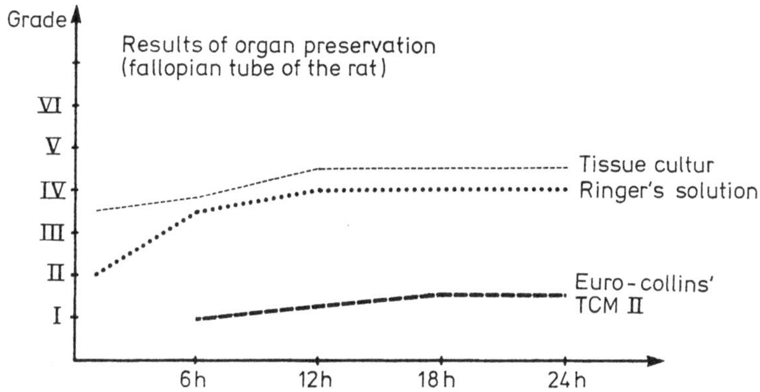

Fig. 2. Preservation of the mucosa in the ampullary part of the fallopian tube – after perfusion with different solutions (see Table 1)

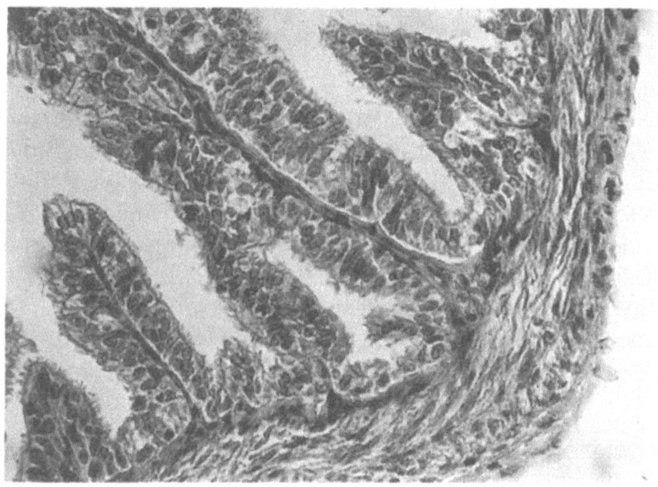

Fig. 3. Microscopic appearance of the rat ampulla after initial perfusion and hypothermic storage (6 h); grade II–III; Masson-Goldner, ×400

After 6, 12, 18, and 24 h storage at 4 °C the organs were examined for morphological alterations by light microscopy. The morphological changes would be observed primarily in the mucosa of the ampulla (Table 2). Therefore the effect of the preservation capabilities of a salt solution can be sufficiently investigated in the fallopian tube (Fig. 1). The hyperosmolarity of the solution seems to be the most important factor (Fig. 2). The exchange of the high-potassium concentration in Collins' solution versus a higher concentration of sodium (TCM II) did not produce more intracellular swellings. After 6 h the differences in preservation were already obvious, when Ringer's solution (Fig. 3) was compared with Euro-Collins' solution (Fig. 4).

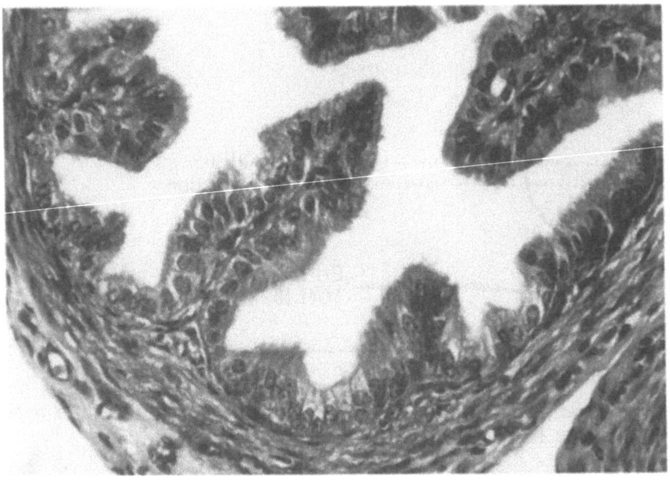

Fig. 4. Intact mucosa of the rat ampulla after 6 h (Euro-Collins); grade I–II; only slight intracellular edema visible; Masson-Goldner, ×400

These results are in close correlation to those observed in renal preservation by Cerra et al. (1977) and Petracek et al. (1981). As it is an undisputed fact that only electron microscopic studies are able to demonstrate alterations of the cellular level (Panigel 1972), we studied the mucosa of the ampulla by transmission electron microscopy and scanning electron microscopy. Preliminary results confirm the findings we have described above. In conclusion, we suggest that up til now the initial perfusion by hyperosmolaric solutions, such as Euro-Collins', and subsequent hypothermic storage is sufficient to preserve even the sensitive structures of the fallopian tube for at least 6 h.

Transplantation Technique

The orthotopic transplantation of the ovary and fallopian tube, simplified by the simultaneous transplantation of a segment of the uterus, was described in detail by Lee and Orloff (1979). They suggested the use of the right adnexe because of the vessels originating from the aorta and the inferior vena cava, thus enabling transplants to be made by end-to-side anastomosis of an aortic and IVC segment. There were two important remarks in this publication:

1. About 80% of Lewis female rats demonstrate variations in the anatomical course of the right ovarian artery and vein.
2. Ideal conditions for a successful ovarian transplantation in the rat include the use of parallel running ovarian vessels of adequate thickness.

In our experience those situations make more than 80% of female rats unacceptable as donors for transplantation of the right adnexe. We have performed approximately 50 transplantations by this technique and have never been able to achieve transplant survival because of thrombosis within 24–48 h. Therefore, we

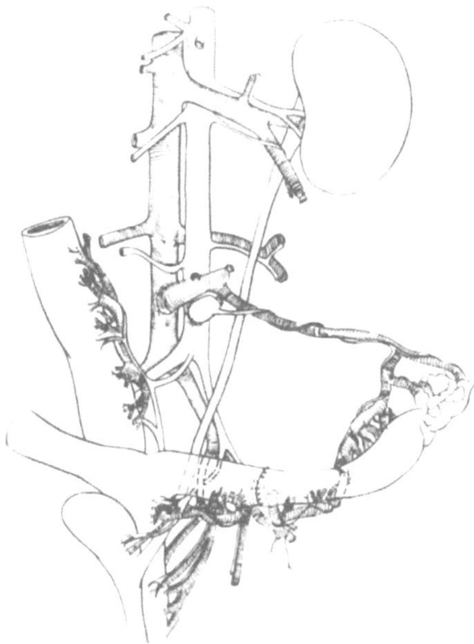

Fig. 5 **Fig. 6**

Fig. 5. Anatomy of the rat: left kidney, ovary, fallopian tube, and uterus; vascular supply

Fig. 6. Transplantation of the left adnexe and uterine segment. (1) End-to-side anastomosis of renal vein segment to inferior vena cava; (2) end-to-side anastomosis of aorta to aorta; (3) end-to-end anastomosis of the uterus

have attempted to use the left adnexe because of the fewer anatomical variations (Fig. 5).

Three different techniques were applied:

1. Transplantation by the use of an aortal and renal vein segment (Fig. 6).
2. Transplantation by the use of a femoral artery and vein segment providing blood supply via the uterine artery as proposed by Cornier and Sibila (1982) (Fig. 7).
3. Transplantation by the use of a renal vein segment and an aortic patch as suggested by Thiede (1983 personal communication) (Fig. 8).

The results of our latest 20 transplants are summarized in Table 3. From our results, the use of an aortic patch and the renal vein segment seems to be the most promising technique. Nevertheless, we have yet to achieve pregnancies after orthotopic transplantation of the genital tract in Lewis rats.

We recently remodified our technique by transplanting the left kidney simultaneously and found absolutely unchanged transplants after 14–16 days postoperatively. We believe that it must be possible to establish a preclinical model of organ transplantation in gynecology with subsequent pregnancies using the rat model. Thus, the interesting questions of the immunological and functional properties of transplanted tubes, ovaries, and uteruses will hopefully be answered in the future.

Fig. 7

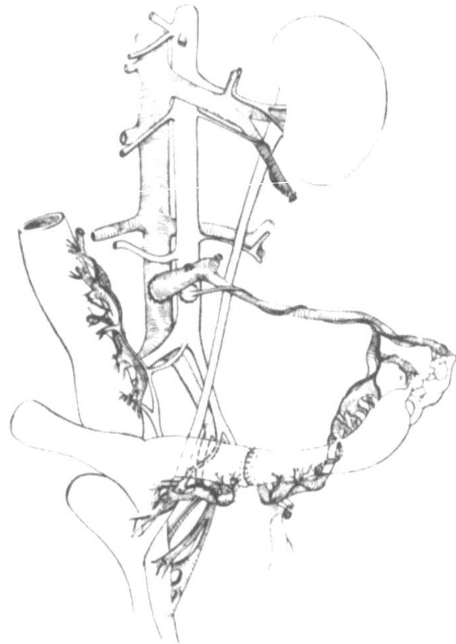

Fig. 8

Fig. 7. Transplantation of the left adnexe and uterine segment. Blood supply via arteria iliaca and arteria uterina: (1) End-to-side anastomosis iliac vein segment to inferior vena cava, (2) end-to-side anastomosis iliac artery to aorta, and (3) end-to-end anastomosis of the uterus

Fig. 8. Transplantation of the left adnexe and uterine segment: (1) End-to-side anastomosis of renal vein segment to inferior vena cava; (2) end-to-side anastomosis of aortal patch to aorta; (3) end-to-end anastomosis of the uterus

Table 3. Results of microsurgical orthotopic transplantation of ovary, fallopian tube, and uterus segment in inbred rats (all anastomosis performed with continuous sutures, 9–0 nylon)

Technique	*n*	Anastomosis patent (7 days postoperatively)		Viable transplant
		Artery	Vein	
Iliac artery segment	9	3	4	2
Renal artery segment	6	3	2	2
Aorta patch	5	4	3	3
Total	20	10	9	7

References

Baird DT, Land RB, Scaramuzzi RJ, Wheeler AG (1976) Functional assessment of the autotransplanted uterus and ovary in the ewe. Proc R Soc Lond [Biol] 192:463–474

Cantor B (1983) Transplantation and replantation of the fallopian tubes and ovaries: a technique for patients undergoing pelvic irradiation. Fertil Steril 39:231–234

Cerra FB, Raza S, Andres GA, Siegel JH (1977) The endothelial damage of pulsatile renal preservation and its relationship to perfusion pressure and colloid osmotic pressure. Surgery 81:534–541

Cohen BM (1974) Vascularised homograft transplantation of the oviduct in the pig. S Afr Med J 48:162

Collins GM, Bravo-Shugarman MB, Terasaki PI (1969) Kidney preservation for transplantation. Lancet 2:1219–1222

Cornier E, Sibila A (1982) Technique microchirurgicale de la transplantation tubo-ovarienne chez le rat (Film). 7th International Congress of the International Microsurgical Society, Lyon

Lee S, Orloff MJ (1979) Microvascular surgery in the rat. In: Silber SJ (ed) Microsurgery. Williams and Wilkins, Baltimore

Panigel M (1972) Morphological evaluation of perfused tissues. Acta Endocrinol [Suppl] (Copenh) 158:74–90

Paterson P (1981) Experimental tubal transplantation. Aust NZ J Obstet Gynaecol 21: 164–166

Petracek MR, Lawson JD, Johnson HK (1981) Three simple solutions for renal flush and preservation. South Med J 73:321–324

Ritala AM (1946) Tubentransplantation bei Sterilität der Frau, verursacht durch Occlusio tubae. Acta Obstet Gynaecol Scand 25:493–527

Scott JR, Keye WR, Poulson AM, Reynolds WA (1981) Microsurgical ovarian transplantation in the primate. Fertil Steril 36:512–515

Thiede A, Brieler HS, Deltz E, Engemann R, Poser H (1978) Zur experimentellen Mikrochirurgie an Ratten. Zentralbl Chir 103:1397–1409

Trounson A, Wood C (1981) Extracorporal fertilization and embryo transfer. Clin Obstet Gynaecol 8:681–717

Winston RML, McClure Brown JCV (1974) Pregnancy following autograft transplantation of fallopian tube and ovary in rabbit. Lancet 2:494–495

Wood C (1981) Ovarian transplantation. Aust NZ J Obstet Gynaecol 21:107–110

Wood C, Downing B, McKenzie I, O'Brien B, Paterson P (1978) Microvascular transplantation of the human fallopian tube. Fertil Steril 29:607–613

Allogeneic and Xenogeneic Biografts in Reconstructive Surgery of the Trachea

R. Ascherl, M. A. Scherer, K. Geißdörfer, and *G. Blümel*

There is a broad range of factors inducing stenoses and defects of the trachea. The most frequent ones are complications due to tracheostomy and iatrogenic injuries. The location of acquired tracheal stenosis is usually in the subglottic region and in the cervical part of the trachea. There are three main types of congenital obstructions according to Meyer (1982) and De Lorimier and Harrison (1981): (1) generalized hypoplasia; (2) fusiform narrowing, usually tapering into a tight stenosis just above the carina; and (3) segmental stenosis, occurring at any level.

Although ample investigations on tracheal prostheses have been performed, today's method of choice is still end-to-end anastomosis using different plastic reconstructive procedures. In adolescent and adult Wistar rats of both sexes preparations of allogeneic and xenogeneic tracheas were investigated and their usefulness as bioprostheses was tested.

Material and Methods

Under general anesthesia with ketamine (100 mg/kg body weight) and xylazine (16 mg/kg body weight) the cervical trachea was dissected following a midline incision of the anterior collar region. After resection of tracheal segments of three, six, or nine cartilages in length, intubation was performed of both the oral and aboral part of the trachea. The animals breathed spontaneously. The anastomoses were performed with single sutures using 8-0 silk or 10-0 nylon. Postoperatively the animals were examined radiologically. The animals were killed within 2-week periods, the longest observation period being 16 weeks. There were thus 14 groups with allogeneic and xenogeneic tracheal implants, the latter derived from guinea pigs. Radiological examinations and microangiographs, micromorphological examinations, scanning electron microscopy, and biomechanical tests were performed. Autologous reimplants of a length of two or six cartilages served as controls.

Microsurgical Models in Rats
for Transplantation Research
Eds. A. Thiede et al.

Preparation Procedure

The preservation process mainly consists in linkage of the free amino groups of the proteins by dicarbonic acids. Ficin is used to destroy and remove all proteins other than collagen. The length of the dicarbonic acids can control the biomechanical properties of the bioprostheses. In a study performed by Oertli (personal communication, 1982) on vessels that were preserved in the same manner as the tracheas, a strongly diminished antigenicity in comparison to unprepared vessels was demonstrated.

Results

In the normal anatomy of the rat trachea there are varying proportions of ciliated, basal and secretory cells. More aborally, near the carina, one can also find the so-called goblet cells. In the control group a nearly perfect recovery of the tissue could be demonstrated, independent of the length of the autologous reimplant. There were no signs of necrosis either in the submucosal region and the cartilages or in the mucosal layer. The cilia were as dense as in normal, nonoperated tissue. The only pathological finding was a reduction of the mucous and seromucous glands. No disordered wound healing in the anastomoses could be observed (Fig. 1).

In the bioprostheses a good incorporation with an ingrowth of epithelium was found within the first 4 weeks. The reepithelialization originated from both anastomotic sites. In the vicinity of the anastomoses a normal ciliated epithelium could be shown. Using longer implants that were say of a length of nine cartilages,

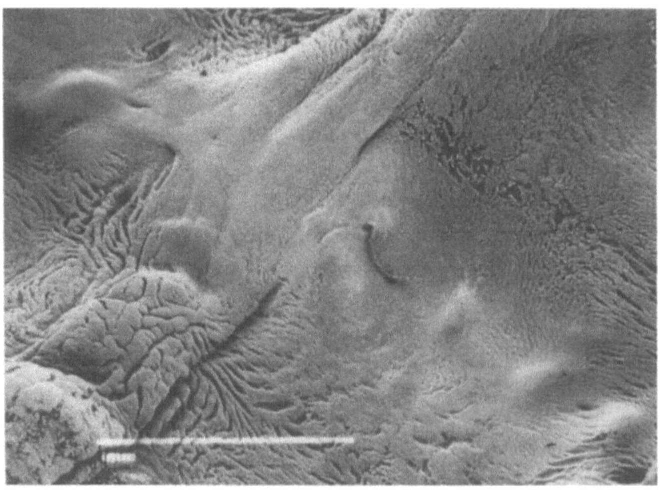

Fig. 1. Inner surface of an anastomotic area of an autologous reimplant, 8 weeks postoperatively, showing excellent wound-healing and epithelial overgrowth of the suture line; scanning electron microscope, × 50

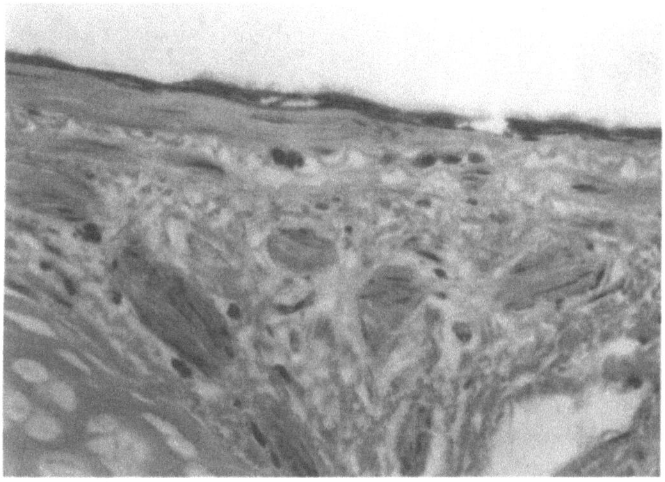

Fig. 2. Ciliated epithelium in the middle of a short allograft (three cartilages in length); no sign of destruction of the prosthetic cartilage; 12 weeks postoperatively; Elastika-Ladewig stain, × 120

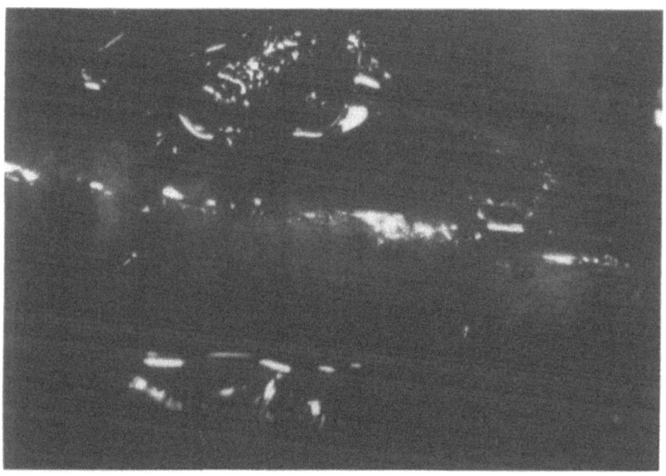

Fig. 3. Macroscopic aspect of a tracheal xenograft six cartilages in length at time of killing; 8 weeks postoperatively, × 4

corresponding to at least one-fourth of the rat's trachea, a multilayer epithelium with a reduction of ciliated cells was sometimes found (Fig. 2).

After 3 months there was a tight and close contact between the implant and host trachea. Using stains according to von Kossa, calcification of the cartilages was dependent on the animal's age; this only occurs in normal cartilages and not in transplanted ones. The preserved cartilages seemed to remain stable with no signs of destruction (Fig. 3). In the long-term groups fibrous infiltration of the

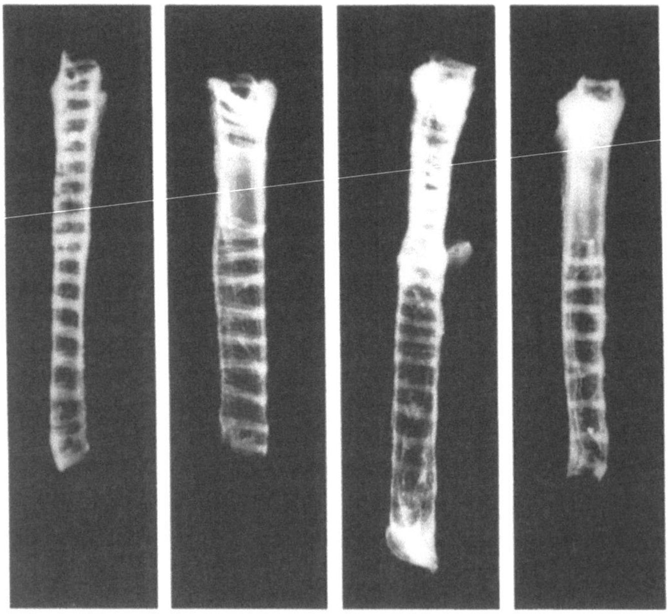

Fig. 4. Microangiographies of four different rat tracheas. *From left to right* Autologous reimplant, 14 weeks; xenograft, 4 weeks postoperatively; allograft, 16 weeks postoperatively, moderate stenosis and slight signs of calcification and fracture of the prosthetic cartilage; xenograft, 8 weeks postoperatively. Note that there is no radiodensity in the cartilages of the prothesis, no alterations – the cartilage was easily demonstrable macroscopically and histologically

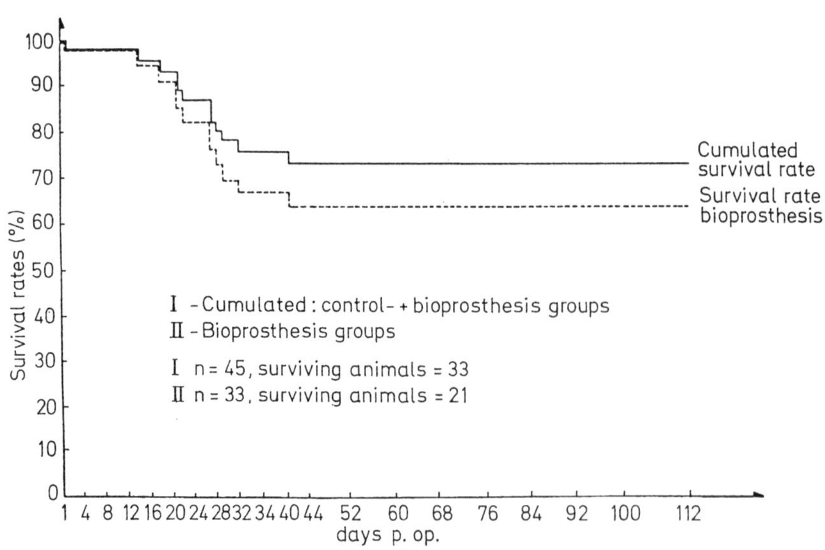

Fig. 5. Survival rates; note that no animals died after the 6th week

submucosa of the implant was visible with deposition of new collagenous fibers. In some animals of the 12- and 16-week groups sometimes stenoses occurred with no obvious respiratory failure. The obstruction was usually located in the posterior pars membranacea of the trachea. This dorsal region, however, seems to serve as a guideline for the ingrowth of epithelium. At least we found the best and mostly ciliated cell-layers in these parts of the tracheas, sometimes even in the mid-portion of the longer implants.

The pattern of revascularization never resembles that of normal tracheas; instead of circular vessels around the intercartilagenous space, an irregular net of vessels can be demonstrated in the microangiographs (Fig. 4).

In the case of paratracheal, heterotopic implantation we also never found any signs of immunogenity of the biografts. In the lumina loose connective tissue was usually found. Fracture of the cartilages, however, resulted in the invasion of foreign body giant cells just at the fracture sites. All these findings reported so far refer to both the allogeneic and xenogeneic implants. In the biomechanical investigations the bioprostheses were less extensible than the host tracheas, with a higher force/lengthening ratio. The breaking strength was nearly the same in both the prostheses and the normal trachea.

None of the animals in the control groups died (autologous reimplants) whereas there was a death rate of about 30% in the bioprostheses groups. These deaths were due to prosthetic failure – stenosis and acute obstruction – in only two cases; the remaining animals died of tracheo pneumonial affection, 90% of these in the interval between the 2nd and 5th weeks (Fig. 5). Implantation of the tracheal prostheses in juvenile rats revealed – measured by gain in body weight – no adverse effects and no functional impairment of normal growth and development.

From these experiments we conclude that the bioprostheses for tracheal replacement seem to be a promising material and method in reconstructive surgery in the trachea. Nevertheless, only further experiments in other mammals will show whether the material is clinically useful. These experiments are now in progress.

References

Borrie JCM, Redshaw NR, Dobbinson TL (1973) Silastic tracheal bifurcation prosthesis with subterminal Dacron suture cuffs. J Thorac Cardiovasc Surg 65/6:956–962

Bottema JR, Feijen J, ten Hoopen HWM, Molenaar J, Wildevuur CRH (1980) Microporous tracheal prosthesis: incorporation and prevention of infection. Trans Am Soc Artif Intern Organs 26:412–417

De Lorrimier AA, Harrison MR (1981) Pediatric surgery. In: Dunphy JE, Way LW (eds) Current surgical diagnosis and treatment. Lange, Los Altos, pp 998–1043

Foet K (1981) Untersuchungen zum Trachealersatz durch homologe Segmentimplantate. In: Cotta H, Martini AK (eds) Implantate und Transplantate in der Plastischen und Wiederherstellungschirurgie. Springer, Berlin Heidelberg New York, pp 310–315

Grillo HC (1979) Surgical treatment of post-intubational tracheal injuries. J Thorac Cardiovasc Surg 78:860

Meyer R (ed) (1982) Reconstructive surgery of the trachea. Thieme, Stuttgart

Neville WE, Bolanowski PJP, Soltanzadeh H (1976) Prosthetic reconstruction of the trachea and carina. J Thorac Cardiovasc Surg 72/4:525–538

Staindl O, Lemetschwandtner A (1979) Ergebnisse heterologer Trachealtransplantation im Tierexperiment. Teil II. Licht-, raster- und transmissionselektronenmikroskopische Befunde am Trachealtransplantat. HNO 27:211–226

Staindl O, Lemetschwandtner A, Zimmermann G, Adam H (1979 a) Ergebnisse heterologer Trachealtransplantationen im Tierexperiment. Teil I. Lichtraster- und transmissionselektronenmikroskopische Befunde am Trachealepithel. HNO 27:7–13

Immunological Mechanisms in Clinical Organ Transplantation

W. Müller-Ruchholtz

This paper will not review mechanisms in transplantation immunobiology but will rather introduce the subject of clinical organ transplantation with some remarks from an immunologist's point of view. As mentioned earlier in "Theoretical Aspects of Rat Organ Transplantation" (p. 77), the basic immunobiological laws appear to apply to all mammals; in other words there is nothing special about the human immune system. It may be assumed that the biological need for immune reactivity is really vital to all mammals and that, therefore, there was no special evolutionary pressure on the human different from that, e.g., on laboratory rodents. Thus: what rat and mouse models teach, the human usually confirms in principle. If this were not so, surgeons should quit their interest in experimental transplantation models.

Nevertheless, there is quite some need for a careful consideration of the clinical relevance of each particular laboratory model.

1. What about the in vivo relevance of in vitro data? We may differentiate between (a) in vitro experiments which include in vitro evaluation and yield the highest level of analytical precision, (b) in vitro manipulations of cells which are followed by in vivo evaluation of the consequences, and (c) in vivo experiments which are followed by in vitro evaluation of one change or the other in immune reactivity. Altogether, we should be aware of something I would like to call a "biological imprecision relationship": A single reaction can be analyzed better the more completely it has been dissociated from the biological context that enables its biological role to be studied.
2. Is the experimental setup adequate with regard to the clinical problem in question? All of us know the experience of reporting something exciting which is then questioned by a clinician with the comment that this finding refers to something clinically impertinent.
3. What is the functional role of a given immune reactivity? This question applies to at least two quite different aspects: (a) Any particular reaction is part of a regulatory network, and the resultant of all reactions may be down regulation rather than rejective sensitization, as, e.g., observed in the immunotolerant rat liver recipients with "locked-in" cytotoxic T-lymphocytes. (b) A well-documented sensitization may not damage a particular tissue to such an extent that

Microsurgical Models in Rats
for Transplantation Research
Eds. A. Thiede et al.

its function breaks down (independent of the damaging effect of the very same sensitization on another tissue graft in the same recipient), as, e.g., reported for rat cornea, trachea, or veins.

Histocompatibility antigens are becoming increasingly perplexing to both immunologists and transplantation surgeons. It is long ago that it was anticipated that the transplantation immunological problems would be solved as easily as blood transfusion immunological problems were solved by donor/recipient matching. Instead, the major histocompatibility complex (MHC) has turned out to be by far the most polymorphic system of genetically determined cell surface structures – and our present understanding is still far from complete. As pointed out in Günther's contribution to this volume (p. 83), there are differences between rat, mouse, and human with regard to the order and number of genes. Significant recent progress has come and will continue to come from gene cloning and transfection experiments and from a discovery explosion regarding the group of human MHC class II antigens. We have now good evidence that the latter, formerly called D/DR antigens, are governed by (at least) three loci, termed *DR*, *D*, and *DP*. But we do not know the number of allelets at each of them and we are still far from routine typing of them.

There are more problems in tissue typing: We have insufficient information about the relative role of the many HLA (human leukocyte antigen) structures for the immunogenic, i.e., the sensitizing potential of a graft. It would be very helpful to know better which antigens we may neglect under which conditions – as it is helpful in blood transfusion to know that in the rhesus blood group system the antigen D is by far the most immunogenic whereas the antigen e is without practical importance. Furthermore, in addition to HLA (the human's MHC) many non-MHC, genetically independently determined cell-surface structures are known or can be postulated. The most impressive evidence for their potential role comes from bone marrow transplantation between HLA-identical siblings, which is followed by a 40%–50% graft versus host reaction incidence in spite of immunosuppressive treatment. On the other hand, not all antigen systems (be they cell-type restricted, cell-group restricted, or common) need to be immunogenic – as the majority of blood group systems consist of practically nonimmunogenic antigens. This certainly is not the place to describe what is known, e.g., about lymphocytic, neutrophil-specific, monocytic, and endothelial-cell antigens – it is still without practical importance.

Does all this (together with the application of today's immunosuppressive wonder drug cyclosporin) mean that tissue typing for graft donor/recipient matching should be given up? Certainly not: it is simply wrong to conclude that matching is worth nothing because it may happen that a match of those strongly immunogenic antigens which can be identified may be overruled by those which we cannot yet identify. And it is not justifiable to generalize from the immunosuppressive potential of a drug given in sufficiently large amounts to override some major histoincompatibilities when grafting a particular organ, to grafting of other tissues with different immunogenic potential, irrespective of the dose-dependent risks of antibiotic-resistant infections and/or drug intoxications. We should not forget that most of the impressive data in support of the role of tissue typing have

Table 1. Antigen-presenting cells (APC)

Cell type	Dendritic — Interdigitating (Langerhans) Dendritic — [Follicular][a] Macrophage subtype
Tissue distribution	Widespread (in every organ)
Properties	Strongly class II antigen (Ia) positive, Thy 1 negative, FcR negative, not or weakly adherent, mobile, rapidly turning over
Origin	Bone marrow Allogeneic (donor) → more effective Autologous (host) → less effective

[a] Confined to germinal centers, Ia negative, B memory generation (Humphrey)

been obtained from the more homogeneous clinical material of few-center or one-center studies in certain geographical areas (such as Scandinavia) – not to mention the analytically superior experimental data obtained from optimally controlled variations of inbred strain combinations.

Many details of the development of immune reactivity against histocompatibility antigens may be claimed to be understood, but not so the entire phenomenon. The antigen-presenting cells (APCs) have recently attracted particular interest and, therefore, some of their major characteristics are summarized, mainly from experimental studies, in Table 1. It appears to be of importance to distinguish clearly between APCs of graft and recipient origin, for two reasons: The former are (1) more potent in presenting "their own" antigens and they are (2) more easily accessible to inactivation by manipulation. Probably such inactivation will be achieved more effectively and faster by incubation with adequate antibodies than by extended in vitro cultures. Thus, bone marrow and pancreatic islet transplantation may well be among the first clinical approaches toward grafting APC-deprived, i.e., immunogenicity-reduced tissues.

The T-cell subset evaluation in clinical transplantation immunology represented one of the major controversial issues at the Ninth International Congress of the Transplantation Society (1982) and continues to do so. It may be recalled again that the marker OKT4 is common to T-helper and T_{DTH} cells whereas OKT8 ist common to T-suppressor and cytotoxic cells; thus, the ratio of $OKT4^+$ /$OKT8^+$ cells can only describe an undefined mixture of ratios of regulator and effector functions. Obviously, such description is of less value than many clinical publications tried to make us believe. Furthermore, this problem is overlapped by the fact that "the peripheral blood represents only a tiny and somewhat atypical sample of the body's lymphoid system, and changes in blood may or may not reflect events within the secondary lymphoid organs or at sites of immune destruction" (Nossal 1982). Thus, a large part of T-cell heterogeneity is still hidden and better markers and more immunohistological studies are urgently needed.

For inhibition of functionally deleterious sensitization against the grafted cells, nonspecific immunosuppression continues to be one of the main pillars, together with tissue typing and, for some organs (particularly for kidney, not at all for bone marrow), pretransfusion of blood. It does not appear appropriate to discuss here such "old hats" as corticosteroids and azathioprine, methotrexate, or cyclophosphamide, especially in view of the present trend toward replacing them in transplantation surgery with the fungal peptide cyclosporin. It may be noted, however, that a similar trend has not developed in those areas of clinical medicine where the vast majority of so-called immunosuppressive drugs is given, namely for treatment of autoimmune diseases. This may, at least in part, be due to a preferential effect of cyclosporin on ongoing cell-mediated sensitization rather than established and largely humoral sensitization.

One may discuss the question why antilymphocyte serum or globulin never became a really "big hit," except for rejection reversion (because of the ratio of effect to side effects in long-term application and the related problems of underdosage and price?). One may ask how it could happen that total lymphoid irradiation was of uppermost interest at the Eighth International Congress of the Transplantation Society (1980) but was "out" 2 years later. I prefer to restrict this discussion of nonspecific immunosuppression to three more general remarks which are based on recent experimental evidence: (1) No immunosuppressive regimen acts selectively on lymphocytes; a parallel unspecifically antiphlogistic effect may simply conceal specific immunological sensitization. (2) The so-called immunosuppressive drugs clearly suppress ongoing sensitization under certain biological conditions but may well act immunostimulatively under other conditions, which may differ from drug to drug. (3) Certain of these drugs (presumably, e.g., ifosfamide and cyclosporin) have a still poorly evaluated immunotolerance-facilitating potential which may be clearly dissociated from their immunosuppressive potential.

Immunotolerance may be the last key word of this overview. Undoubtedly, establishment of specific tolerance remains the ultimate goal of transplantation immunology. The literature is full of reports on induction of this particular state, many of which lack adequate specificity controls. If, e.g., a graft recipient is moderately immunodepressed and a third party control graft is more immunogenic than the test graft, only the third party graft may be rejected, seemingly (!) indicating specificity. Thus, the control graft should be less immunogenic than the test graft – a postulate which is rarely met.

Induction of immunotolerance must be followed by its maintenance, and this cannot be achieved without transplantation antigen persistence in systemically accessible form. Such is optimally provided by hemopoietic chimerism (as briefly discussed in my paper "Theoretical Aspects of Rat Organ Transplantation," p. 77) but rarely provided in skin graft recipients. Recipients of larger organs may be seen as located on a scale between these extremes, probably closer to the former.

A discussion of the mechanisms of immunotolerance may fill volumes or it may be reduced to the simple statement that we now know of three principal alternatives: (1) functional noneffectivity because of strong dominance of suppression-type regulator reactivity, (2) antigen-mediated reversible blockade of immune reactivity, and (3) deletion/abortion of specifically reactive immunocompetent cells. These alternatives are not mutually exclusive, not even within the same organism at different times. And it may well be assumed that prevention of immunological autoaggression is so vitally important that nature had to provide more than just one mechanism to guarantee this.

Finally, transplantation immunological research and clinical follow-up require quite some psychological tolerance, since teamwork of people as different as surgeons and immunologists is required. The surgeon is expected to tolerate that the immunologist asks basic questions, and the immunologist is expected to tolerate that the surgeon asks practical questions.

Immunohistological Analysis of Late Cellular Rejection and Cyclosporin-A-Resistant Acute Rejection in Human Kidney Allograft: Evidence of Delayed-Type Hypersensitivity Reaction Mechanisms

H. K. Müller-Hermelink, R. Schoel, and *F. Strahlendorf*

The investigation of biopsy material in renal allografts focuses attention on several questions, which are summarized in Table 1. Some of these questions, in particular recurrence of glomerular disease, preservation damage, ischemic changes, and infectious complications, are too complex to be considered here. Different immune mechanisms are thought to be involved in kidney allograft rejection. They may change clinical and histopathological findings. According to Porter (1967) the types of rejection phenomena in Table 2 are well accepted and discussed in standard texts (Meadows 1978). Recent studies of cellular infiltrates in human kidney allografts cast doubt on the importance of some of the described immune phenomena and stress the importance of cellular immune mechanisms (Hancock 1984; Mason 1983).

This paper aims to discuss some of these recent findings of intragraft effector mechanisms associated with renal allograft rejection as revealed by immunohistological studies using a set of monoclonal antibodies. The administration of cy-

Table 1. Pathogenic consideration of biopsy material in renal allografts

1. Rejection	4. Interstitial nephritis
2. Glomerular disease	Bacterial
De novo	Viral
Recurrent	Unknown
Treatment related	5. Treatment damage (e.g., cyclosporin A)
3. Circulatory changes	6. Combinations
Arterial	
Venous	
Lymphatic	

Table 2. Types of kidney allograft rejection (Porter 1967)

Type 1:	Acute immediate rejection "hyperacute"
Type 2:	Acute early rejection "cellular"
Type 3:	Acute late rejection "vascular"
Type 4:	Insidious late rejection "glomerular"

Microsurgical Models in Rats
for Transplantation Research
Eds. A. Thiede et al.

closporin A (CyA) as a major immunosuppressive agent in the prevention of graft rejection was started in December 1982 in the Transplantation Unit in Kiel. Rejection periods previous to that date and those occurring since are compared in order to elucidate whether the effector mechanisms and phenotype have changed in CyA-resistant rejection as compared with the rejection seen under conventional immunosuppressive treatment.

Materials and Methods

Patients

Needle-and-drill biopsies of 26 patients were studied at different times after kidney allografting. From 13 patients only one biopsy was available; from 13 pa-

Table 3. Polyclonal antisera (rabbit to human) and monoclonal antibody (mouse to human)

Antibody	Specificity	
Polyclonal antisera (rabbit to human)		
Anti IgA	α-Chain	(Dako)
Anti IgM	μ-Chain	(Dako)
Anti IgG	γ-Chain	(Dako)
Anti C_{1q}	Complement Component C_{1q}	(Behring-Werke AG)
Anti C_{5b9}	C_{5b9} complex	(Dr. Bhakdi, Dept. of Microbiology, University of Giessen, FRG)
Monoclonal antibody (mouse to human)		
Anti C_{3b}	C_{3b} complement Component	(Bethesda Research Laboratories, United States)
Anti Leu_1	Peripheral T-lymphocyte	(Becton Dickinson, United States)
Anti Leu_4	Peripheral T-lymphocyte	
Anti Leu_{2a}	Cytotoxic/suppressor T-lymphocyte	
Anti Leu_{3a}	Helper/inducer T-lymphocyte, some phagocytic mononuclear cells, some dendritic cells	
Anti Leu_7	Killer cells, natural killer cells (HNK-1 cells)	
Anti HLA-ABC	Common HLA-ABC determinant	(Becton Dickinson)
Anti HLA-DR	Common HLA-DR determinant	(Dr. Anderson, Dept. of Immunology, University of Kiel, FRG)
Pan B (To 15)	B-cell normal and neoplastic	(Dako)
OkM_1	Monocyte, granulocyte, and null cells	(Ortho Diagnostic Systems, United States)
Ki M_1	Monocyte, connective tissue macrophages, peritoneal macrophages	(Dr. Radzun and Prof. Parwaresch, Dept. of Pathology, University of Kiel, FRG)
Ki M_2	Monocyte, macrophages	
Ki M_6	Monocyte, macrophages	

tients two and more biopsies were studied. Immunohistological analysis on cryostat sections was performed on 14 patients, and 18 biopsies with the histological diagnosis of graft rejection were carried out. Six patients were treated with conventional immunosuppression (azathioprine and prednisolone) and eight patients were treated with CyA and low-dose methylprednisolone.

Histological Analysis

Biopsy material was fixed with 10% buffered formalin and embedded in JB-4 resin (Dupont); 2-µm-sections were stained with hematoxylin and eosin, Giemsa, Silver stain, and PAS.

Immunohistological Methods

Second biopsies in the case of needle biopsy or one-half of the cylinder of drill biopsies were immersed in Jung embedding for cryostat sections (Jung GmbH, Heidelberg) and shock frozen in liquid nitrogen. Cryostat sections of this material were prepared for the immunohistological study as published elsewhere in detail (Stein et al. 1980). The specific monoclonal antibodies and heterologous antisera, which were used in the primary step, are listed in Table 3.

As a second step the sections were incubated with a peroxidase-conjugated rabbit antibody to mouse immunoglobulins (Dako) for detection of the specific-bound primary monoclonal mouse antibody. This was followed by a third step using peroxidase-conjugated goat antibody to rabbit immunoglobulins (Medac), subsequently revealing the enzyme label by incubating with a substrate, which consisted of diaminobenzidine tetrahydrochloride plus hydrogen peroxide dissolved in Tris buffer. Slides were counterstained with hematoxylin and mounted in glycerol gelatin. A more detailed description is published elsewhere (Stein et al. 1980; Bolton et al. 1983). When heterologous antisera were used as the primary antibody, the peroxidase-antiperoxidase technique according to Sternberger was performed (Sternberger 1979).

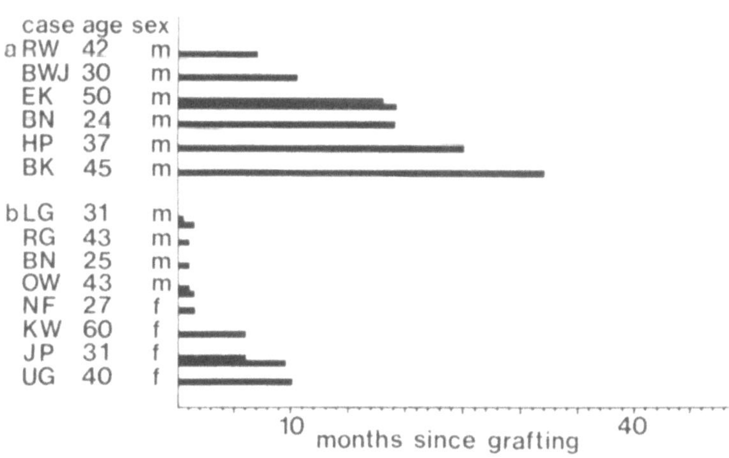

Fig. 1. Periods between grafting and biopsy (*a*) with immunosuppression by azathioprine and prednisolone and (*b*) with immunosuppression by CyA and methylprednisolone

Table 4. Immunological identification of interstitial infiltrates in cellular rejection episodes under conventional immunosuppression and cyclosporin A treatment. The percentage is calculated as immunoreactive cells from total counterstained interstitial infiltrate. The succession of cases is the same as in Fig. 1

Case	Type of rejection	Leu_{3a}/ Leu_{2a}	T cell (Leu_4)	B cell (To 15)	Macro- (Ki $M_{1,6}$)	Special features
R. W., 42, m	Late cellular and vascular rejection	2:1	75	10	15	
B. W.-J., 30, m	Late cellular and vascular rejection	2:1	70	10	20	
E. K., 50, m	Acute cellular rejection	1:5	80	5	15	
Same patient, follow-up after 5 weeks	Less acute cellular rejection	1:1	75	5	20	Beginnings of interstitial fibrosis
B. N., 24, m	Late cellular and vascular rejection	3:1	75	10	15	Interstitial fibrosis
H. P., 37, m	Late cellular and vascular rejection	4:1	70	5	25	
B. K., 45, m	Late cellular and vascular rejection	2:1	70	10	20	
L. G., 31, m	Acute cellular rejection	3:1	70	10	20	Many Leu_{2a}-positive cells in vessels
Same patient, follow-up after 4 weeks	Acute cellular rejection	3:1	70	n.d.	20	Interstitial fibrosis
R. G., 43, m	Acute cellular rejection	4:1	70	10	20	
B. N., 25, m	Acute cellular rejection	1:1	70	n.d.	20	5% of cellular infiltrate is Leu_7^+ cells
O. W., 43 m	Acute cellular rejection	8:1	60	20	30	Macrophage granulomas around tubules many eosinophil granulocytes
Same patient, follow-up after 2 weeks	Acute cellular rejection	3:1	60	20	30	Increase in peritubular macrophages granulomas
N. F., 27, f	Acute cellular rejection	2:1	65	20	15	Macrophage granulomas at tubules
K. W., 60, f	Acute cellular rejection	3:1	60	15	25	
J. P., 31 f	Acute cellular rejection	2:1	60	20	10	Many eosinophil granulocytes

Case	Type of rejection	Leu_{3a}/Leu_{2a}	T cell (Leu_4)	B cell (To 15)	Macro- (Ki $M_{1,6}$)	Special features
Same patient, follow-up after 17 weeks	Late cellular rejection	1.5:1	65	5	25	Macrophage granulomas at tubules, 5% of cellular infiltrate Leu_7^+ cells
U. G., 40, f	Acute cellular rejection	2:1	60	15	20	5% of cellular infiltrate is Leu_7^+ cells

n.d., not done

The periods between grafting and biopsy are shown for both treatment groups in Fig. 1. The percentages of immunohistologically identified cell types in the infiltrates are given in Table 4. These results will now be shown in relation to the histological diagnosis and treatment schedule.

Results

Acute Cellular Rejection in Patients with Conventional Immunosuppression

Only one patient and two biopsies were investigated immunohistologically which showed the typical features of acute cellular rejection. The interstitial infiltrate is mainly found in the periglomerular and intertubular areas as well as around small- and medium-sized blood vessels. Many activated lymphoid cells are seen. Eighty percent of infiltrating cells were T cells (Leu_4^+, Leu_1^+). The ratio of Leu_{3a}^+/Leu_{2a}^+ T cells was 1:5. Five weeks after treatment of rejection it was 1:1. Macrophages are mainly accumulated in the periglomerular intertubular areas as seen with the monoclonal antibodies Ki M_1, Ki M_2, and Ki M_6. Some T-lymphocytes of Leu_{2a}^+ phenotype also infiltrate the medium-sized arterial vessels and the subendothelial region of small capillaries. The relative amount of Leu_{3a}^+ and Leu_{2a}^+ cells differs to some extent in different regions. The predominance of Leu_{2a}^+ cells is always clear cut. The distribution of the HLA-DR$^+$ reaction is far more widespread than in the normal kidney. Tubular epithelial cells, vascular endothelial cells, interstitial dendritic cells, macrophages, and activated T cells exhibit HLA-DR reactivity as judged from the distribution of HLA-DR$^+$ immune reactions.

Late Cellular Rejection in Patients with Conventional Immunosuppression

Five patients and five biopsies which were studied immunohistologically were in this group. The interstitial infiltrate is more pronounced in the deeper cortex, mainly around the larger blood vessels. Tubular atrophy and some interstitial fibrosis was always found, showing a predilection for the distal convoluted parts of the renal tubules. The main features are the arterial vessel changes. A considerable thickening of the arterial intimal layer leads to a subcomplete obstruction

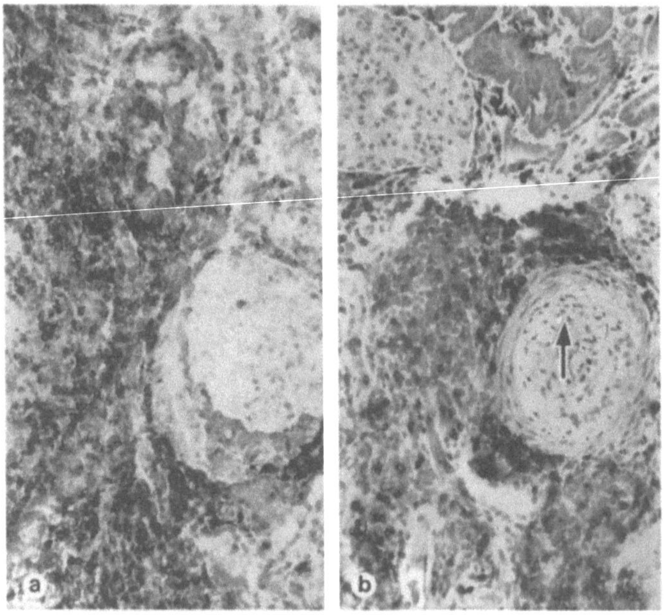

Fig. 2a, b. Late cellular and vascular rejection after conventional immunosuppression. **a** Leu_4-reactive cells; **b** Leu_{2a}-reactive cells: The infiltrate mainly consists of Leu_{2a}-negative Tcells. Only a minority show the Leu_{2a} antigen. Small artery completely obstructed by intimal proliferation (*triangle*)

of medium-sized and larger arterial vessels at the corticomedullary junction and the lobular arteries. Immunohistologically the main infiltrating cell type is again the T cell (70% Leu_+ cells). Leu_{3a}^+ cells are more numerous than Leu_{2a}^+ cells, the ratio being 2:1 up to 4:1 (Fig. 2). Macrophages are seen in the cortical intertubular area showing dendritic or plump cellular projection. The immunological phenotype is that of mature tissue macrophages (Ki M_1^+, Ki M_2^+, Ki M_6^+). As in the acute rejection episodes renal tubular epithelial cells and some T-lymphocytes exhibit HLA-DR reactivity in addition to the interstitial dendritic cells, macrophages, and vascular endothelial cells. The thickened intimal layer of larger arteries contains some Leu_{2a}^+-activated T cells, macrophages, and an increase in endothelial cells and subendothelial myofibroblasts. It thus appears as a consequence of T-cell-mediated endothelial damage in the larger arteries (Fig. 3).

Acute Cellular Rejection in CyA-Treated Patients

This group consisted of 8 patients and 11 biopsies, studied immunohistologically. The interstitial infiltrate showed a predilection for the intertubular areas among the distal convoluted tubules and the periglomerular region. Small blood vessels of the cortex were infiltrated and often at least partially destroyed. The tubules were infiltrated by activated lymphocytes (Fig. 4). Tubular epithelial cells were destroyed. Mitoses in tubular epithelial cells were frequent. Among lymphocytes and macrophages eosinophils were also seen. Immunohistologically the main in-

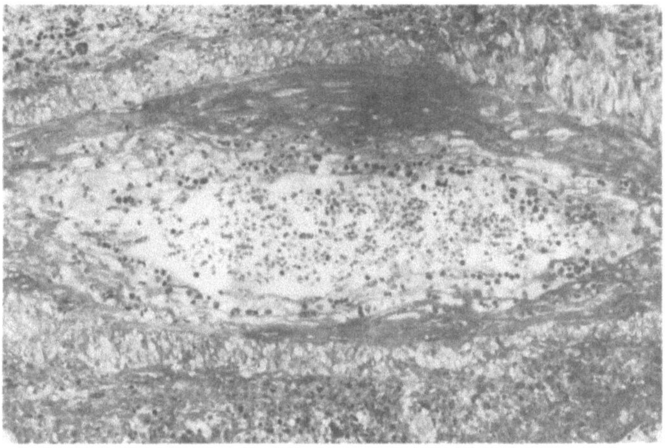

Fig. 3. Large artery in acute rejection showing infiltration of endothelial and subendothelial layer by lymphoid cells and macrophages. Early intimal proliferation. Giemsa stain, ×200

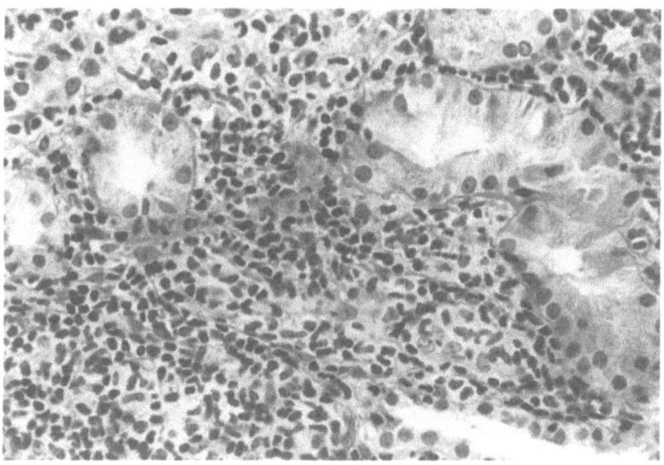

Fig. 4. Strong infiltration and destruction of distal cortical tubules in acute cellular rejection under CyA treatment

filtrating cells were T cells although their relative amount was lower (60%–70% Leu_4^+ T-lymphocytes). Leu_{3a}^+ cells were much more numerous than Leu_{2a}^+ cells. The ratio of Leu_{3a}^+/Leu_{2a}^+cells was 2:1 up to 8:1 (Fig. 5). In addition to T-lymphocytes many macrophages were found. They constitute up to 30% of infiltrating cells. Their immunological phenotype is that of mature macrophages (Ki $M_{1'2'6}^+$). Interestingly they form granulomas consisting of activated rounded macrophages around infiltrated and damaged renal tubules (Fig. 6). Equally B-lymphocytes and plasma cells were always found and accounted for 20% or slightly more of infiltrating cells. The main tissue damage was seen in the distal cortical tubules

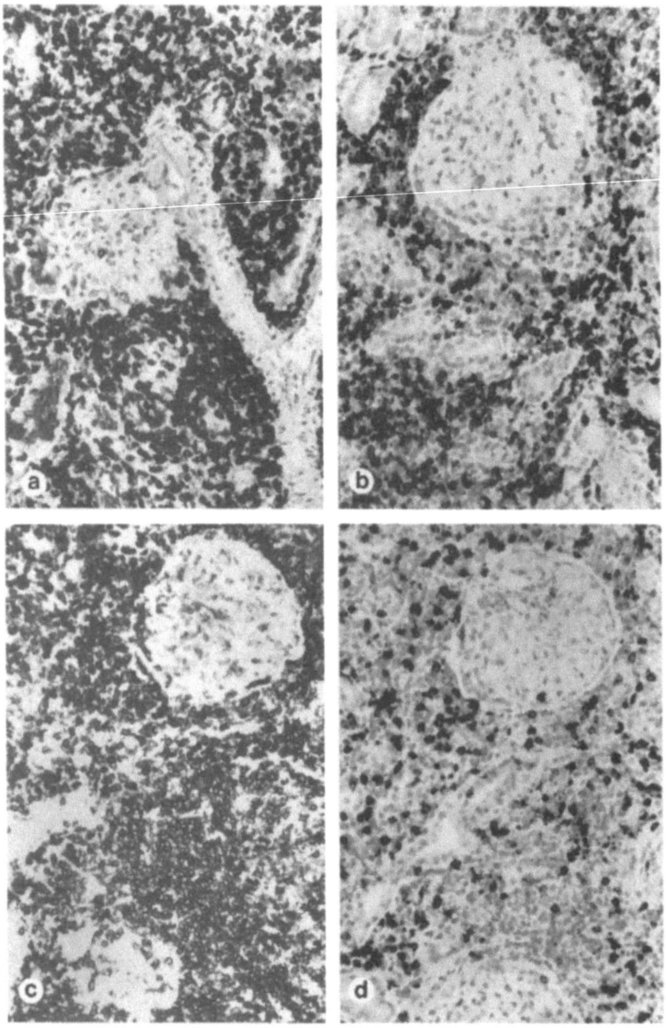

Fig. 5 a–d. Acute cellular rejection under CyA treatment: demonstration of phenotypes of the infiltrating cells a–d in the same case: **a** Leu_4 (pan T); **b** To 15 (pan B); **c** Leu_{3a} (T helper/inducer); **d** Leu_{2a} (T cytotoxic/suppressor), ×200

and in the small- to medium-sized arterial vessels. HLA-DR is often expressed on tubular epithelial cells to a considerable extent.

Discussion

For obvious reasons allograft rejection of the human cadaveric kidney is always modified by more or less specific attempts at its prevention or treatment. Thus, in contrast to the unmodified situation in experimental organ transplantation in rodents therapeutic approaches may cause a selection of effector mechanisms

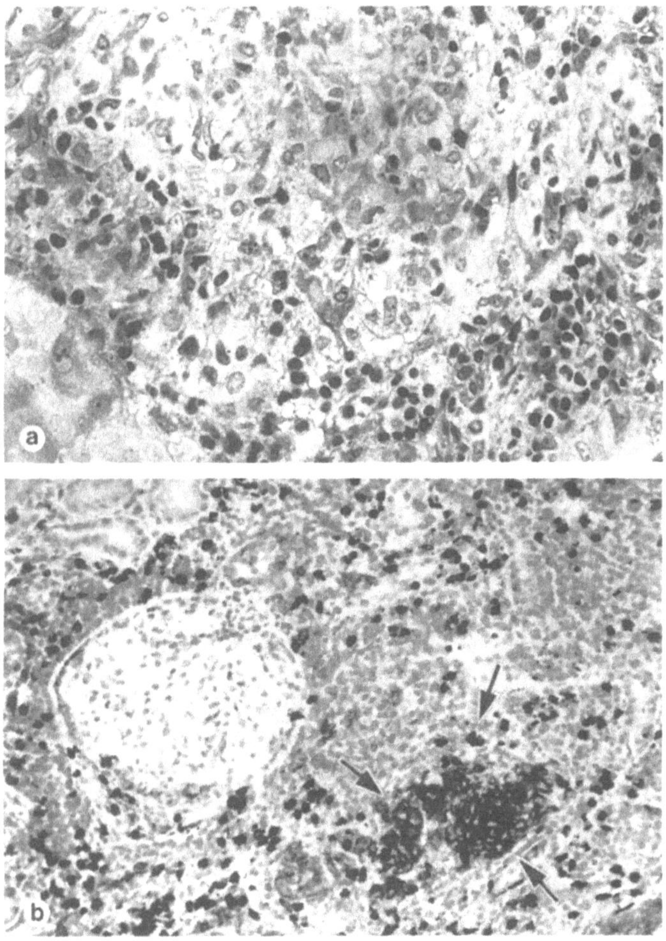

Fig. 6a, b. Macrophage granulomas in acute cellular rejection under CyA treatment: **a** Macrophages infiltrating and destroying renal tubules. Giemsa, ×500 **b** Immunological reaction of Ki M_1 antigen. Dispersed macrophages in the interstitial infiltrate and reacting granuloma (*arrows*)

leading to graft destruction from the well-known orchestra of different mechanisms in allograft rejection (Mason 1983; Ascher et al. 1983; Gurley et al. 1983). Changes in therapeutic schedules may lead to differences in the course of time, the histological expression, and the intragraft target of rejection, which are difficult to judge in their clinical and pathological implications. Recent reports of the early cellular rejection of human renal allograft have shown that under azathioprine-prednisolone treatment the acute rejection is mediated by T cells exhibiting the phenotype of the cytotoxic/suppressor subtype acting most likely as cytotoxic T cells (Platt et al. 1982; Hancock 1984) and by macrophages. Only a small proportion of infiltrating T cells exhibit the phenotype of T-helper cells. In view of cross-reactivities of antibodies to T-helper cells with macrophages, however, the number of T-helper cells is difficult to evaluate. Other immuno phenomena during renal allograft rejection especially antibody-mediated damage or DTH reactions seem to play a minor role, although their relative importance is hard to evaluate in situ (Hancock 1984).

The present study shows that in the late cellular rejection Leu_{3a}^+ cells predominate among the infiltrating-cell types and in the azathioprine-prednisolone group as well. Only one acute cellular rejection was investigated in which Leu_{2a}^+ cells proved to be the main type of infiltrating cell. The distribution of the inflammatory infiltrate is different in acute and late cellular rejection crisis. In the acute phase lymphoid cells are localized in the interstice around small vessels and the distal convoluted tubules of the renal cortex. B-cells in this group never play a significant role. Total numbers rarely exceed 10% of the interstitial infiltrate. Macrophages are present in a significant number up to 20% of cells. Damage of these distal convoluted tubules is a major effect since in the later stages interstitial fibrosis and dropout of whole areas of tubules are found in the regions where interstitial infiltrates are seen in the acute phase. This effect appears not to be only the consequence of vascular injury to endothelial cells, for infiltration of activated lymphoid cells and cytological signs of cell death are seen in this part of the renal tubules. It well may be that the expression of HLA-DR and also of class I antigens by renal tubules in allograft rejection are of importance for these findings. In the late cellular rejection the main damage is seen in the larger arteries of the corticomedullary junction. Although endothelial cell swelling and cytolysis of endothelial cells is also seen in the early rejection phase, the pathological changes in the late cellular rejection are far more pronounced.

The immunohistological study shows that the arterial occlusion which is so characteristic for this type of rejection is caused by endothelial proliferation and accumulation of macrophages. Only few lymphoid cells are found in the inner part of the arteries although many cells are accumulated in the surrounding and outermost part of the arterial wall.

The immunological phenotype (Ki M_1^+, Ki M_2^+, Ki M_6^+) and in particular the amount of Ki M_6, a lysosomal antigen (Radzun and Parwaresch 1983), expression show that most macrophages are mature. But they are not as activated as seen in early rejection in the CyA-resistant rejection.

Cyclosporin-A-resistant rejection has been observed mainly in the early post-transplant period. In contrast to most of the reports dealing with the composition and phenotypes of the interstitial infiltrate after treatment with azathioprine and prednisolone, the main cell type is the Leu_{3a}^+ cell. The ratio of Leu_{3a}^+/Leu_{2a}^+ cells varies from 2:1 to 8:1. As judged from the cell size and the corresponding histological reaction many of these cells are activated lymphoid cells and immunoblasts. In addition the amount of macrophages is significantly higher than in the patients treated with azathioprine and prednisolone. In the most severe phases of rejection, granulomas consisting of accumulation of activated macrophages are formed around distal convoluted tubules. Some eosinophil granulocytes are seen. The number of B-lymphocytes although small is higher than reported in the other treatment group. The composition of the interstitial infiltrate in the early CyA-resistant cellular rejection shows all the features of DTH reaction rather than a cytotoxic T-cell response as seen in early rejection of azathioprine-treated patients. Earlier studies on interstitial infiltrates in transplant kidneys treated with CyA have been attributed to CyA nephrotoxicity. Platt et al. (1983) reported a predominance of OKT_4^+ cells and a small content of macrophages in CyA nephropathy compared with acute rejection. The same group, however, also showed that vasculitis and "tubulitis" are manifestations of interstitial inflammation attributed to rejection rather than to CyA nephropathy (Sibley et al. 1983). All cases included in our study exhibited strong interstitial inflammation mainly focused on small vessels and renal tubules as is characteristic in acute rejection. The "ex juvantibus" method of distinction between CyA nephrotoxicity and rejection in the study of Sibley et al. (1983) (prednisone sensitivity = rejection; response to reduction of CyA dosage = CyA nephrotoxicity; inefficacy of both = irreversible rejection) appears questionable in view of the selectivity of CyA action and prednisolone treatment as well as of the lack of similar features in CyA-treated bone marrow transplant patients (Shulman et al. 1981). The modification of inflammatory infiltrates in kidney allograft rejection has also been shown experimentally (Nemlander et al. 1982).

Recent studies in rodents have proved the importance of DTH reactions for the rejection of skin allografts (Loveland et al. 1981). Similar mechanisms have also been postulated for the experimental renal allograft rejection (Gurley et al. 1983). For discussion of these findings, however, see Mason (1983). The present study shows that DTH effector mechanisms may well be of major importance in the CyA-resistant rejection crisis.

Cyclosporin A is able to prevent the effector function of primary T-helper cells but not the priming of helper cells. Adoptive transfer of primed T-helper cells to CyA-treated animals gave levels of help comparable to that in control animals (Klaus and Kunkl 1983). The early appearance of CyA-resistant acute cellular rejection and its cellular composition indicate the presence of primed alloreactive T-helper cells in the host which are not detected by the usual pretransplant cross-matches. These particular cells are most likely to lead to DTH reaction in the renal allograft. It is of interest that Hall et al. (1978) found the memory cells mediating second set heart graft rejection among long-lived, nonrecirculating, small Ig^- lymphocytes. Thus, cross-matches using peripheral blood lymphocytes will not necessarily show this type of memory. The effector phase of this CyA-resistant rejection is determining renal tubular damage that presumably is induced by activated macrophages, and by endothelial damage in small vessels. Only in later stages are vascular lesions of larger arteries observed as seen in the late cellular rejection of azathioprine-treated patients. Thus, although the results of CyA as an immunosuppressive drug are superior to those of conventional immunosuppression, the rate of graft loss due to acute rejection seems to be higher and the incidence of rejection episodes is not reduced. Graft losses were most often reported when the CyA levels were persistently low in blood. The majority of rejection episodes, however, are less severe under CyA and more easily handled (Merion et al. 1984). Prompt success of treatment of these rejection episodes by intravenous methylprednisolone may also speak in favor of an intragraft DTH reaction as expected from the composition of the interstitial infiltrate.

References

Ascher NL, Hoffman R, Hanto DW, Simmons RL (1983) Cellular events within the rejecting allograft. Transplantation 35:193–197
Bolton EM, Thompson JF, Wood RFM, Morris PJ (1983) Immunoperoxidase staining of fine-needle aspiration biopsies and needle core biopsies from renal allografts. Transplantation 36:728–731
Gurley KE, Lowry RP, Forbes RDC (1983) Immune mechanisms in organ allograft rejection. T helper cells delayed-type hypersensitivity, and rejection of renal allografts. Transplantation 36:401–405
Hall BM, Dorsch S, Roser B (1978) The cellular basis of allograft rejection in vivo. The nature of memory cells mediating second set heart rejection. J Exp Med 148:890–902
Hancock WW (1984) Analysis of intragraft effector mechanisms associated with human renal allograft rejection: immunohistological studies with monoclonal antibodies. Immunol Rev 77:61–84
Klaus GGB, Kunkl A (1983) Effects of cyclosporine on the immune system of the mouse. II. Cyclosporine inhibits the effector function of primary T-helper cells, but not helper cell priming. Transplantation 36:80–84
Loveland BE, Hogarth PM, Ceredig R, McKenzie IFC (1981) Cells mediating graft rejection in the mouse. I. Lyt_{-1} cells mediate skin graft rejection. J Exp Med 153:1044–1057
Mason DW (1983) The mechanism of allograft rejection – progress and problems. Transplant Proc 15:264–268
Meadows R (1978) Renal histopathology. A light, electron, and immunofluorescent microscopy study of renal disease, 2nd edn. Oxford University Press, Oxford
Merion RM, White DJG, Thiru S, Evans DB, Calne RY (1984) Cyclosporine: five years' experience in cadaveric renal transplantation. N Engl J Med 310:148–154
Nemlander A, Soots A, von Willebrand, Tallqvist G, Häyry P (1982) Effect of cyclosporine A on the in situ inflammatory response of rat renal allograft rejection. Scand J Immunol 16:91–102
Platt JL, Le Bien TW, Michael AF (1982) Interstitial mononuclear cell populations in renal graft rejection. Identification by monoclonal antibodies in tissue sections. J Exp Med 155:17–30
Platt JL, Ferguson RM, Sibley RK, Gajl-Peczalska KJ, Michael AF (1983) Renal interstitial cell population in cyclosporine nephrotoxicity. Identification using monoclonal antibodies. Transplantation 36:343–346
Porter KA (1967) Rejection in treated renal allografts. J Clin Pathol 20:518–534
Radzun HJ, Parwaresch MR (1983) Differential immunohistochemical resolution of the human mononuclear phagocyte system. Cell Immunol 82:174–183
Sibley RK, Rynasiewicz J, Ferguson RM, Fryd D, Sutherland DER, Simmons RL, Najarian RS (1983) Morphology of cyclosporine nephrotoxicity and acute rejection in patients immunosuppressed with cyclosporine and prednisone. Surgery 94:225–234
Shulman H, Striker G, Deeg HJ, Kennedy M, Storb R, Thomas ED (1981) Nephrotoxicity of cyclosporine A after allogenic marrow transplantation. Glomerular thromboses and tubular injury. N Engl J Med 305:1392–1395
Stein H, Bonk A, Tolkdsdorf G, Lennert K, Rodt H, Gerdes J (1980) Immunhistologic analysis of the organization of normal lymphoid tissue and non Hodgkin's lymphomas. J Histochem Cytochem 28:746–760
Sternberger LA (1979) Immunocytochemistry, 2nd edn. Wiley, New York, p 104

Immunological Monitoring of Heart Transplant Patients: Clinical and Experimental Studies *

M. H. Yacoub, M. L. Rose, J. H. Cox, J. A. Gracie, P. M. Chisholm, R. J. Griffin, and *M. I. Coles*

Introduction

Between January 1980 and October 1983, 60 orthotopic and 17 heterotopic cardiac transplants were performed at Harefield Hospital. During this period three regimens of immunosuppression were used. Prior to September 1982, patients were treated by antithymocyte globulin (ATG), corticosteroids and azathioprine. Between January 1980 and July 1981, high-dose steroids were used, and from July 1981 to September 1982 low-dose steroids were used. From this date onwards, cyclosporin (CyA), azathioprine and minimal or no steroid therapy has been used.

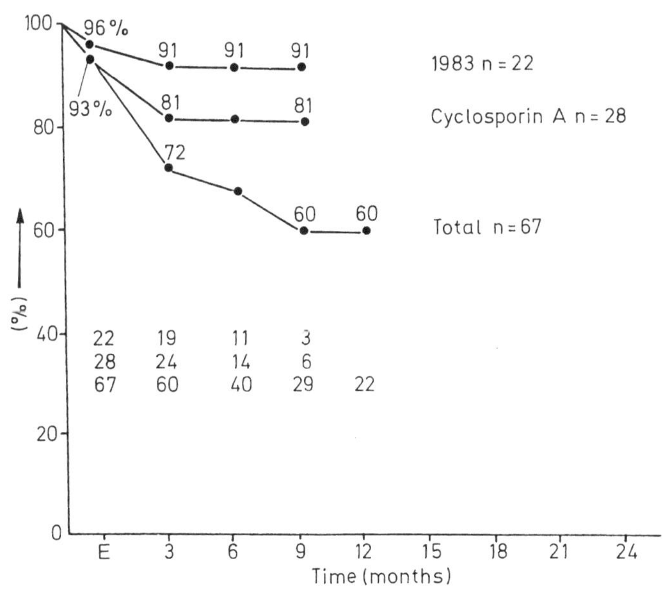

Fig. 1. One-year survival of cardiac transplant patients at Harefield Hospital

* This work was supported by a grant from the British Heart Foundation (81/86) and the Harefield Heart Transplant Trust

Microsurgical Models in Rats
for Transplantation Research
Eds. A. Thiede et al.

The 1-year survival of patients receiving CyA is 81% (Fig. 1). The overall actuarial 1-year survival is 60% (Fig. 1). The 8-month actuarial survival rate of patients operated upon this year (1983) is 91% (Fig. 1).

In this series heterotopic transplantation with or without revascularization of the native heart was used in patients who satisfied the following criteria:

1. Salvagable areas of myocardium with normal or revascularable coronary arteries
2. Absence of significant valve disease
3. Absence of prostethic valves
4. Absence of clots in the ventricle
5. Presence of pulmonary hypertension [pulmonary vascular resistance (PVR) $\geqq$ 8 units]

Further improvement in the results of cardiac transplantation depends on the early diagnosis of rejection and effective specific antirejection therapy based on a thorough understanding of the process of acute rejection. Microsurgical techniques provide a very useful model for the study of acute cardiac allograft rejection in rats. The clinical usefulness of experimental studies, however, depends on the ability to link them to specific clinical problems and on critical comparison between parallel clinical and experimental results. The following clinical and experimental studies were designed to discover new methods for detecting rejection and to understand the cellular interactions within the allograft which lead to graft destruction.

Monitoring of T-Cell Subpopulations in Peripheral Blood

Clinical

The total number of T cells in the peripheral blood of cardiac transplant patients has been followed using the E-rosette technique. This has proved to be useful in monitoring the efficacy of ATG immunosuppressive therapy but, in our experience, has been found to be an overall poor predictor of rejection episodes. The advent of monoclonal antibodies (MABS) which identify subpopulations of T-lymphocytes, helper/inducer T cells (T4), and suppressor/cytotoxic cells (T8), has stimulated investigations into the role of T-cell subpopulations in causing allograft rejection. The first study from this centre investigated the ratios of T4/T8 cells in the blood of five cardiac transplant patients who had been operated upon 0-4 months earlier and were receiving corticosteroids/azathioprine immunosuppression. Blood samples were taken 1 or 2 days before and on the same day as cardiac biopsy. The lymphocytes were separated using the density-gradient Ficoll-Hypaque technique. The ability of the cells to bind MAB against T-cell subpopulations was tested using fluoresceinated MABs against OKT4- and OKT8-positive cells (Ortho Diagnostic Ltd.). The ratios of helper (T4) to suppressor cells (T8) were plotted against the grade of rejection found in the biopsy. There was some correlation between increasing T4/T8 ratio and increasing severity of rejection (Yacoub et al. 1981).

In a more recent study we have investigated the ratios of the T4/T8 subpopulations in a different group of patients, i.e., those who are now 6–12 months post-transplant and who were receiving azathioprine corticosteroid immunosuppression (conventional immunosuppression) and those who were 4 months post-transplant receiving CyA immunosuppression (in combination with small does of corticosteroids and azathioprine). In this study the blood samples were taken on the same day as the biopsy.

The technique used to assess the number of T cells and T-cell subpopulations was similar to that described above except that MABs of the Leu series (Becton Dickinson Co. Ltd.) were used. Leu 4 binds to all T cells, Leu 2a to suppressor/cytotoxic T cells and Leu 3a to helper/inducer T cells. All groups of patients were found to have reduced numbers of T cells compared with the normal controls (Table 1). The decrease in numbers affected both cells of suppressor/cytotoxic phenotype and those of the helper/inducer phenotype. However, the cells of the helper/inducer phenotype were affected to a greater degree than those of the suppressor/cytotoxic phenotype. All patients were found to have a low T4/T8 ratio compared with normal controls. Patients receiving CyA had more T cells and a higher T4/T8 ratio than did patients receiving conventional immunosuppression. There was no significant difference in the blood T-cell profile of patients whose biopsies were showing histological signs of rejection to those who were not.

Experimental

Monoclonal antibodies against T-cell subpopulations in the rat are now available (Mason et al. 1980) and the subpopulations thus distinguished are functionally equivalent to human T-cell subpopulations. The MABs used in the following study and their equivalent for detecting human lymphocytes are described in Table 2.

The availability of such MABs makes possible the monitoring of T-cell subsets in cardiac-allografted rats, and thus enables investigation of:

1. Whether the onset of untreated rejection is heralded by significant changes in the ratios of T-cell subsets found in the peripheral blood
2. Whether drug treatment alone produced significant changes in the ratios of peripheral blood T cells which express different phenotypes

The phenotypes of peripheral blood lymphocytes from CyA-treated, as well as from untreated, rats grafted with either an allograft or isograft were compared.

Five rats of the highly inbred PVG/C strain were grafted with a syngeneic heart (mean survival time >70 days) and 12 rats were grafted with an allogeneic A0 heart (mean survival time 10 ± 1 day) using standard microsurgical techniques. Three of the isografted rats and eight of the allografted rats were treated daily with CyA, 15 mg/kg intramuscularly, from the day of grafting (day 0) until day 10. In CyA-treated rats the heart graft maintained strong ventricular contraction for >70 days.

Rats were bled before transplantation; 1, 3, 5, 7, 9, and 14 days after transplantation, and weekly thereafter. Lymphocytes were separated by density-gradient centrifugation on Ficoll-Hypaque. Prepared cytospins were incubated with the

Table 1. Binding of monoclonal antibodies to mononuclear cells in peripheral blood of cardiac transplant patients and normal controls

Group	No. of Patients	Histology	% cells binding MAB Mean (SD)			T4/T8 ratio	No. cells binding MAB $\times 10^6$/ml Mean (SD)		
			Leu 4 (T3)	3a (T4)	2a (T8)		Leu 4 (T3)	3a (T4)	2a (T8)
Normal	14		57.7 (10.4)	41.6 (7.3)	26.6 (5.2)	1.7	0.87 (0.25)	0.70 (0.21)	0.44 (0.15)
Immunosuppression	4	R	60.3 (10)	29 (7.2)	33.7 (10.3)	0.9	0.36 (0.24)	0.12 (0.007)	0.17 (0.17)
	13	NR	50.4 (28.1)	25 (13.1)	34 (16.8)	0.9	0.39 (0.35)	0.12 (0.11)	0.21 (0.21)
Cyclosporin	14	R	53.3 (10.4)	33.4 (7.7)	32.1 (9.2)	1.1	0.6 (0.4)	0.25 (0.24)	0.21 (0.21)
Immunosuppression	14	NR	53 (18.7)	34.3 (8.6)	33 (10.7)	1.2	0.46 (0.25)	0.27 (0.19)	0.28 (0.23)

R, biopsy showed histological signs of rejection; NR, biopsy did not show histological signs of rejection

Table 2. MABs used to identify lymphocyte subpopulations in rats and humans

Designation of rats[a]	Designation of humans			Specificity
	Standard	Ortho[b]	Becton-Dickinson[c]	
W3/13	T3	OKT3	Leu 4	All T cells
W3/25	T4	OKT4	Leu 3a	Helper/inducer
MRC 0x8	T8	OKT8	Leu 2a	Suppressor/cytotoxic

[a] From Sera-Lab Ltd
[b] From Ortho Diagnostics Ltd
[c] From Becton-Dickinson Monclonal Centre Ltd

monoclonal antibodies (MABs) W3/13 (Pan T cell), W3/25 (T helper), MRC 0x8 (T suppressor/cytotoxic) and MRC 0x6 (B-lymphocytes) (kindly donated by Dr. A. Williams, MRC Cellular Immunology Unit, Oxford). Positive (MAB binding) cells were detected by incubation with horse radish peroxidase (HRP)-conjugated rabbit anti-mouse followed by HRP-conjugated goat anti-rabbit (Miles-Yeda Ltd.). After washing, the peroxidase was revealed by reacting with diaminobenzidene tetrahydrochloride (Sigma). At least 200 cells in each of two cytospins were scored by two independent observers.

The results are expressed as the percentage positive lymphocytes which bound the different MABs. W3/13-positive polymorphonuclear leucocytes and 0x6 and W3/25-positive macrophages, identified by morphological criteria, were excluded. In all cases changes in the percentage positive lymphocytes were reflected by changes in the absolute numbers of lymphocytes in the peripheral blood. The results are presented in Fig. 2.

There was an early drop in the proportion of total T cells (W3/13) in allograft and isografted rats whether untreated or CyA treated. This may have been due to operative trauma. In the two CyA-treated groups and in the isografted untreated rats the porportion of total T cells was elevated compared with ungrafted rats at 3, 5, and 7 days after grafting. In the allografted untreated rats the proportion of total cells was not significantly different from normal up to day 7 but was significantly depressed on days 9 an 14.

There was no significant difference in any of the groups in the proportion of T-helper (W3/25) cells from day 0 up to 3 days after grafting; the proportion of T-helper cells in untreated isografted rats did not change up to 3 weeks after grafting. In CyA-treated rats, however, irrespective of the presence of an allograft or isograft, there was an increase in the proportion of T-helper cells in the blood. This CyA effect has disappeared by day 14, i.e., 4 days after stopping CyA treatment. In untreated allografted rats the proportion of T-helper cells declined from day 5 onwards and by day 14 was significantly depressed.

In untreated allografted or isografted rats there was no difference from normal in the proportion of T suppressor/cytotoxic (0x8) cells in the peripheral blood at any time after grafting. However, in CyA-treated rats again, whether allografted or isografted, there was a significant increase in the proportion of these cells at all times tested.

There was an increase in the proportion of B-lymphocytes (0x6-positive cells) in untreated rats with an allograft from day 5 until graft rejection. In contrast, in untreated rats with an isograft, the proportion of 0x6-positive lymphocytes was reduced throughout the period tested. Unlike W3/25 and 0x8 there was no evidence of a drug effect, i.e., in neither allografted nor isografted CyA rats was the proportion of 0x6-positive cells different from normal.

It is concluded from these studies that in patients changing ratios of T-cell subsets in peripheral blood may have some diagnostic value in a limited (and increasingly rare) group of patients, namely during the first 3 months after operation on patients receiving conventional immunosuppression. However, even in this group, results should be interpreted with caution; a number of studies have now shown that the T4/T8 ratio falls well below 1.0 during a subclinical infection with cytomegalovirus. All new patients now receive CyA immunosuppression and it

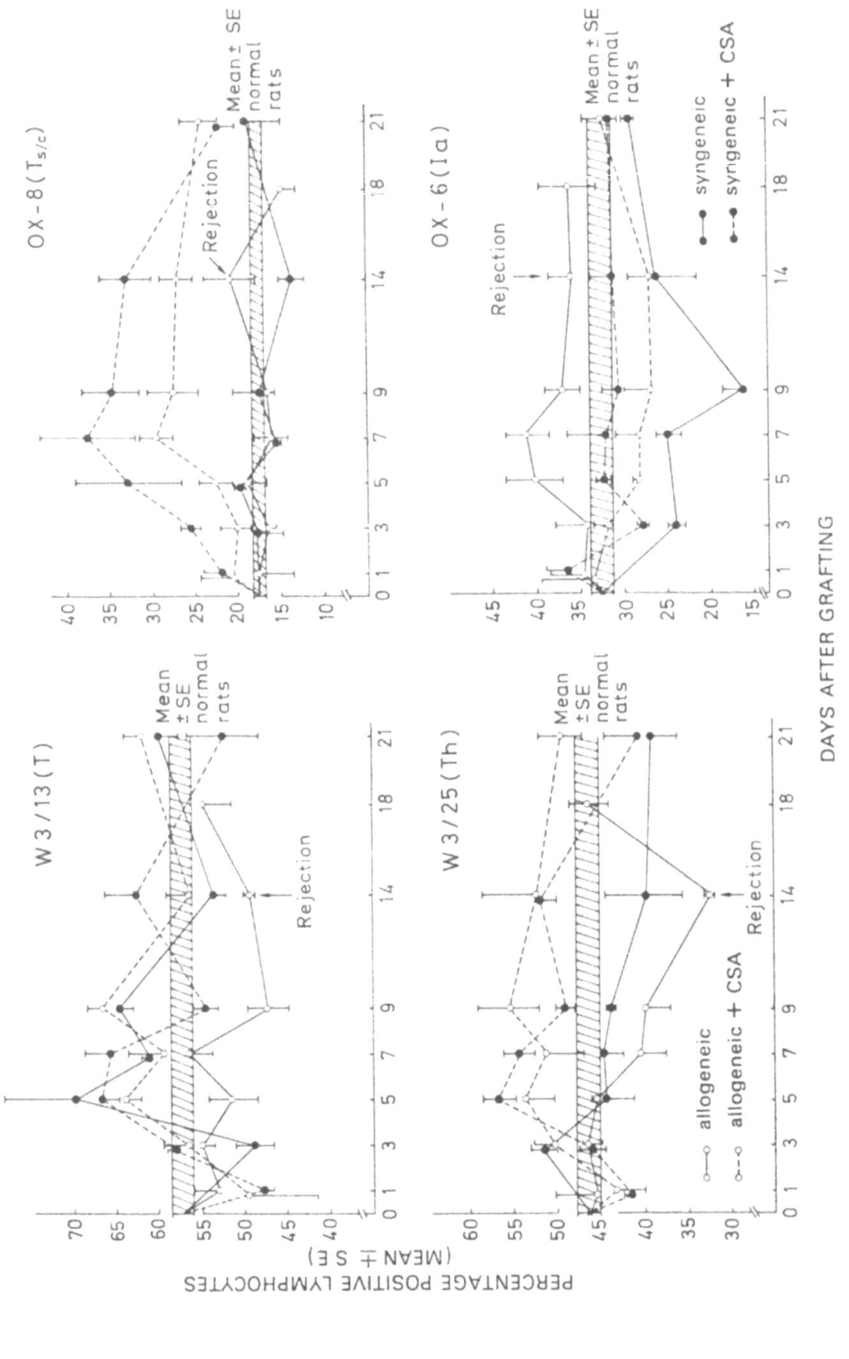

Fig. 2. Subpopulations of peripheral blood lymphocytes in cardiac-grafted rats

would be clinically very useful to find a cell marker in the blood which was diagnostic of rejection.

The experimental studies confirm the clinical observations that changing ratios of T-cell subsets are not reliably predictive or diagnostic of rejection. The experimental results also show that cyclosporin produces a substantial rise in both suppressor/cytotoxic (0 x 8) and helper (W3/25) T cells after grafting and that this is a drug effect.

Monitoring of T-Cell Subpopulations in Patient Cardiac Biopsies

Acute rejection of cardiac allografts is characterized by a mononuclear infiltrate (Billingham 1979). The nature of the cells composing the infiltrate has been investigated using monoclonal antibodies against T cells and T-cell subsets using an immunoperoxidase technique (Yacoub et al. 1983). Cryostat sections of cardiac biopsies were treated with MABs Leu 4 (all T cells), Leu 2a (suppressor/cytotoxic) and Leu 3a (helper/inducer) followed by biotinylated goat anti-mouse antibody (from TAGO, tissue culture services) and by avidin-biotin-horseradish peroxidase (ABC complex or Vecterstain from Sera laboratories). The sites of peroxidase fixation were identified using diaminobenzidene tetrahydrochloride (DAB from Sigma). The numbers of cells binding MABs were counted and results are expressed as percentage infiltrating leucocytes binding MAB (Fig. 3a, b).

Biopsies from patients receiving conventional immunosuppression which were showing histological signs of rejection contained many more T cells than biopsies not showing signs of rejection (mean $47.0 \pm 6.8\%$ of leucocytes compared with $9.0 \pm 13.5\%$ of leucocytes) and, moreover, 86% of the infiltrating T cells in rejecting biopsies were of the suppressor/cytotoxic phenotype (Fig. 3a). Accumulation of cells of suppressor/cytotoxic phenotype have also been reported during mild rejection of renal allografts (Platt et al. 1982). Biopsies from patients receiving CyA immunosuppression contained more T cells when rejecting than when not rejecting (Fig. 3b), but there was no predominance of any particular phenotype associated with rejection. It is not easy to diagnose rejection by conventional histological criteria in biopsies from patients receiving CyA because these biopsies often contain large cellular infiltrates which are not necessarily associated with other histological or clinical signs of rejection. Studies using MABs have demonstrated that many of the cells in the localized accumulation are T cells (Yacoub et al. 1983) but the fact that no particular phenotype predominates has not helped the diagnosis of rejection. We are now using other cell surface markers to identify the leucocytes present in biopsies from patients receiving CyA.

Infiltration of Heart Grafts by Indium-111-Labelled Leucocytes

Indium-111 is a powerful gamma-emitting radioisotope with a half-life of 2.8 days, which is suitable for clinical studies. The observation that it can be used at non-damaging doses to label blood leucocytes makes it of potential use as a marker for clinical investigations into the migration of blood leucocytes into allografted organs. The experimental studies described below demonstrate that in-

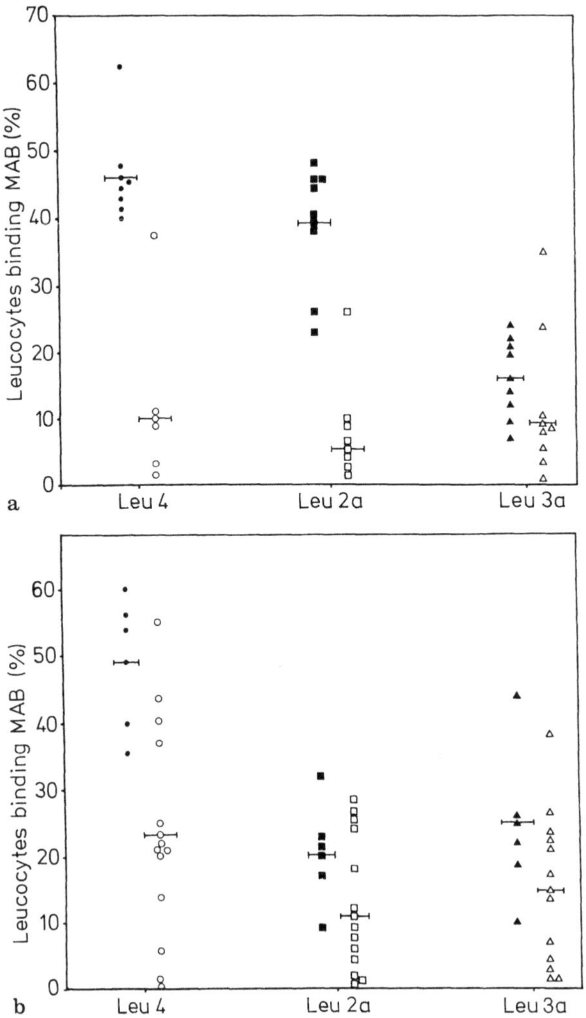

Fig. 3. a Percentage of leucocytes infiltrating cardiac biopsy which bind Leu 4 (all T cells), Leu 2a (suppressor/cytotoxic) or Leu 3a (helper/inducer) MABs. Each *point* is the result of one biopsy. *Bar* is the mean value. Patients receiving conventional immunosuppression. **b** Percentage of leucocytes infiltrating cardiac biopsy which bind Leu 4 (all T cells), Leu 2a (suppressor/cytotoxic) or Leu 3a (helper/inducer) MABs. Each *point* is the result of one biopsy. *Bar* is the mean value. Patients receiving CyA immunosuppression

dium-111-labelled lymphocytes accumulate in allografted hearts and form the rationale for the subsequent clinical investigation (p. 336).

Experiments

We have investigated lymphocyte infiltration of rat cardiac allografts using pulses of isotope-labelled lymphocytes. Syngeneic thoracic duct lymphocytes were labelled with 5–10 $\mu Ci/10^8$ cells of [^{111}In] oxine (Amersham International) for

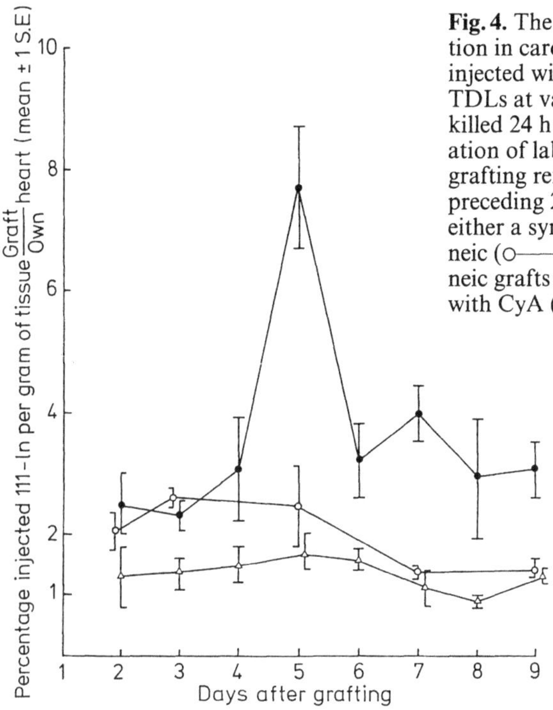

Fig. 4. The kinetics of lymphocyte accumulation in cardiac grafts. Graft recipients were injected with indium-111-labelled syngeneic TDLs at various times after grafting and were killed 24 h later. The figure given for localization of labelled cells at each period after grafting refers to the accumulation in the preceding 24 h. Graft recipients had received either a syngeneic (△———△) or an allogeneic (○———○) graft. Recipients of allogeneic grafts were either untreated or treated with CyA (○———○)

15 min at room temperature. The labelled cells were injected into PVG/C rats which had received an allogeneic (AO) or syngeneic heart graft 1, 2, 4, 6, or 8 days previously. In each experiment all recipients received the same number of labelled cells. The recipients of allografts were either untreated or injected with CyA daily, 15 mg/kg intramuscularly, from the day of grafting, the last injection of CyA being given on the same day as the labelled cells. In the allogeneic strain combination used the mean survival time was 10 ± 1 day. Syngeneic heart grafts survived indefinitely.

Twenty-four hours after cell injection the animals were killed and their tissues removed and counted for indium-111. The localization of the labelled cells in the grafts was expressed as the ratio of the percentage injected indium-111/g in the grafted heart divided by the percentage injected indium-111 in the recipient's own heart. The pooled results of 13 separate experiments involving a total of 99 grafts are included in Fig. 4.

Localization of labelled cells in isografts was only slightly greater than in the recipient's own heart and at no time after grafting was the difference between the two statistically significant. In contrast, the labelled lymphocytes accumulated in the allografts to a greater extent than in the isografts during each of the 24-h periods tested, from as early as 24–48 h after grafting. The extent of lymphocyte accumulation in each 24-h period was not the same during the course of rejection but showed a sharp increase between day 4 and 5 after grafting and then declined. Nevertheless, from day 5 onwards the accumulation in the allografts was always significantly greater than the accumulation in the isografts.

On days 2 and 3 after grafting labelled lymphocytes accumulated in the allografts of cyclosporin-treated and untreated recipients to the same extent. However, the substantial increase in the extent of infiltration which occurred in untreated allografts between days 4 and 5 was completely abolished in cyclosporin-treated animals. From day 5 until day 9 the accumulation of labelled cells in cyclosporin-maintained grafts was no greater than in syngeneic non-rejecting grafts and was significantly less than in the untreated rejecting grafts.

Clinical

Thirty-two patients have received intravenous injections of autologous leucocytes labelled with indium-111. The patients fell into three groups: (1) thirty-two cardiac transplant patients; (2) two patients with coronary artery grafts, i.e., had recently had open heart surgery; and (3) eight patients with pyrexia of unknown origin.

Patients received $(200–400) \times 10^6$ cells containing 130–400 μCi. They were scanned immediately upon injection and then for the next 40 min; during this time radioactivity was seen to disappear from the lungs and accumulate in the liver and spleen. They were scanned again at 2 h and 24 h postinjection. Patients received indium-111 leucocytes on the same day as their cardiac biopsy.

The amount of activity in the heart was estimated visually by four independent observers; it was also quantitated and compared with the radioactivity in the spleen, lungs and liver. None of the patients from groups 2 and 3 had activity in the heart region. However, 14 of the 22 transplant patients showed positive scans. There were varying degrees of activity in the heart region of these patients which did not correlate with the histological assessment of rejection. Seventy-five percent of the patients with rejecting biopsies had positive scans. However, 80% of the patients apparently not rejecting also had positive scans. It is thought this technique may reflect in a very sensitive manner the migration of host leucocytes into the allografted organs. Further studies are currently in progress to define further the causes and significance of positive scans in an attempt to establish the possible future role of this method in diagnosing and grading acute rejection in these patients, making use of both further clinical and experimental methods.

References

Billingham M (1979) Some recent advances in cardiac pathology. Hum Pathol 10:365–380

Mason DW, Brideau RJ, McMaster WR, Webb M, White KAH, Williams AF (1980) Monoclonal antibodies that define T-lymphocyte subsets in the rat. In: Kennett RH, McKearn TJ, Becktol KB (eds) Monoclonal antibodies. Plenum, New York, p 251

Platt JL, Le Bien T, Michael AF (1982) Interstitial mononuclear cell populations in renal allograft rejection. J Exp Med 155:17–30

Yacoub MH, Froud DJR, Ashraf MH, Campalani G, Forsyth AT, Pomerancel A (1981) The use of monoclonal antibodies to monitor T-cell subpopulations after human cardiac transplantation. In: Peeteis H (ed) Protides of the biological fluids, 29th colloquium. Pergamon, Oxford, p 691

Yacoub MH, Rose ML, Fraser AK, Gracie JA, Cox J, Chisholm PM (1983) T-cell phenotype characterisation in human cardiac allografts. Heart Transplant 2:259

Intestinal Transplantation: Laboratory Experience and Clinical Consequences

M. A. Hardy, Ch. Iga, and *H. Lau*

Introduction

It has been estimated that several thousand patients die each year in the United States, because of lack of small-intestine secondary to mesenteric vascular occlusion, strangulation-obstruction due to volvulus, congenital malformations, regional enteritis, abdominal trauma, or extensive involvement by tumor, malignant or benign. Although it is well known that various compensatory changes occur in the small intestine following resection, which increase the absorptive capacity of the intestine (Flint 1912; Porus 1965; Weinstein et al. 1969; Dowling 1967; Wilmore 1969), total resection of the jejunum and ileum has been incompatible with life, except for the recent increased use of total parenteral hyperalimentation.

Although progress in parenteral nutrition has been extremely rapid in the past 5 years, the problems with this approach continue to be (1) maintenance of adequate access without recurrent infection, which is frequently life-threatening; (2) the cost, which is estimated to be \$ 70 000–\$ 90 000/year per patient; (3) the necessity for frequent physiological adjustments; and (4) the inconvenience and interference with quality of life. With the advent of better immunosuppressive methods and the maintenance of recipients with total parenteral nutrition (TPN) increasing, continued investigations in the laboratory are necessary to make small-intestine transplantation clinically successful.

In this paper we will very briefly review the clinical, and particularly the experimental, work performed in intestinal transplantation and attempt to relate this to the difficulties that might be encountered in clinical small-intestinal transplantation.

Clinical Intestinal Transplantation – A Historical Glance

The few attempts at clinical transplantation of the small intestine have so far been unsuccessful (Lillehei et al. 1967; Olivier et al. 1969; Okumura et al. 1969; Fortner et al. 1972; Detterling 1971; Alican et al. 1971; Butt, personal communication), and all patients have died. The first reported intestinal transplant was done at the University of Minnesota in 1967 by Lillehei et al. (1967). The patient died 12 h

Microsurgical Models in Rats
for Transplantation Research
Eds. A. Thiede et al.

after surgery with recurrent occlusions of the venous system. Okumura's case (Okumura et al. 1969) was performed because of superior mesenteric arterial thrombosis, and the graft was rejected. Olivier et al. (1969) performed an intestinal transplant for a type of Gardner's syndrome by anastomosing mesenteric vessels of the graft to those of the host (portal venous drainage) after resection of the small bowel. The patient was immunosuppressed with corticosteroids, azathioprine, and antilymphocyte globulin (ALG) and survived for 26 days. An intestinal transplant performed by Alican et al. (1971) was the first one done in a child, a 10-year-old boy who required a total small-bowel resection for gangrene secondary to strangulation by a mesenteric band. The child was maintained by intravenous hyperalimentation, but had many problems with catheter sepsis. This patient received a one-haplotype matched 3-foot (0.9-m) segment of ileum from his mother and was immunosuppressed. He developed a perforated ulcer and was reexplored 9 days after intestinal transplantation, and the ischemic but nonrejected graft was removed. Another intestinal transplant performed by Fortner et al. (1972) in 1970 involved an allograft of 170 cm lower jejunum and upper ileum from an identically matched sister. The mesenteric vessels were anastomosed to the iliac artery and vein of the recipient and the intestine was placed in continuity with the recipient's duodenum and colon. The recipient was maintained on intravenous hyperalimentation for 15 months prior to receiving the transplant (Shils et al. 1970) and tolerated a soft diet and had formed stools 3 weeks after transplantation. This patient had an early rejection episode which may have been reversed with increased use of immunosuppression. The patient died 76 days after transplantation with *Escherichia coli* septicemia. An auxiliary jejunal pouch from the uppermost portion of the graft was constructed at the time of transplantation and was used for frequent biopsies to diagnose rejection. Other sporadic, unpublished, reports of intestinal transplantation include two cases of Detterling (1971) in very young children, one from the mother and another from a cadaveric donor; our own case where we transplanted a segment of adult intestine from a cadaver donor to an infant who died 12 h after transplantation of various medical conditions, with the graft showing no signs of rejection; and the most recent case that I am aware of, performed by Doctor K. Butt, using a segment of small intestine from a living related donor where the recipient died of various associated problems within a week of transplantation.

This brief glance at the history of intestinal transplantation has been most discouraging in that only one patient survived longer than 2 months and appeared to have function of the graft. The cases described were desperate attempts at prolonging life in patients who were usually either septic or metabolically unstable, or were not maintained adequately with parenteral hyperalimentation.

Anatomical and Technical Considerations

Arteries and Veins

A problem that has not been fully addressed is whether systemic venous drainage, rather than portal venous drainage, will have major clinical implications on intestinal absorption or hosts' immunological reactions. Clinically, only Olivier's patient had intraportal venous drainage. In other cases there have been problems with thromboses of the systemic veins. It would seem logical that any prospective recipient of intestinal transplantation should have a venogram prior to surgery. Although systemic venous drainage seems acceptable, intraportal venous drainage may offer a physiological advantage in absorption of nutrients as well as an immunological advantage, as has been suggested in transplantation of other organs (Kort et al. 1973; Sakai 1970). As this volume clearly shows, the issue of anastomoses of small vessels is no longer a major problem in expert hands. Common surgical sense will be required in eventual operations on recipients of small-intestinal allografts, since these patients will probably have undergone multiple previous procedures, may have previously compromised venous and arterial systems, and may have chronic metabolic derangement secondary to long-term parenteral hyperalimentation.

Lymphatics

The transplantation of large segments of small intestine requires division of main intestinal lymphatic channels, while transplantation of short segments requires division of smaller lymphatic channels. There are three options in the management of lymphatic channels: (1) ligation of the channels leading to lymphatic stasis and decreased rate of lymphatic regeneration; (2) leaving the lymphatic channels untied with the risk of development of chylous ascites, but with greater probability of lymphatic regeneration; and (3) reconstruction of the lymphatic channels by lymphovenous anastomoses, a most difficult and time-consuming procedure with a relatively low success rate in the best of hands. The importance of intestinal lymphatics and intestinal transplantation has been emphasized by Olivier et al. (1972). The danger of lymphatic stasis may complicate the problem of rejection in that it leads to thickening and edema of the small intestine and may have a harmful effect on the intestinal flora with the subsequent possibility of infection in an immunosuppressed host. That regeneration of lymphatics occurs within 6 weeks after intestinal transplantation in dogs has been elegantly shown by Kocandrle et al. (1966). We have previously shown (Hardy et al. 1970; Stamford and Hardy 1974), by lymphangiography, that lymphatic regeneration and subsequent fatty acid absorption occur earlier (at approximately 3 weeks following allotransplantation) when short segments of jejunum are used. More thorough studies of this problem are needed, both at the experimental and clinical levels since absorption of complex fats, as well as many medications, including cyclosporin, requires reestablishment of lymphatic connections between the donor and the recipient.

Length of Intestinal Segment and Reestablishment of Intestinal Continuity

The critical length of small intestine which allows a child to survive is approximately 20–30 cm. The length of the segment to be transplanted will depend partially on the anatomical considerations in the donor (short segments in all cases of living-related donors), but particularly on the size of the recipient. It is obvious that a small child will require a much shorter segment of adult intestine with its large absorptive surface area than will an adult recipient. The issue of whether the proximal and distal stoma should be placed in continuity with the recipient's small and large bowel has not been answered. The method proposed by Kunlin et al. (1971) to use a small auxiliary allograft pouch for repeat biopsies, while the rest of the small intestine is anastomosed to the recipient's gastrointestinal tract, seems the most attractive and most efficient way of following the morphological changes in the graft. The problems that might be encountered with the management of cutaneous enterostomies in an immunosuppressed patient discourage me us from this approach, which, initially, may appear to be the safest. Further experimental studies are required on the healing of intestinal anastomoses in the face of immunosuppression and in relation to the anatomical segments used.

Physiology

Intestinal Denervation

Total denervation of the canine small intestine, carefully studied by Ballinger et al. (1962), led to decreased absorption of fats and glucose, as well as to morphological changes which lasted from 3 to 6 weeks after denervation. Interestingly, the same group (Ballinger et al. 1964) showed that total vagotomy in dogs resulted in similar changes in the small intestine, but to a lesser degree. Our own studies with short isolated segments of jejunum, both in autotransplantation and allotransplantation (Hardy et al. 1970; Stamford and Hardy 1974; Cohen et al. 1969), showed no significant effect of denervation on the absorption of glucose, amino acids, and even some short chain fatty acids. No morphological differences were appreciated between the denervated autotransplants and the enervated Thierry-Vella loops. No significant alteration in peristalsis and small-intestinal motility has been noted in clinical or experimental allografts. Reenervation and the intrinsic pacemaker of small intestine need to be studied further in the appropriate experimental models, as well as clinically.

Absorption of Nutrients and Medications

The major, if not the only, reason for transplantation of the small intestine would be to provide an adequate balanced nutrition to the recipient. There is little doubt that the absorptive capacity of the small bowel is seriously reduced immediately after transplantation, as has been shown by many workers in autotransplants as well as allotransplants. Such impairment is probably due to denervation, interruption of lymphatic drainage, and particularly the transient ischemia, to which the mucosa of the small intestine is extremely sensitive. In replacement of the whole small bowel, it has been suggested that several months are necessary for the return of normal intestinal function (Ruiz et al. 1972). With short segments of the transplanted intestine, we have previously shown that absorption of glucose, amino acids, and simple fatty acids (Hardy et al. 1970; Stamford and Hardy 1974) resumes promptly, and that both passive and active absorption of nutrients recovers quickly both in autografts and in nonrejecting allografts. The absorption of complex fatty acids does not occur until 3–4 weeks after transplantation and is related to the regeneration of the lymphatics. We have previously suggested that some of the absorption studies, such as glucose absorption, may be utilized as an indicator of rejection of allografts. This was recently confirmed by Nordgren et al. (this volume, p. 131) in studies using cyclosporin as the immunosuppressive agent. Holmes et al. (1971 a), on the other hand, felt that the changes in absorption of glucose are late changes and that rejection can and should be diagnosed earlier by serial biopsies of the allograft. The Minnesota group (Toledo-Pereyra and Najarian 1973) examined the intestinal absorption of D-xylose and vitamin A in dogs that had whole small-bowel replacement, both as autografts and allografts, and found significantly diminished absorption of these substances in the first weeks after transplantation. Clinical recovery and weight gain in the autotransplanted dogs took 2 months after surgery. An important observation by Ruiz and Lillehei (1972), which needs to be confirmed experimentally, was that there is a 25% reduction in the absorption of D-xylose in normal dogs treated with azathioprine and prednisone, i.e., the use of immunosuppression alone, particularly with the cytotoxic agents, may cause problems in absorption.

With the recent interest in cyclosporin, the problem of absorption of cyclosporin by the small intestine has been of major interest and importance. The Toronto group (MacKenzie et al. 1982) performed both auto- and allotransplants of small intestine in dogs and observed a 40%–60% reduction in cyclosporin absorption from control values. They were further able to demonstrate that interruption of lymphatic drainage during intestinal transplantation may well cause malabsorption of cyclosporin and have suggested that systemic administration of cyclosporin and other drugs must be used in intestinal transplantation during the first few months and particularly during rejection episodes when absorption by the oral route is completely uncertain. It is clear from their studies that the careful monitoring of cyclosporin levels in plasma will be necessary and appropriate adjustments of the oral and/or parenteral cyclosporin dosage will need to be made to maintain the survival of intestinal allografts. Further studies are necessary to evaluate the ability of small-intestinal allografts and autografts to absorb various other medications.

Alterations of Mucosal Metabolism and Bacterial Flora

The pathophysiology of mucosal metabolism in intestinal transplantation is extremely complex and beyond the scope of this discussion. However, disturbances secondary to ischemia alone, such as depression of oxidative phosphorylation and nucleotide synthesis in the cells of the intestinal mucosa which lead to marked alterations in active absorption of nutrients as well as in secretion of enzymes and hormones, need to be studied further. Another problem that needs to be evaluated is the loss of the continuous coat of mucin, as well as intracellular mucin, which may lead to the loss of the barrier effect against bacteria and toxins, a normal function of the intestinal mucosa (Gurd 1965). The question of how to alter or eliminate the bacterial flora of the intestinal allograft also requires a thorough experimental evaluation, in relation to immunosuppression, as well as to metabolic and absorptive functions of the graft.

Immunological Considerations

Allograft Rejection

The best method of diagnosing intestinal allograft rejection is still uncertain. Studies by Holmes et al. (1971 b) and Fortner et al. (1972) suggested that serial biopsies of intestinal allografts are most accurate and practical in diagnosing early rejection. Previous work by my group (Hardy et al. 1970; Cohen et al. 1969) and the more recent work by the Toronto group (Reznick et al. 1982) suggest that the use of absorption studies, particularly with radiolabeled isotopes, may give an early answer in regard to rejection that correlates closely with the morphological appearance of intestinal allograft rejection. It is likely that in clinical situations both methods will need to be used, and that the morphological analysis, particularly with precise staining for lymphocyte subpopulations using monoclonal reagents, may give the most accurate determination of rejection in the future.

Since immunological rejection is still the major problem in intestinal transplantation, recent reports from the Toronto group (MacKenzie et al. 1982; Reznick et al. 1982; Craddock et al. 1983) on the use of cyclosporin in prolonging the functional survival of intestinal transplants in dogs are most encouraging. Previously, the use of standard immunosuppressive agents, including azathioprine, steroids, and antilymphocyte globulin, was relatively ineffective experimentally and clinically. Cyclosporin, given both intramuscularly and then orally, led to a medium survival time (MST) of intestinal allografts in dogs of 103.8 ± 39.4 days, with the longest survivor dying after 432 days. Dogs treated only with oral cyclosporin had an MST of 30.4 ± 7.6 days, because the absorption of orally administered cyclosporin was relatively ineffective. Important observations in several long survivors included (1) the demonstration of lymphatic regeneration, (2) reversal of a rejection episode by an increased dose of cyclosporin, (3) the observation of chronic rejection of small intestine in the longest-surviving allograft, and (4) no evidence of clinical or histological graft versus host (GVH) disease. The prolongation of survival achieved with cyclosporin is greater than any previously achieved with conventional immunosuppression. Further studies are required to document the best method of administration and dose of this drug and possible synergistic effect of this agent with other immunosuppressive or immunomanipulative approaches.

Graft Versus Host Disease

Although the short intestinal segments in dogs are usually rejected within 6–10 days, when nonimmunosuppressed, and have the histological appearance of rejection, Lillehei et al. (1959) and many authors since then have noted that transplantation of the entire small intestine frequently leads to the death of the host without histological evidence of rejection in the graft. Since the quantity of lymphoid tissue transplanted in the mesentery of the small intestine varies with the length of the segment transplanted, and plays an important role in initiating rejection and/or graft versus host disease, attempts at manipulation of this lymphoid tissue have been suggested by Deltz et al. (in this volume, p. 145) and have been tried by Toledo-Pereyra and Najarian (1973) and Cohen et al. (1976). The evidence for GVH disease in intestinal transplantation continues to be controversial, since no evidence for this has ever been seen clinically. There is no doubt that a generalized GVH reaction may occur, particularly when the number of "passenger" leukocytes in the graft is large, as it is in the intestine or in the spleen. It also appears clear from preliminary studies that this process may be modified by allograft pretreatment with cytotoxic agents and/or various forms of radiation. The elegant studies of Monchik and Russell (1971), in an F_1-hybrid-to-parent rat bowel transplantation model, showed that the presence of immunologically competent lymphocytes in the allograft does result in a lethal GVH reaction. This cannot be ignored, particularly in view of the possibility of using living related parents as donors of intestinal allografts to children.

General Discussion

Many issues that are important for clinical intestinal transplantation have not been previously discussed and need to be further investigated, both at the experimental and at the clinical levels. The whole question of who should serve as donor remains uncertain, with the choice for allografts being either a living related or cadaveric donor. As with other organs, the advantages of using a living related donor are the possibility of excellent tissue matching, ready availability, and therefore the possibility of a prompt procedure. On the other hand, there is a minimal danger to the donor, and only a short segment of bowel can be used which technically may be more difficult but immunologically may be better tolerated. With cadaveric donors the great sensitivity of the intestinal mucosa to ischemia would appear to be a major disadvantage. However, several studies (Toledo-Pereyra et al. 1975; Toledo-Pereyra and Najarian 1973) have demonstrated that small intestine can be functionally well-preserved with pulsatile and nonpulsatile perfusion and hypothermia, but further preservation studies are needed, both in experimental animals, and particularly in man. The problem of preparation of the intestinal tract in both the donor and the recipient with appropriate antibiotic agents needs to be addressed both experimentally and clinically in intestinal transplantation. The bacterial flora may increase the immunological damage of rejection, while the possibility of infection in immunosuppressed patients must also be avoided. Finally, further attempts at modification of the immunogenicity of small-intestinal transplants, particularly of its large lymphoid tissue antigenic load, must be continued, both at the experimental and clinical levels using perfusion and/or irradiation techniques in matched and mismatched donor-recipient combinations.

Although most of the absorption studies suggest that intestinal allografts function normally as digestive organs, the question of whether the enzyme contents of the mucosa return to normal is unsettled. Although enzyme production by intestinal epithelium appears to be preserved after transplantation, further studies, both histo- and biochemical, are necessary to delineate whether they are normal in quantity and quality.

The technical aspects of intestinal transplantation in regard to vascular anastomoses appear to be solved particularly by those who are expert in microvascular anastomoses. The majority of the investigators favor systemic drainage over portal vein drainage as being more practical and physiologically acceptable. It has, however, been pointed out that portal vein drainage may have both immunological and physiological advantages and this needs to be further studied at the experimental and clinical levels.

Although the prevention of rejection of transplanted intestine appeared initially to be more difficult than the rejection of other organs, it has been recently reemphasized by the Toronto group (Reznick et al. 1982; Craddock et al. 1983), using cyclosporin A as the major immunosuppressive agent, that this problem can be solved. Immunologically, it will be important to evaluate further the relative benefits of using short segments of small intestine for transplantation (a necessity when living related donors are used) from the point of view of rejection and/or possible GVH disease. As previously pointed out by Monchik and Russell (1971) the small-intestine allograft is unique in that it consists of both fixed lymphoid cells of the small intestine (Peyer's patches) and of immunocompetent cells in the mesenteric lymph nodes, both of which are capable of initiating potent immunological reactions. Our future ability to eliminate rejection and/or the GVH-like response in intestinal transplantation, if it exists in man, will partially depend on our further understanding of the role and fate of the transplanted lymphoid tissue in intestinal transplantation.

The growing success of total parenteral nutrition in maintaining many patients with short-bowel syndrome alive for months and years in a reasonable state of health (Greig et al. 1981), and the recent encouraging results with the use of cyclosporin in experimental intestinal transplantation, have led to renewed interest in the experimental and particularly clinical intestinal transplantation in man. Improved understanding of the mechanism of rejection and GVH reaction, as well as of the physiological and anatomical problems in intestinal transplantation, should make clinical intestinal transplantation a safe and efficient method of treatment of patients surviving on prolonged intravenous hyperalimentation.

Acknowledgments. I am grateful to Ms. Renee Grobtuch for secretarial assistance, and to many of my former surgical research fellows for their persistent efforts and stimulating thoughts.

References

Alican F, Hardy JD, Cayirli M, Varner JE, Moynihan PC, Turner MD, Pandeli A (1971) Intestinal transplantation: laboratory experience and report of a clinical case. Am J Surg 121:150–159

Ballinger WF, Christy MG, Asby WB (1962) Autotransplantation of the small intestine: the effect of denervation. Surgery 52:151

Ballinger WF, Iida J, Aponte GE, Wirts W, Goldstein F (1964) Structure and function of the canine small intestine following total abdominal vagotomy. Surg Gynecol Obstet 118:1305–1308

Cohen WB, Hardy MA, Quint J, State D (1969) Absorptive function in canine jejunal autografts and allografts. Surgery 65:440–446

Cohen Z, MacGregor B, Moore KTH, Falke RE, Langer B, Cullen JB (1976) Canine small bowel transplantation. Arch Surg 111:248–253

Craddock GN, Nordgren SR, Reznick RK, Gilas T, Lossing AG, Cohen Z, Stiller CR, Cullen JB, Langer B (1983) Small bowel transplantation in the dog using cyclosporine. Transplantation 35:284–288
Detterling R (1971) Discussion to Alican. Am J Surg 121:159
Dowling RH (1967) Compensatory changes in intestine absorption. Br Med Bull 23:275–278
Flint JM (1912) The effect of extensive resections of the small intestine. Bull Johns Hopkins Hosp 23:127–135
Fortner JG, Sichuk G, Litwin SD, Beattie EJ (1972) Immunological responses to an intestine allograft with HL-A identical donor-recipient. Transplantation 14:531–535
Gurd FN (1965) Metabolic and functional changes in the intestine in shock. Am J Surg 110:334–339
Greig P, Jeejeebhoy KN, Langer B, Cohen Z (1981) Decade of home parenteral nutrition. Gastroenterology 80:1164–1169
Hardy MA, Quint J, State D (1970) Effect of antilymphocyte serum and other immunosuppressive agents on canine jejunal allografts. Ann Surg 171:51–60
Holmes JT, Yeh SDJ, Winawer SJ, Kawano N, Fortner JG (1971 a) Absorption studies in canine jejunal allografts. Ann Surg 174:101–108
Holmes JT, Klein MS, Winawer SJ, Fortner JG (1971 b) Morphological studies of rejection in canine jejunal allografts. Gastroenterology 61:693–706
Kocandrle V, Houttuin E, Prohaska JV (1966) Regeneration of the lymphatics after autotransplantation and homotransplantation of the entire small intestine. Surg Gynecol Obstet 122:587–592
Kort WJ, Westbroek DL, MacDicken I, Lameijer LDF (1973) Orthotopic total small bowel transplantation in the rat. Eur Surg Res 5:81–89
Kunlin A, Shiu MH, Winawer SJ, Fortner JG (1971) The isolated allograft pouch: a useful method for monitoring small bowel allografts. Surg Forum 22:237–239
Lillehei RC, Goott B, Miller FA (1959) Homografts of the small bowel. Surg Forum 10:197–199
Lillehei RC, Idezuki J, Feemster JA, Dietzmann RH, Kelly WD, Merkel FK, Goetz FC, Lyons GW, Manax WG (1967) Transplantation of stomach, intestine and pancreas: experimental and clinical observations. Surgery 62:721–741
MacKenzie R, Nordgren S, Lossing A, Craddock G, Cohen Z, Stiller C, Langer B (1982) Cyclosporine A absorption in canine small intestinal transplantation. Transplant Proc 14:646–648
Monchik GJ, Russell PS (1971) Transplantation of the small bowel in the rat: technical and immunological considerations. Surgery 70:693–702
Okumura M, Fujimura J, Ferrari AA (1969) Transplante de intestine delgado. Rev Hosp Clin Fac Med Sao Paulo 24:39–45
Olivier C, Rettori R, Olivier C, Baur O, Roux J (1969) Un cas de transplantation orthotopique de jejuno-ileon et des colons droit et transverse. Presse Med 77:1275–1278
Olivier C, Rettori R, Olivier C, Camilleri JP (1972) Interruption of the lymphatic vessels and its consequences in the total homotransplantation of the small intestine and right side of the colon in man. Lymphology 5:24–31
Porus RL (1965) Epithelial hyperplasia following massive small bowel resection in man. Gastroenterology 48:753–755
Reznick RK, Craddock GN, Langer B, Gilas T, Cullen JB (1982) Structure and function of small bowel allografts in the dog: immunosuppression with Cyclosporine A. Can J Surg 25:51–55
Ruiz O, Lillehei RC (1972) Intestinal transplantation. Am J Proctol 23:379–393
Ruiz OJ, Uchida H, Schultz L, Lillehei RC (1972) Problems in absorption and immunosuppression after entire intestinal allotransplantation. Am J Surg 123:297–303
Sakai A (1970) Role of the liver in kidney allograft rejection in the rat. Transplantation 9:333–338
Shils ME, Wright WL, Turnbull A, Brescia F (1970) Long-term parenteral nutrition through an external arteriovenous shunt. N Engl J Med 283:341–344

Stamford WP, Hardy MA (1974) Fatty acid absorption in jejunal autograft and allograft. Surgery 75:496–502
Toledo-Pereyra LH, Najarian JS (1973) Preservation of the small intestine for allotransplantation. Surg Gynecol Obstet 137:445–449
Toledo-Pereyra LH, Simmons RL, Najarian JS (1975) Absorption of carbohydrates and vitamins in the preserved and transplanted small intestine. Am J Surg 129:192–200
Weinstein LD, Shoemaker CP, Hersh T, Wright HK (1969) Enhanced intestinal absorption after small bowel resection in man. Arch Surg 99:560–566
Wilmore D (1969) Short bowel syndrome. A comprehensive approach to patient management. I. Pathophysiology following massive intestine resection. J Kans Med Soc 70:233–239

Experience with Segmental Pancreatic Transplantation in Stockholm

C. G. Groth, G. Lundgren, G. Tydén, R. Gunnarsson, and *J. Östman*

Pancreatic transplantation for the treatment of diabetes mellitus in man has proved increasingly successful in recent years. In consequence, the number of patients undergoing such transplantation is rapidly increasing and several new centers have started programs in the past few years (Sutherland 1984). Nevertheless, the results of pancreatic transplantation are much inferior to those of the transplantation of other organs, such as the kidney and the heart. It is our belief that the main reason for this lies in the fact that the surgical technical problems of pancreatic transplantation have not been completely solved. The most common complications in this regard are related to difficulties in the handling of the exocrine juice and to vascular thrombosis. The data so far accumulated indicate that the pancreas is no more prone to rejection than other organs, perhaps even less so. There have, however, been considerable difficulties in making an early diagnosis of pancreatic graft rejection. This paper reviews some of these problems as they have been encountered in 27 segmental pancreatic transplantations in Stockholm. It also reports that those patients who have retained their grafts have excellent control of their glucose homeostasis.

The Patients

Twenty-seven segmental pancreatic transplantations were performed in 24 diabetic patients; thus there were three retransplantations. All of the patients suffered from insulin-dependent diabetes of long standing (19–42 years). Their age range was 25–47 years. The first nine transplantations were performed in patients with complicated juvenile diabetes mellitus but without uremia. The next 18 transplantations were performed in patients who had end-stage diabetic nephropathy. Two of these transplantations were performed in patients with a previous kidney graft, while the remaining 16 were performed in conjunction with kidney transplantation and with both of the organs from the same donor.

Organ Procurement and Preservation

In 26 instances the segmental pancreatic graft was obtained from an ABO-compatible cadaveric donor (age 13–51 years), while in one instance a maternal graft

Microsurgical Models in Rats
for Transplantation Research
Eds. A. Thiede et al.

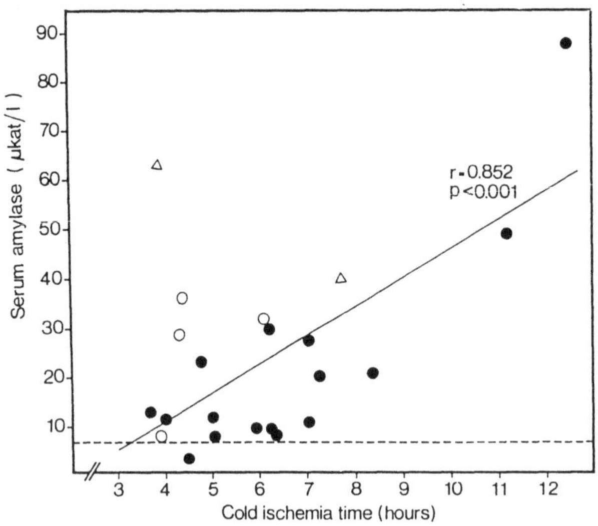

Fig. 1. Effect of cold ischemia time on postoperative serum amylase levels. ●, grafts perfused with Perfadex and with enteric diversion; △, grafts perfused with Sack's solution and with enteric diversion; ○, duct-ligated grafts perfused with Perfadex. The *regression line* relates to 16 grafts with enteric diversion which were perfused with Perfadex

was used. In the cadavers, the kidneys and the pancreas were removed by an evisceration technique, which permitted rapid cooling of the organs by intra-arterial perfusion in situ (Björken et al. 1976). For the perfusion, refrigerated (4 °C) extracellular-type solution (Perfadex, Pharmacia, Sweden) was used in 22 instances, Sack's solution in 2, and Euro-Collins' solution in 2.

The time between cardiac arrest[1] and the initiation of cold perfusion – that is, the warm ischemia time – was 4–18 min (9.2 ± 0.7 min). The cold ischemia time was 3 h 39 min – 12 h 32 min, means 5 h 56 min, including a reanastomosing time of 19–30 min, mean 28 min. In one instance where Euro-Collins' solution was used, the perfusion cannula was mistakenly placed in the superior mesenteric artery. Following revascularization in the recipient, the pancreas became hemorrhagic. This was assumed to be due to preservation injury and the graft was removed. Two grafts, both perfused with Perfadex, thrombosed within 24 h.

Of the remaining 23 grafts, 19 immediately made the recipient insulin independent, while 4 recipients required occasional injections of insulin during the first postoperative weeks. There was no correlation between the postoperative blood glucose levels and the warm or cold ischemia times. For 16 grafts which had been perfused with Perfadex and had exocrine diversion to the bowel a highly significant correlation was found between the highest postoperative serum amylase value and the time of cold ischemia. Recipients of duct-ligated grafts and grafts perfused with Sack's solution had comparatively higher serum amylase levels (Fig. 1). There was no correlation between the warm ischemia time and the serum amylase levels. In all patients the amylase levels returned to normal in 5–7 days.

[1] Under Swedish regulations, organ removal can begin only after cardiac arrest has occurred

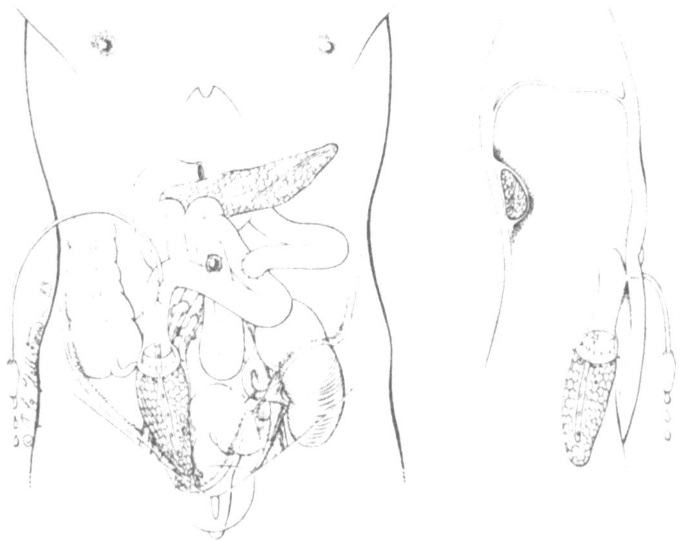

Fig. 2. Technique used for intraperitoneal segmental pancreatic transplantation with pancreaticoenterostomy to a jejunal Roux loop with pancreatic duct tube drainage to the exterior. Note that the pancreas is placed intraperitoneally. The concomitantly transplanted kidney is placed extraperitoneally on the contralateral side

The Transplantation

The graft vessels were anastomosed to the patients' iliac vessels and the grafts were placed either extra- or intraperitoneally. Three methods have been used for handling the exocrine secretion: duct ligature, diversion to the bowel via a pancreaticojejunostomy, or ductoenterostomy (Groth et al. 1976, 1980a, 1983). Since November 1981, only end-to-end pancreatico-Roux-en-Y-loop jejunostomy has been used (Groth et al. 1982). The grafts were placed intraperitoneally instead of extraperitoneally as before. The cut end of the graft was telescoped far into the bowel and a catheter was placed in the pancreatic duct to temporarily divert the secretion to the exterior (Fig. 2).

Anticoagulant and Immunosuppressive Therapy

All patients received a small dose of heparin (2500–5000 IU i.v.) at the time of revascularization. Most of the patients were given 500 ml dextran 70 (Macrodex) daily for 3–4 days after transplantation as prophylaxis against graft thrombosis. Seventeen of the patients were also given anticoagulant therapy with sodium warfarin for some weeks or months after transplantation. Immunosuppressive therapy was given with azathioprine, prednisolone, and horse antilymphocyte globulin or rabbit antithymocyte globuli (RATG). In two patients, thoracic duct drainage was used for the first few postoperative weeks. In some recent patients azathioprine was replaced with cyclosporin one or more months after transplantation. Rejection was treated with 0.25- to 1-g doses of methylprednisolone given i.v. over several days and in one patient RATG was also given.

Technical Complications Related to the Exocrine Secretion (Table 1)

Duct Ligature (n=4). One graft was placed extra- and three intraperitoneally. In all four patients an exocrine leakage developed. One patient formed a pancreatic pseudocyst and one had painful pancreatic ascites. Exocrine fistulae to the exterior occurred in two patients.

Pancreaticoenterostomy to Jejunal Roux Loop (n=5). All grafts were placed extraperitoneally. Two of the patients developed graft vascular occlusion within 24 h, and were never at risk for complications related to the exocrine secretion. The remaining three patients all developed a pancreatic fistula to the exterior, which soon became infected with enteric bacteria. One well-functioning graft had to be removed after 2 weeks because of bleeding caused by the pancreatic fistula. In one patient the fistula was repaired successfully.

Duct Enterostomy to Jejunal Roux Loop or Ileal U-Loop (n=5). All grafts were placed extraperitoneally. Four of five patients developed an exocrine pancreatic fistula with drainage through the wound. Three of the grafts, although still functioning, had to be removed because of a peripancreatic infection with enteric bacteria. In the fourth patient with a leakage an extensive wound drainage was attempted, but to no avail.

Modified Pancreaticoenterostomy to Jejunal Roux Loop (n=13). All grafts were placed intraperitoneally. One graft was removed intraoperatively because of a preservation injury. In 8 of the 12 remaining grafts there was a flow of pancreatic juice via the ductal tube for 3–28 days (mean, 13 days); 24-h volumes were as high as 1500 ml. In four patients no flow occurred, either because the tube had slipped out of the duct or because the juice escaped around the tube into the bowel. Two patients developed exocrine leakage. In one instance this came from the pancreaticojejunal anastomosis. In this patient the catheter had functioned for 3 days. Another patient developed a deep wound abscess which led to purulent peritonitis. Subsequently, a perforation occurred in the jejunal Roux loop; however, the anastomosis to the pancreatic graft was intact. In both of these patients a well-functioning graft had to be removed.

Vascular Occlusions (Table 1)

The second pancreatic graft recipient in the Stockholm series lost his graft because of vascular occlusion within 24 h. The subsequent 18 recipients were treated with sodium warfarin. Four of these patients nonetheless developed vascular occlusion leading to graft destruction. The last six patients were not given this drug. Of these grafts one was lost because of vascular occlusion. Thus 4 of 18 patients (22%) given warfarin had a vascular occlusion compared with 2 of 8 of those not given this drug (25%). The occlusions occurred 1–65 days after transplantation.

Table 1. Incidence of graft exocrine leakage and vascular occlusion in 27 segmental pancreatic transplantations

Handling of exocrine secretion	Technical complications: Exocrine leakage	Technical complications: Vascular occlusion	Functioning at present
All cases 1974–1981	11/12	4/14	0/14
Duct ligature ($n=4$)	4/4	2/4	0/4
Pancreaticoenterostomy to jejunal Roux loop ($n=5$)	3/3[a]	2/5	0/5
Ductoenterostomy to jejunal Roux loop or ileal U-loop ($n=5$)	4/5	0/5	0/5
Modified pancreaticoenterostomy to jejunal Roux loop ($n=14$) 1981–1983	2/12[b]	2/12[b]	6/13

[a] Two patients were not at risk because of early graft removal due to thrombosis
[b] One patient was not at risk because of early graft removal due to preservation injury

Table 2. Parameter used to evaluate graft function (values in parentheses indicate the length of time the patients have had a pancreatic graft

Patient	Fasting blood glucose (<7.0 mmol/liter)	IVGTT ($k>1.0$)	OGTT 2 h blood glucose (<8.0 mmol/liter)	Hb A_1 (<8.5%)	Serum amylase (<6 µkat/liter)	Serum creatinine (<120 µmol/liter)
SF (20 months)	4.4	1.0	7.3	7.6	4.9	196
AW (16 months)	3.9	1.2	6.2	6.6	2.3	128
KD (15 months)	5.7	1.0	6.8	6.8	3.3	169
RJ (4 months)	4.2	1.0/0.7[a]	–	6.7	6.2	250
AMW (3 months)	5.3	1.1/0.8[a]	–	5.8	3.5	246
GA (½ months)	4.6				2.4	249

[a] Recently switched to cyclosporin A with a subsequent decline in the k value. Recovery usually occurs when the cyclosporin A dose is lowered
IVGTT, intravenous glucose tolerance test; OGTT, oral glucose tolerance test

Diagnosis and Treatment of Rejection

Of the nine segmental pancreatic allografts performed in nonuremic diabetic patients, four were lost because of technical complications. The remaining five grafts were all irreversibly rejected after 5–51 days; in four instances this occurred in spite of anti-rejection treatment. Two rejection episodes were successfully treated, but these grafts were later lost because of technical failure and rerejection, respectively. An increase in the postprandial blood glucose preceded the increase in the fasting blood glucose in five rejection episodes. The increase in the fasting blood glucose was always abrupt, necessitating insulin therapy within 24 h. There was no consistent change in serum amylase or body temperature before the deterioration of the fasting blood glucose.

Of the two pancreatic grafts performed in patients with previous kidney grafts one was lost because of rejection after 36 days. In the latter case the pattern of rejection was similar to that in isolated pancreatic grafts, with an increase in the postprandial blood glucose preceding the increase in the fasting blood glucose by 2 days. The serum creatinine remained normal.

Of the 15 combined pancreatic and kidney grafts, 9 pancreatic grafts were lost because of technical complications. In the remaining seven patients there was one irreversible re-rejection of both grafts on day 4. This patient had a simultaneous abrupt rise in the postprandial blood glucose, fasting blood glucose, and serum creatinine over 24 h, while the serum amylase remained unchanged. Six patients have both grafts functioning after ½–20 months. There have been five renal graft rejection episodes and one rejection, all occurring 10–30 days after transplantation. The diagnosis was based on a deterioration in renal function and changes in the radioisotope renogram with no other cause at hand. In three instances the diagnosis was verified by findings on renal biopsy. The postprandial blood glucose and fasting blood glucose remained stable in all but two cases in whom an increase occurred when anti-rejection therapy with steroids was instituted. In one patient there was a moderate increase in the serum amylase on the day after the increase in the serum creatinine, but in the others the serum amylase remained unchanged (Fig. 3).

Metabolic Control

Six pancreatic grafts with modified pancreaticoenterostomy are presently functioning 0.5–20 months after transplantation. At this time all six patients have normal fasting blood glucose levels (Table 2). The serum amylase levels also are normal, or almost normal, in all patients. The serum creatinine levels are between 128 and 250 μmol/liter. The moderate elevations in the serum creatinine reflect a mild cyclosporin toxicity or are the result of rejection. Studies of the intravenous and oral glucose tolerance tests in the three patients who have had their grafts more than 1 year have revealed normal or near-normal values. Three and 4 months after transplantation two other patients had normal k values on the intravenous glucose tolerance test until they were switched from azathioprine to cyclosporin

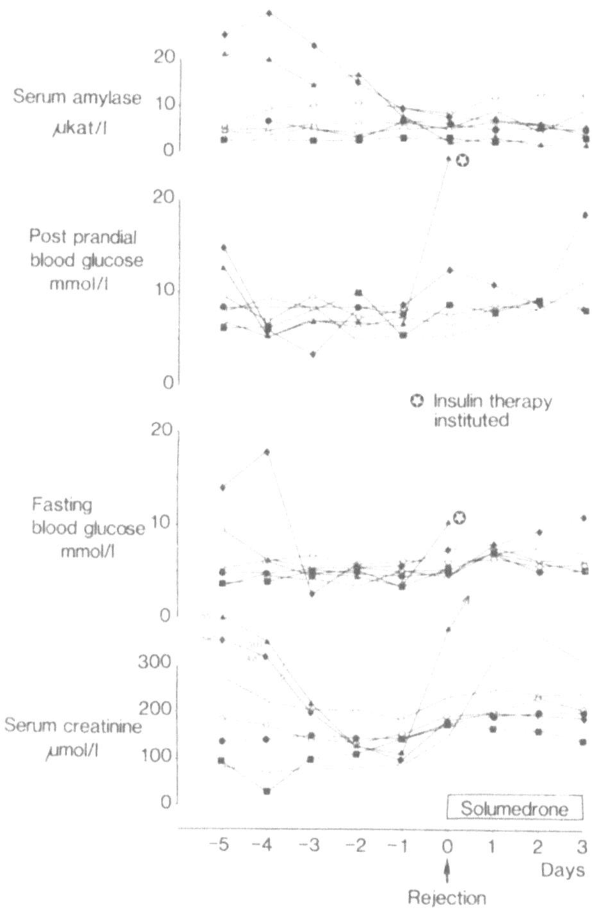

Fig. 3. Laboratory findings during rejection episodes in diabetic patients who had undergone combined pancreatic and renal transplantation. When a rise in the serum creatinine signaled renal graft rejection, parameters indicative of pancreatic graft function were usually little affected

therapy; then the *k* values declined. A recovery is expected when the cyclosporin dose is lowered. HbA_1 levels were normal in all five patients (Table 2).

In two of the three patients who had well-functioning pancreatic grafts for more than 1 year an improved nerve conduction velocity was found on examination with electroneurography (Fig. 4). In the third patient no such improvement was noted. Examinations of retinopathy failed to reveal any improvement, but one of the patients was blind and the two others had advanced vascular changes at the time of transplantation. Renal transplant biopsy carried out 15 months after transplantation in the patient with the longest follow-up showed no thickening of the glomerular basement membranes; such a finding would have suggested a recurrence of diabetic nephropathy.

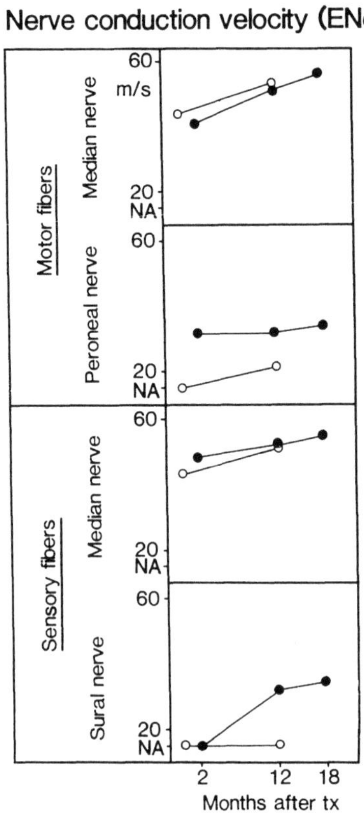

Fig. 4. Nerve conduction velocity, measured with electroneurography, in two pancreatic transplant patients who had a graft for more than 1 year. Note the improvement in both patients

Overall Patient and Graft Survival

None of the 14 pancreatic grafts transplanted prior to November 1981 functioned beyond 2 months. The causes of failure were irreversible rejection (6), vascular occlusion (4), and complications related to exocrine leakage (4). Two of the patients died following loss of the pancreatic graft, the causes of death being gastrointestinal bleeding and sudden cardiac arrest. Seven of the 13 pancreatic grafts that have been transplanted since November 1981 with modified pancreaticoenterostomy to a jejunal Roux-Y-loop were lost because of preservation injury (1), rejection (1), vascular occlusion (2), complications related to exocrine leakage (1), and peritonitis (2). Six of the grafts are presently functioning 20, 16, 15, 4, 3, and 0.5 months after the transplantation (46%). The patient in whom a damaged pancreas was immediately removed never received a kidney; among the 12 patients who received a renal graft 9 retain a functioning graft at this time (75%). One patient died of septicemia. She developed a deep wound abscess that resulted in a purulent peritonitis followed by perforation of the jejunal Roux loop. Thus the patient survival rate is 12/13 (92%).

Discussion

Since the purpose of pancreatic transplantation is to prevent the secondary complications of diabetes, the procedure should preferably be undertaken before multiple organ damage has occurred. However, the experimental nature of the operation and the hazard of the immunosuppressive treatment has prompted most transplantation groups to transplant the pancreas only to uremic diabetics eligible to receive renal grafts, for these patients require immunosuppressive therapy in any case for the renal graft. The renal and pancreatic transplantations can be performed simultaneously, thus avoiding two separate operative procedures. Another advantage of this approach is that the high-dose immunosuppression usually given immediately after transplantation does not have to be repeated. In some cases, however, the operations have been performed at an interval of some weeks or months, using organs from two donors.

In a few cases nonuremic diabetic patients have been given only pancreatic grafts, the indication for transplantation then being either rapidly progressing angiopathy with threatening eye or kidney damage or hyperlabile diabetes. It is obvious that in the future it would be hoped to perform pancreatic transplantation at this early stage of the disease.

It is clear from our findings that human pancreatic grafts tolerate warm ischemia for at least 18 min. Simple cold storage, using an extracellular type solution (Perfadex), permits preservation for up to 10 h. Postoperative serum amylase levels correlated with the duration of cold storage thus seem to represent an index of the degree of ischemic injury. When storage times were in the 10- to 12-h range the serum amylase levels were quite high and it may be advisable to avoid ischemic times beyond 8 h with this technique.

The handling of the exocrine secretion has been a stumbling block in segmental pancreatic transplantation. Obviously, the simplest solution is duct ligature and in animal experiments this has been used without too many complications. However, in man, this procedure has invariably resulted in exocrine leakage in our hands as well as in others. Moreover, findings in several experimental studies indicate that duct ligation may gradually impair the endocrine function of the graft. On the whole, it seems inadvisable to continue to use duct ligation in human pancreatic transplantation.

Pancreaticoenterostomy and ductoenterostomy are both standard procedures after pancreaticoduodenectomy for malignancy (Whipple's procedure). In some animal experiments they have also been used with reasonable success in conjunction with pancreatic transplantations. However, initial clinical experience with segmental pancreatic transplantation and enteric diversion was discouraging: Pancreatic fistula almost invariably occurred and when the fistula came from a pancreaticoenteric anastomosis it became infected with enteric bacteria. In some cases this dangerous complication necessitated the removal of a well-functioning graft; in others it made the treatment of an acute rejection episode too risky and thus indirectly caused the loss of the graft.

However, following modifications of various details, we have recently been able to use pancreaticojejunostomy with some success. We believe that the use of intra- instead of extraperitoneal placement of the graft has provided better conditions for anastomotic healing. Pancreatic duct drainage to the exterior via a catheter was introduced in the hope that this would allow the anastomosis to heal without being exposed to the digestive action of the pancreatic secretion. However, in some patients the catheter did not function and in others it functioned for only a few days. The only patients who developed an anastomotic leak had had good drainage to the exterior through the catheter for 3 days.

A high incidence of graft vascular occlusions has been reported in all large series of pancreatic transplantations. One explanation of this may well be the abnormal hemodynamic situation of the segmental pancreatic vessels since the large splenic artery is left to drain via the small pancreatic vessels. Such a poor runoff must increase the risk of thrombosis formation. Unfortunately, it seemed that treatment with the powerful anticoagulant drug, warfarin, did not avert this problem. Therefore we have abandoned the use of this agent. We plan, however, to continue to use a small dose of heparin during surgery and postoperative prophylaxis with dextran.

The present data again underline the fact that it is difficult to diagnose rejection in the pancreatic grafts when transplanted alone (Groth et al. 1976, 1980 b). There was no increase in serum amylase to indicate damage to the exocrine gland. An increase in the postprandial blood glucose preceded that in the fasting blood glucose by a few days but was still not an early enough sign since the institution of anti-rejection therapy, even at that stage, proved effective in only two of seven rejection episodes. In patients receiving combined grafts no isolated pancreatic rejection episodes occurred. Furthermore, of the six rejection episodes recorded in these patients there were signs of pancreatic involvement only in two, as judged by changes in the fasting and postprandial blood glucose and in the serum amylase. It is of interest that only one pancreatic graft in the latter series was lost due to rejection, as compared with several in the nonuremic patient group. It seems likely that the kidney signaled the immunological attack at such an early juncture that pancreatic rejection could be averted.

It is encouraging that the patients in whom technical complications and graft rejection were avoided had satisfactory control of the blood glucose homeostasis, as reflected by the glucose tolerance tests and glycosylated hemoglobin levels. The recent finding that cyclosporin may cause a deterioration in glucose control is worrisome in this context (Gunnarsson et al. 1983). The role of cyclosporin in pancreatic transplantation will have to be reassessed with this finding in mind.

The real benefit of pancreatic transplantation consists in the prevention or reversal of the secondary complications of the disease. So far the number of patients who have had pancreatic grafts for any length of time has been small. Moreover, many of the patients have had very severe secondary lesions at the time of transplantation which made it difficult to assess an effect on the lesions. Two of our patients showed regression of their neuropathy. Since these patients had both uremia and diabetes at the time of transplantation, their recovery might have been due to the improvement in their uremic state as well as the improvement in blood glucose control. Further studies in this regard are needed. The fact that no increase in basement membrane thickness has occurred in the one patient who underwent late biopsy is also encouraging.

In our last 13 patients the morbidity and mortality rates have been acceptable. The use of combined pancreatic and renal transplantation seems not to have jeopardized the results of renal grafting. Six patients have well-functioning pancreatic grafts with good or excellent glucose control.

References

Björkén C, Lundgren G, Ringdén O, Groth CG (1976) A technique for rapid harvesting of cadaveric renal and pancreatic grafts after circulatory arrest. Br J Surg 63:517–519

Groth CG, Lundgren G, Arner P, Collste H, Hårdstedt C, Lewander R, Östman J (1976) Rejection of isolated pancreatic allografts in patients with diabetes. Surg Gynecol Obstet 143:933–940

Groth CG, Lundgren G, Gunnarsson R, Hårdstedt C, Östman J (1980a) Segmental pancreatic transplantation, with special reference to the use of ileal exocrine diversion and to the hemodynamics of the graft. Transplant Proc [Suppl 1] 12:62–67

Groth CG, Lundgren G, Gunnarsson R, Arner P, Berg B, Östman J (1980b) Segmental pancreatic transplantation with duct ligation or drainage to a jejunal Roux-en-Y loop in non-uremic diabetic patients. Diabetes [Suppl 1] 29:3–9

Groth CG, Collste H, Lundgren G, Wilczek H, Klintmalm G, Ringdén O, Gunnarsson R, Östman J (1982) Successful outcome of segmental human pancreatic transplantation with enteric exocrine diversion after modifications in technique. Lancet 2:522–524

Groth CG, Collste H, Lundgren G, Ringdén O, Thulin L, Wilczek H, Gunnarsson R, Östman J (1983) Surgical techniques for pancreatic transplantation. A critical appraisal of methods used and a suggested new modification. Horm Metab Res [Suppl 1] 13: 37–41

Gunnarsson R, Klintmalm G, Lundgren G, Tydén G, Wilczek H, Östman J, Groth CG (1983) Deterioration in glucose metabolism in pancreatic transplant recipients after conversion from azathioprine to cyclosporin. Lancet 2:571

Sutherland DER (1984) International pancreas and islet registry statistics. Transplant Proc 16(3):593–598

Clinical Experience in Islet Transplantation

D. Lorenz, H. Wolff, H. Lippert, H.-J. Hahn, O. Abri, and *E. Zander*

The objectives of transplantation of the whole pancreas or of the islets of Langerhans are the same – to establish normal carbohydrate metabolism in patients with diabetes mellitus to prevent or to reduce secondary complications (Sutherland 1981).

Islet transplantation was originally conceived as an alternative to pancreas transplantation because of the high rate of complications of the latter procedure. Some of the problems are being solved, but islet transplantation remains attractive for some reasons (Lorenz et al. 1981). Islet transplantation should be simpler and safer for the patient; one pancreas theoretically could provide islet tissue for more than one diabetic recipient and the immunogenicity of islets might be reduced in vitro prior to transplantation by special culture conditions.

In 1977 the first attempts at clinical application were made, but they were largely unsuccessful, if not premature (Sutherland 1981). Procurement of a sufficient quantity of viable islet tissue from human pancreases is a major problem and the development of a high-yield isolation technique for human islets is necessary.

Isolation Technique

Collagenase digestion remains the most effective and practicable method for islet isolation. The problem consists in solving the technical problem of separating isolated islets from undigested tissue during the collagenase digestion. We have tried to do it by the following means. Human pancreases, harvested from brain-dead donors, were perfused via Arteria lienalis and Arteria gastroduodenalis for 20 min by Collins'-B solution. After removing the pancreas was stored up to the beginning of isolation in Hank solution at 4 °C. The time between removal and isolation ranged from 30 min to 2 h. For the isolation of islets, the tissue was freed from large vessels, pancreatic duct tissue, and surrounding connective tissue. After cutting the pancreas by a tissue chopper the microfragments were put into a special incubator containing two filters. As a result of continuous digestion by 0.25% collagenase at 37 °C the microfragments can pass these two filters; the first has a pore size of 1 mm and the second a pore size of 0.2 mm. All the fragments

Microsurgical Models in Rats
for Transplantation Research
Eds. A. Thiede et al.

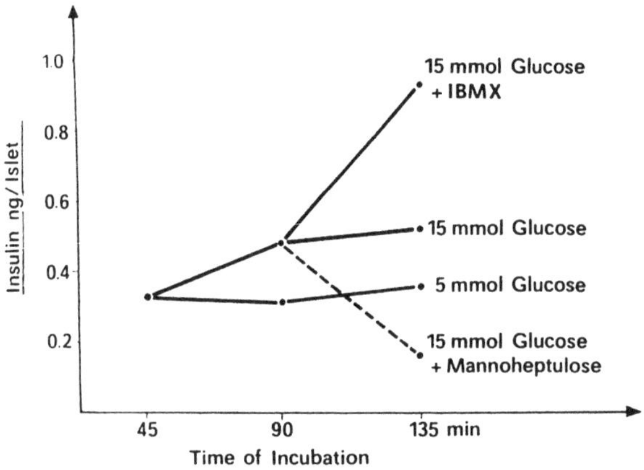

Fig. 1. Estimation of stimulated insulin released from isolated human islets. The glucose-stimulated effect can be potentiated by isobutylmethylxanthine (IBMX) and is inhibited by mannoheptulose

smaller than 0.2 mm are conveyed by a roller pump to a separator which contains a third filter with a pore size of 0.05 mm. On this filter the microfragments are caught while the collagenase can pass and flows back to the incubator.

In periods of 2–4 min the flow is interrupted and by a counterflow of Hank's solution the microfragments stored on the filter of the separator are conveyed to a sample container. To estimate the yield of islets from one donor pancreas we determined the insulin and amylase content of the tissue before and after isolation. The insulin/amylase ratio could be increased three- to four-fold in comparison to the ratio before collagenase digestion. Calculating the insulin content and tissue weight before and after isolation we reached a yield of endocrine tissue of between 15% and 25% of the endocrine tissue of the pancreas prior to isolation.

In some cases there were completely isolated islets but in most preparations they were enclosed into exocrine tissue. The viability of the isolated islets was proved by the estimation of the insulin secretion in vitro. Undamaged islets have to respond with an enhanced insulin secretion in the presence of 15 mmol glucose in comparison to 5 mmol glucose. This effect can be potentiated by glucose and isobutylmethylxanthine (IBMX), whereas the glucose-stimulated insulin secretion is inhibited by mannoheptulose (Fig. 1).

Autologous Islet Transplantation

Rejection is probably largely responsible for the failure of islet allografts, since the technical feasibility of the approach could be demonstrated by the autologous islet transplantation in pancreatectomized patients. In some cases of chronic pancreatitis it is necessary to remove either the entire pancreas or to carry out a subtotal pancreatectomy (95%). The indication for pancreas removal is given in serious changes of the pancreas parenchyma and duct system (cysts, tumors), stenoses of common bile duct and duodenum, ascites, and severe and resistant pain seizures. The indication for autologous islet transplantation can be made if the endocrine reserve is more than 20% in comparison to normal controls and if the

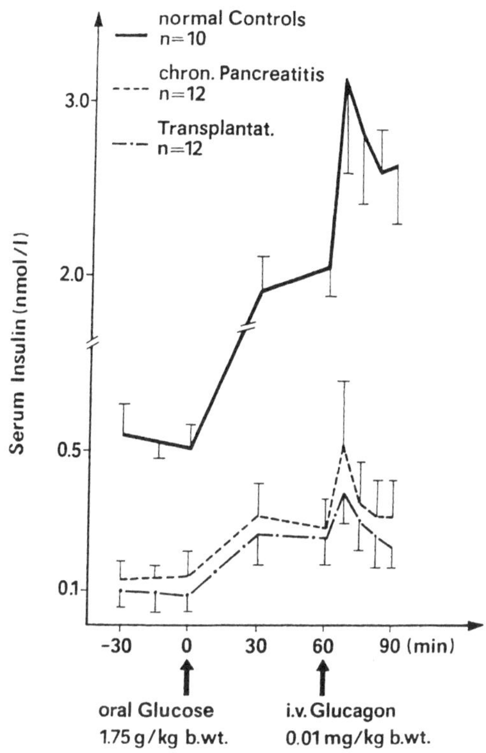

Fig. 2. In comparison to normal controls the glucose- and glucagon-stimulated serum insulin concentration is remarkably reduced in patients with chronic pancreatitis in whom a pancreas resection is indicated. Three months after intraportal autologous islet transplantation the glucagon-stimulated serum insulin concentration rose to about 40%–50% in comparison to the preoperative levels

liver and the portal system show normal conditions. Then the islets can be isolated during pancreatectomy and after this and under continuous measuring of the portal pressure and portal flow the islets are infused into the portal vein via the vena lienalis within a time of 30 min. If there is an increase of portal pressure to more than 12 mmHg it is necessary to finish the islet transplantation. So we never observed signs of liver disturbances or portal hypertension as a result of our careful isolation technique, using precise indications and accurate preoperative investigations of the liver and the portal system.

We have performed autologous islet transplantation via the portal system in 12 cases: seven times after duodenopancreatectomy and five times after subtotal pancreatectomy. The first transplantation was done in 1979. Two patients with duodenopancreatectomy were not submitted to insulin substitution for 3 and 6 months and then, like the other five patients, they needed between 8 and 18 units insulin per day. None of the five patients with subtotal pancreatectomy received insulin substitution for at least 12 months (Lorenz et al. 1981).

Discussing the results of autologous islet transplantation in pancreatectomized patients it is important to take into consideration a reduced endocrine function of the pancreas in cases of severe chronic pancreatitis. In all the 12 recipients the glucose- and glucagon-stimulated insulin concentration of the serum was reduced in contrast to normal controls to about 20%–30%. That is why it is remarkable that 3 months after autologous intraportal islet transplantation the glucagon-

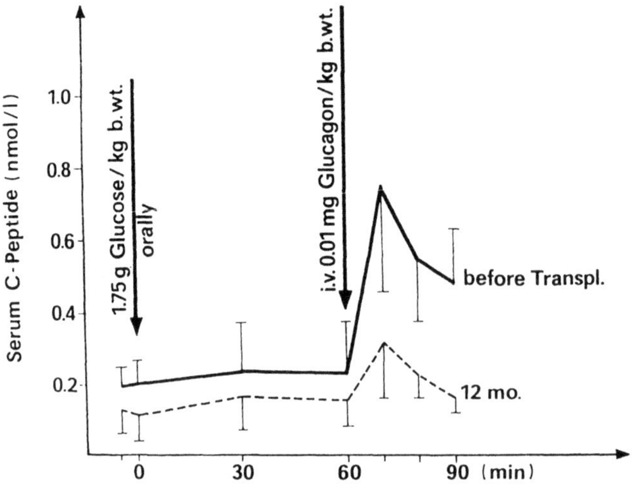

Fig. 3. In the estimation of the glucose- and glucagon-stimulated serum C-peptide levels 12 months after intraportal autologous islet transplantation the values rose in 12 patients from about 0.1 nmol/liter to about 0.3 nmol/liter. Thus the islets responded 12 months postoperatively with a C-peptide concentration of about 50% of that of the preoperative levels

stimulated serum insulin concentration rose to about 50% of that of the preoperative level (Fig. 2).

In the estimation of the glucose- and glucagon-stimulated serum C-peptide levels we obtained the same results. Twelve months after autologous islet transplantation the C-peptide level rose in 12 recipients from about 0.1 nmol/liter to about 0.3 nmol/liter by the stimulation of glucagon. Thus the transplanted islets responded 12 months postoperatively with a C-peptide concentration of about 50% in comparison to the preoperative C-peptide levels (Fig. 3).

Altogether we have to take into consideration a continuous decrease in the function of the transplanted autologous islets. If we assume a requirement of 0.4 units insulin/kg body weight per day we reached this limit of insulin substitution in one patient 1 year after duodenopancreatectomy and islet transplantation and in three other patients 3 years postoperatively. According to these findings we had to start with insulin substitution in two patients with subtotal pancreatectomy 12 months after islet transplantation (Wolff et al. 1981).

Seventy-seven cases of intraportal islet autotransplantation after total or near total pancreatectomy for treatment of chronic pancreatitis have been reported to the Human Pancreas and Islet Transplant Registry (Table 1). Thirty-five recipients of intraportal islets were reported to be insulin independent after the operation, but only four of these had a total duodenopancreatectomy.

Thus it is likely that the small pancreatic remnant is partially responsible for maintaining glucose homeostasis in some of the other patients. Direct evidence of function of the transplanted islets comes from the estimation of serum C-peptide and from insulin measurement. Signs of function of intraportal transplanted islets have been obtained by Cameron et al. (1980). In one patient serum insulin concentrations were higher in hepatic veins than in the portal vein during the in-

Table 1. Seventy-seven cases of intraportal islet transplantation after pancreas resection reported to the Human Pancreas and Islet Transplant Registry. Thirty-five of them were reported to be insulin independent

Team	Patients
Valentes (Genoa)	18
Wolff, Lorenz, Lippert (Berlin)	12
Sutherland, Najarian (Minneapolis)	11
Cameron (Baltimore)	8
Nardi (Boston)	5
Altman (Paris)	5
Dobroschke (Giessen)	4
Traverso (Los Angeles)	4
Five other teams	10
Total	77

travenous glucose tolerance test. Ten patients of the Minneapolis group (Sutherland 1981) received an intraportal autologous islet transplantation and three of them with near total pancreatectomies are insulin independent. Nine of ten partially pancreatectomized patients of Valente et al. (1980b) did not need insulin substitution. Other groups (Dobroschke et al. 1980; Traverso et al. 1981; Cameron et al. 1981; Hinshaw et al. 1981; Grodsinsky et al. 1981; Altman et al. 1983) reported their cases in autologous islet transplantation and the results were dependent on the preoperative endocrine function and on the extent of the pancreas resection.

Allogenic Islet Transplantation

According to published cases and transplantations submitted to the Islet Transplant Registry over the past 5 years, there have been 92 islet allotransplants in 85 diabetic patients. Almost all these attempts have failed. Three patients became insulin independent for sustained periods after islet transplantation. An overall interpretation of the results is extremely difficult, because a variety of sources, techniques, and sites have been used for tissue preparation and transplantation (Table 2).

Histocompatibility data have often been lacking and some patients have not even been immunosuppressed. The reason for uncontrolled application of the allogenic islet transplantation could be that this approach is free of complications. No patients have died as a result of an islet transplant.

We have performed the allogeneic islet transplantation in six diabetic patients. The age of the recipients was between 16 and 50 years and they suffered from diabetes Type I for more than 5 years. They were free from serious angiopathies and islet cell-surface antibodies. All the brain-death donors were less than 30 years old and the pancreases of the donors were perfused during the removal by Hanks' solution for 30 min at 4 °C. The islets were transplanted via the recanalized vena umbilicalis into the portal system. The immunological regime included a preoperative allogenic blood transfusion (500 ml) within 6 months preopera-

Table 2. Ninety-two islet allotransplants submitted to the Human Pancreas and Islet Transplant Registry. Only three patients became insulin independent for a definite period after transplantation

Team	Adult islets	Neonatal fetal	Total
Valentes/Tossati (Genoa)	13	12	25
Najarian, Sutherland (Minneapolis)	13	6	19
Shumakov (Moscow)		17	17
Groth (Stockholm)		9	9
Largiader, Kolb (Zürich)	7		7
Wolff, Lorenz, Lippert (Berlin)	5	1	6
Six other teams	6	3	9
Total	44	48	92

tively and the HLA and DR typing. Immunosuppressive treatment by 2 mg azathioprine (Imuran) and 1.5 mg prednisone/kg body weight was given up to the 14th postoperative day. In all the recipients there was a small increase of the serum C-peptide concentration in comparison to the preoperative levels. Because the effect of the allogenic islet transplantation was too small the immunosuppressive treatment was ended 14 days after transplantation and 6 weeks later there were no signs of graft function. We observed no complications in our patients, but it is to be presumed that while allogenic islet transplantation is not dangerous, it is not an effective approach at the moment.

At the University of Minnesota there were 19 islet allotransplants performed up to 1978. No patients could be completely withdrawn from insulin substitution (Najarian et al. 1977). Largiader et al. (1980) have transplanted islet tissue to seven diabetic patients and the islet transplants were carried out simultaneously with kidney transplants. The kidneys came from the same donor and the recipients were immunosuppressed. Four recipients received the islets into the portal system without success. Three others received the islets into the spleen and in one patient there was a full islet function. The diabetic patient was independent of insulin substitution from 9 up to 20 months postoperatively (Kolb and Largiader 1980). Groth et al. (1980) transplanted minced fetal pancreas to nine patients with long-standing diabetes. No patients were withdrawn from insulin substitution. Shumakov et al. (1983) transplanted fetal islet tissue into the rectus abdominis muscle in 16 diabetic patients and in one case into the pulp of the spleen. After intrasplenic transplantation and in most cases following intramuscular transplantation the recipient's requirement for insulin substitution was reduced.

Valente et al. (1980b) reported their experience with two intraperitoneally and ten intramusculary transplanted islets. In four cases the allogenic fetal pancreatic fragments were cultured for 4 days and in eight instances they were uncultured. One recipient of noncultured tissue and without immunosuppression was withdrawn from insulin substitution 5 months postoperatively. In another group of diabetic patients 13 patients were implanted with one to five diffusion chambers filled with collagenase-dispersed, cultured adult pancreatic tissue. In one patient the chamber was implanted intraperitoneally and in 12 other recipients the chambers were implanted subcutaneously. In two patients it was possible to end the insulin substitution 3 months after implantation of the diffusion chambers. All the results of Valente's group are in contrast to animal experiments and further analysis is necessary.

Future Prospects for Islet Transplantation

Total endocrine replacement to treat insulin-dependent diabetes mellitus is possible by the transplantation of either vascularized pancreatic grafts or free grafts of the islets of Langerhans. The basis for efforts in this field is the hypothesis that secondary complications can be prevented by B-cell replacement (Steffes et al. 1982).

Clinical application of pancreas transplantation has been difficult (Sutherland 1981). First attempts were associated with a high morbidity and mortality and new hopes came from the segmental pancreas transplantation. Islet transplants can ameliorate or treat experimental diabetes in different animal models (Ballinger and Lacy 1972; Sutherland et al. 1974; Sharp et al. 1975; Lorenz et al. 1979).

For clinical islet transplantation an important precondition is a sufficient quantity of viable islets from one donor pancreas. It is possible the islet isolation technique by Velcro, described by Lacy et al. (1982), could solve this problem. Other problems are immunological. The vulnerability of islets to immune destruction was demonstrated by Naji et al. (1979). In a series of experiments Lacy's group found that the influence of islet tissue by special culture conditions was able to reduce the immunogenicity of adult rat islets (Faustman et al. 1982). Possibly the application of special antisera such as antilymphocyte and antimacrophage sera could improve the results of allogenic islet transplantation further (Ulrichs et al. 1983).

Autologous islet transplantation is an acceptable approach for preventing severe deterioration in carbohydrate metabolism resulting from pancreas resections. In patients with severe chronic pancreatitis without insulin dependence before removal of the pancreas it is possible to isolate the islets and to infuse them during the operation into the portal system. An important precondition for preventing complications like disseminated intravascular coagulation consists in a sufficient isolation technique and a careful investigation of the portal system and liver function (Mehigan et al. 1980). Walsh et al. (1982) reported portal hypertension, hepatic infarction, and liver failure after pancreatic islet autotransplantation.

These experiences have resulted in a visible decrease in islet autotransplants in the past 2 years in favor of Whipple's procedures combined with duct occlusion of the remaining left part of the pancreas. At the moment we have to conclude that allogenic islet transplantation is a safe procedure but at this time it is clearly less efficient than pancreas transplantation. Possibly the improvement of islet isolation techniques combined with an effective reduction of immunogenicity by antisera and culture conditions could make islet allotransplantation more attractive in the future.

References

Altman JJ, Houlbert D, Bruzzo F, Desplanque N, Manoux A, Galletti PM (1982) Implantation of semi-permeable hollow fibers to prevent immune rejection of transplanted pancreatic islets. In: Federlin K (ed) Islet-pancreas-transplantation and artificial pancreas. Thieme, Stuttgart, pp 43–45

Ballinger WF, Lacy PE (1972) Transplantation of intact pancreatic islets in rats. Surgery 72:175–186

Cameron JL, Mehigan DG, Harington DP, Suidema GD (1980) Metabolic studies following intrahepatic autotransplantation of pancreatic islet grafts. Surgery 87:397–400

Cameron JL, Mehigan DG, Broe PJ, Zuidema GD (1981) Distal pancreatectomy and islet autotransplantation for chronic pancreatitis. Ann Surg 193:312–317

Dobroschke J, Langhoff G, Kunze HH, Laube H, Schwemmle K (1981) Intraportal transplantation of human islets of Langerhans. In: Federlin K, Bretzel RG (eds) Islet isolation, culture and cryopreservation. Thieme, Stuttgart, pp 168–175

Faustman D, Lacy PE, Davie JM (1982) Transplantation without immunosuppression. Diabetes 31:11–16

Grodsinsky C, Malcom S, Goldman J, Dienst S, Oh HK, Westrick P (1981) Islet cell autotransplantation after pancreatectomy for chronic pancreatitis. Arch Surg 116:511–516

Groth CG, Andersson A, Bjorken C, Gunnarsson R, Hellerström C, Lundgren G, Petersson B, Swenne I, Ostman J (1980) Attempts at transplantation of fetal pancreas to diabetic patients. Transplant Proc [Suppl] 2:208–212

Hinshaw DB, Jolley WB, Hinshaw DB, Kaiser JE, Hinshaw K (1981) Islet autotransplantation after pancreatectomy for chronic pancreatitis with a new method of islet preparation. Am J Surg 142:118–122

Kolb E, Largiader F (1981) Transplantation of pancreatic microfragments in patients with juvenile diabetes. In: Federlin K, Bretzel RG (eds) Islet isolation, culture and cryopreservation. Thieme, Stuttgart, pp 188–194

Lacy PE, Lacy ET, Finke EH, Yasunami Y (1982) An improved method for the isolation of islets from the beef pancreas. Diabetes 31:109–111

Largiader F, Kolb E, Biswanger U (1980) A long term functioning human pancreatic islet allotransplant. Transplantation 39:76–77

Lorenz D, Lippert H, Tietz W, Worm V, Hahn HJ, Dorn A, Koch G, Ziegler M, Rosenbaum KD (1979) Transplantation of isolated islets of Langerhans in diabetic dogs. I. Results after allogeneic intraportal transplantation. II. The influence of histoincompatibility. III. Donor selection by mixed lymphocyte reaction and immunosuppressive treatment. J Surg Res 27:181–213

Lorenz D, Wolff H, Lippert H, Sutherland DER, Hahn HJ (1981) Transplantationsregister für Pankreas und Langerhanssche Inseln. Chirurg 52:590–594

Mehigan DG, Bell WR, Zuidema GD, Eggleston JC, Cameron JL (1980) Disseminated intravascular coagulation and portal hypertension following pancreatic islet autotransplantation. Ann Surg 191:287–293

Najarian JS, Sutherland DER, Matas AJ, Steffes MW, Simmons RL, Goetz FC (1977) Human islet transplantation: a preliminary report. Transplant Proc 9:233–236

Naji A, Barker CF, Silvers WK (1979) Relative vulnerability of isolated pancreatic islets, parathyroid, and skin allografts to cellular and humoral immunity. Transplant Proc 11:560–562

Sharp DW, Murphy JJ, Newton WT, Ballinger WF, Lacy PE (1975) Transplantation of islets of Langerhans in diabetic rhesus monkeys. Surgery 77:100–105

Shumakov VI, Blumkin VN, Ignatenko SN, Skaletsky NN, Kaurechiva NI, Seid-Guseinov AYA, Slovesnova TA (1983) Transplantation of islet cell cultures of the pancreas of human fetuses to diabetes mellitus patients. Klin Med (Mosk) 61:48–51

Steffes MW, Vernier RL, Brown DM, Basgen JM, Mauer SM (1982) Diabetic glomerulopathy in the uninephrectomized rat resists amelioration following islet transplantation. Diabetologia 23:347–353

Sutherland DER (1981) Pancreas and islet transplantation. I. Experimental studies. II. Clinical trials. Diabetologia 20:161–185, 435–450

Sutherland DER, Steffes MW, Mauer SM, Najarian JS (1974) Isolation of human and porcine islets of Langerhans and islet transplantation in pigs. J Surg Res 16:102–111
Traverso LW, Alou-Zamzam AN, Longmire WP (1981) Human pancreatic cell autotransplantation following total pancreatectomy. Ann Surg 193:191–195
Ulrichs K, Schang T, Keller R, Müller-Ruchholtz W (1984) Reactivity of pancreas islet cells with antisera of known unspecificity. Transplant Proc 16:857–860
Valente U, Ferro M, Campisi C, Parodi F, Tosatti E (1980a) Allogenic pancreatic islet transplantation by means of artificial membrane chambers in 13 diabetic recipients. Transplant Proc [Suppl 2] 12:223–228
Valente U, Ferro M, Campisi C, Cataldi L, Traverso C, Fontana I, DeRosa E, Cama A, Tosatti E (1980b) Report of clinical cases of islet transplantation. Transplant Proc [Suppl 2] 12:202–204
Valente U, Ferro M, Barocci S, Campisi C, Parodi F, Cataldi L, Arcuri V, Tosatti E (1980c) Report of clinical cases of human fetal pancreas transplantation. Transplant Proc [Suppl 2] 12:213–217
Walsh TJ, Eggleston JC, Cameron JL (1982) Portal hypertension, hepatic infarction, and liver failure complicating pancreatic islet autotransplantation. Surgery 91:485–487
Wolff H, Lorenz D, Lippert H (1981) Der gegenwärtige Stand der Pankreasinsel-Transplantation. Dtsch Med Wochenschr 106:506–511

Present Status of Clinical Liver Transplantation and Potential Developments Based on Experimental Microsurgical Models

C. E. Brölsch

Clinical liver transplantation has proceeded to a new stage of clinical importance based upon three major developments: acceptance of clinical indications for a variety of end-stage hepatic diseases, more experience in surgical management, and the introduction of cyclosporin A for immunosuppressive therapy (Brölsch et al. 1984; Calne et al. 1979b, 1981; Pichlmayr et al. 1983; Starzl et al. 1982). The first development has derived from consensus among hepatologists and surgeons, announced in the National Institutes of Health Consensus Development Conference Statement of 1983, accepting liver transplantation primarily as therapy for benign end-stage liver diseases which have been treated in the past by management of their complications without substantially changing the prognosis of the patients. Hepatic tumors are rarely considered as an indication because of the high probability of recurrent tumor within the first postoperative year (Calne et al. 1981; Fortner et al. 1970; Krom et al. 1983; Pichlmayr and Brölsch 1983). Selection criteria possibly based upon histological differentiation need to be determined (Pichlmayr et al. 1981). The number of patients requiring liver transplantation by far exceeds the capacity of the transplantations centers currently undertaking. Consequently, new centers for liver transplantation need to be established to treat an increasing number of patients suffering from progressive liver disease for which there is no alternative therapy.

Clinical acceptance is based upon the achievements of the major transplantation groups in Pittsburgh (Starzl, Iwatzuki, Shaw), Cambridge (Calne, Rolles), Hannover (Pichlmayr, Brölsch, Neuhaus), and Groningen (Krom, Slooff, Gips). Several other institutions have launched programs for liver transplantation, emphasizing the need for more transplantations to be performed particularly in pediatric patients (Ascher and Najarian 1984; Brölsch et al. 1984; Iwatsuki et al. 1984; Krom et al. 1983). Improved surgical technique and postoperative patient management have resulted in increased numbers of transplanted patients and better survival rates (Brölsch et al. 1984; Calne 1969; Calne et al. 1981; Starzl et al. 1982). In particular reinvention of the porta-cava-jugular bypass for decompression during the cross-clamping phase has resulted in less postoperative complications (i.e., renal failure, respiratory insufficiency) because sequelae of bowel congestion and hemodynamic disturbances could be avoided. Even successful performance of hepatic retransplantation has become possible, since recognition

Microsurgical Models in Rats
for Transplantation Research
Eds. A. Thiede et al.

Table 1. Centers and numbers of transplantations per year

Orthotopic liver transplantation 1983/84		
Center	Number of transplantations	
	Currently	Aimed per year
Denver/Pittsburgh	ca. 400	ca. 100
Cambridge	145	20–30
Hannover	117	30
Groningen	34	15
Minnesota	32	5–10
Berlin Charité	22	3–5
Villejuif	13	2–4
Innsbruck	12	10
Hamburg	2	10
Berlin	2	3–5

of problems of early graft failure due to preservation damage, vascular occlusion, or chronic rejection have become accepted (Pichlmayr and Brölsch 1983; Starzl et al. 1982). Today it is reported that 1-year survival rates following liver transplantation are in the order of 60% and in selected groups of patients, i.e., children, in the range of between 75% and 83%. Nevertheless the risk of dying while waiting to be transplanted is high in the pediatric recipient group (Iwatsuki et al. 1984) (25/64) and fairly high in the adult group (18/68) according to van Thiel et al. (1982).

The Hannover experience (Pichlmayr et al. 1983) with a limited number of patients is clearly in support of this view, which has been mainly put forward by the leading transplant group in Pittsburgh. However, there is still particular interest in separating the groups of indications rather than calculating overall results. It has been shown (Pichlmayr and Brölsch 1983) that, depending on the time of indication as well as the underlying disease, different results can be obtained. With patients transplanted in the presence of coma of "urgent clinical indication" the outcome was poor mainly due to multiorgan failure in the postoperative course. In contrast, of patients transplanted in an elective situation, 68% survived the operation and the subsequent course, despite being in the final stage of their disease. It is therefore concluded that liver transplantation should be performed in patients in the best possible situation at an appropriate stage of their disease, with the prospect of regaining normal quality of life without anticipating repeated complications and eventual sudden death. The proposed differentiation is not always accepted but it is essential to maintain it at least temporarily until objective and comparable guidelines for indications have been developed. Liver transplantation will continue to be performed on critically ill and patients in coma since there is no alternative therapy available. It is conceivable that an operation of such magnitude has a high operative mortality when performed in desperate situations. The life-saving benefit can therefore be masked by its application to patients requiring transplantation earlier than they finally received it (Brölsch et al. 1982; Calne 1969; Starzl et al. 1982).

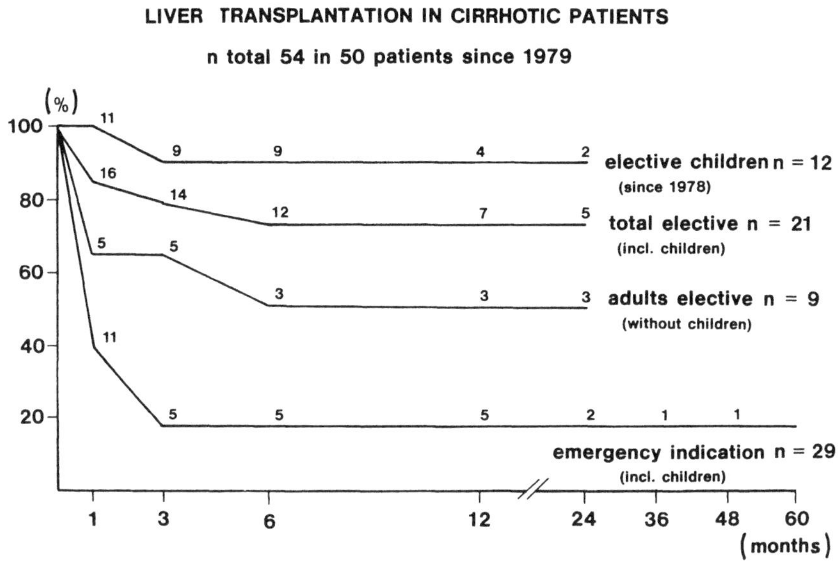

Fig. 1. Survival rates according to the time of indication. (Brölsch et al. 1984)

Additional major progress in liver transplantation has been achieved by the introduction of cyclosporin A (CyA) for immunosuppressive therapy. Clinical reports are increasing in number (Ascher and Najarian 1984; Brölsch et al. 1984; Iwatsuki et al. 1984; Starzl et al. 1982) although the mechanism of action and pharmacodynamic properties still need to be elucidated. In a series of 37 patients treated with conventional immunotherapy after liver transplantation 24 developed recognizable rejection episodes and 4 died from uncontrollable rejection. Introduction of CyA reduced the total number of rejections and basically avoided acute uncontrollable rejection and graft loss (Pichlmayr et al. 1983). As a result of this improvement survival figures in other transplant groups increased significantly (Iwatsuki et al. 1984; Krom et al. 1983; Pichlmayr and Brölsch 1983; Starzl et al. 1976, 1982).

Despite this clinical development and improvement the frequency, clinical course, and prognosis of rejection have only altered since CyA has been available but the basic problems of rejection, its diagnosis, and its treatment remain the most important issue in clinical liver transplantation. Hence, there are very few experimental models available where the variety of immunological problems of liver grafting can be precisely evaluated.

One of the most important contributions microsurgical models can make to clinical progress is in the area of evaluation of rejection phenomena. In the past, the immunological effects have been regarded generously due to the shortage of donor organs, urgency of indications, and some experimental data mainly from pig experiments (Balner 1980; Bockhorn et al. 1974; Calne et al. 1969, 1970; van Es and Balner 1979). Compared with skin, kidney, heart, and pancreas transplants, liver grafts showed fewer episodes of rejection and indefinite survival in unrelated strain combinations. Investigations in tissue-typed animals demon-

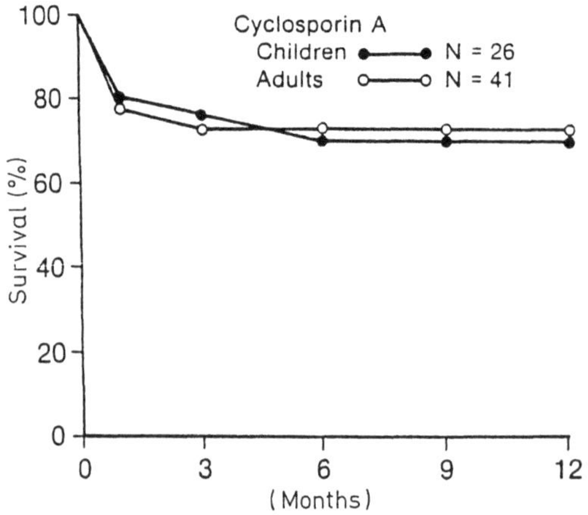

Fig. 2. Survival rates in the Pittsburgh group. (Starzl et al. 1982)

strated spontaneous tolerance even in major histocompatibility complex (MHC)-different animals. If two organs from the same donor are transplanted simultaneously, the liver as well as the other organ will be accepted. It was concluded by Calne et al. (1969) that this phenomenon is a modification of the immune response by the liver transplant. Nevertheless, the underlying mechanism requires further investigation.

Differentiation of the allograft functional disturbances has proved to be of importance, since therapeutic approaches to rejection, perfusion damage, vascular occlusion or biliary tract obstruction, and infection are distinctly different. Since the introduction of the technique of orthotopic liver grafting in the rat by Lee et al. (1972), long-term survival rates of the graft have rarely exceeded 60% (Lee et al. 1972). Improved survival was achieved by Engemann et al. (1982) and by Zimmermann et al. (1983) with an advanced technique of reanastomosing the arterial supply of the liver graft. Improvements in observation of the graft morphology were achieved and specific transplantation tolerance in the fully allogeneic strain combination was observed (Engemann et al. 1982; Zimmermann et al. 1983).

Technically, the findings of Engemann et al. (1982) pointed clearly toward the necessity of arterial blood supply to a liver graft. This corresponds with clinical and morphological features in liver homografts which are even more susceptible to hepatic arterial obstruction, which inevitably results in either acute or progressive graft loss by necrosis of the intrahepatic biliary tract system.

Of major importance to the microsurgical models is immunological evaluation of liver grafts. Although the model is limited because of the extraordinary skill required to perform the operation, some aspects can readily be transferred to the human situation.

In the Cambridge laboratory Zimmermann et al. (1983) started to use the rat orthotopic model for consecutive immunological evaluations. They were the first to take advantage of the availability of strains of known major histocompatibility types and could clearly demonstrate that despite major differences in the MHC spontaneous survival of liver allografts could be observed. As a consequence of observations in many fully allogeneic combinations, they were able to demonstrate that prolonged survival of orthotopic liver allografts in rats is not a general

Table 2. Current problems under investigation in the rodent animal model

Diagnosis of graft functional disturbances	Differentiation of allograft functional disturbances caused by rejection, cholangitis, circulation
Immunology	Significance of histocompatibility antigens Improvement of immunosuppression Drug mechanism of action and metabolism

Table 3. Immunology in liver grafting

	Clinical relevance	Results in experimental rat model	Further significance
Rejection			
Tolerance phenomena	Questionable	Depending on species observed	Open
Hyperacute rejection	Never observed	Questionable, on species inductable depending	Open
Acute / Chronic rejection	Frequently observed	Established in defined strain combinations	Evaluation of characteristic features and mechanisms

property of this organ but is strictly dependent on the strain combinations involved. This points toward evaluation of the significance of histocompatibility antigens in the genetically well-defined rat strains and the possible genetic control of a certain responsiveness to organ allografts.

Wonigeit et al. (1979) described post-transplant changes in specific precursor T-cell reactivity inducing a status of hyporesponsiveness after transplantation. This phenomenon was studied by precursor-cell reactivity against MHC alloantigen differences in humans by the mixed leukocyte culture (MLC) and the cell-mediated cytotoxicity (CMC) assay. In the rat orthotopic model, the same phenomenon recently was described by Kamada et al. (1982) in fully allogeneic rat strains.

Tolerance phenomena and recently discussed features of hyperacute rejection (Andrzejewski and Brölsch 1982; Bockhorn et al. 1974; Calne et al. 1970; Jaffe et al. 1967) are still open for further investigation. The rat orthotopic liver model provides a unique opportunity to study both phenomena under genetically controlled biomodels. The frequently used heterotopic rat liver model produces controversial results as well as new information. Heterotopic grafts are rejected in strong allogeneic strain combinations (DA-LE and PVG) whereas orthotopic grafts are accepted. Additional heart allografts are also rejected if transplanted in a second fashion set (i.e., 30 days after heterotopic liver grafting with no tolerance to be seen). It is therefore concluded that the host liver remnant exerts an immunostimulating effect mediated by the reticuloendothelial system which is left unchanged. Removal of the host liver leads to less immunogenicity, thereby explaining donor-specific tolerance phenomena (Andrzejewski and Brölsch 1982; Müller et al. 1983).

For clinical work, phenomena of acute and chronic rejection will be the key points of research in the near future. It will be necessary to evaluate characteristic features and mechanisms of rejection in order to obtain information of the underlying mechanism itself and its particular influence by immunosuppression. It is no longer justifiable to think about an immunological privilege of the liver graft. It is rather important to clarify the variety of factors involved, such as HLA typing and immunosuppressive therapy. Because of the complexity of the clinical situation even in long-term surviving patients, observations will not be conclusive. Evaluations have to be made in animal experiments, basically in the rat microsurgical model and possibly, if precise findings can be elucidated, in a preclinical situation such as the primate model.

For the animal species investigated thus far, discussion is undoubtedly required of specific immune mechanisms, which are seen as specific tolerance, hyporesponsiveness, pecularities of the antigen content, or release of substances with immune specificity. The question remains of whether the described phenomena are exceptional findings specific aspects of immunoregulation, or key models similar to the homograft situation (Andrzejewski and Brölsch 1982; Balner 1980; Bockhorn et al. 1974; Calne et al. 1970). Analysis of the experience in homografting would definitely require research in the rat animal model with emphasis of the rejecting combinations. The course of rejection, the underlying morphological and functional disturbances in the graft, and the effectiveness of immunosuppression can be exclusively and conclusively studied in a model of rejecting combinations. Therefore immunogenetically determinable species such as the rat and primates provide the optimal basic research models (Balner 1980; Calne et al. 1970; van Es and Balner 1979).

Starzl et al. (1961, 1966, 1967) has repeatedly emphasized the importance of transplant rejection according to his experiments in dogs. His subsequent search for effective immunosuppressive therapy by employment of antilymphocyte globulin (ALG), thoracic duct drainage, and recently CyA supports his opinions about the significance of immunosuppressive therapy. Use of the latter immunosuppressive regimen points toward the currently most relevant problems in clinical liver transplantation such as optimizing of immunosuppression. Since the use of CyA possibly opens up a new era of transplantation (heart, heart-lung, pancreas) many problems still have to be answered.

Dosage of the drug, its metabolism, and its mode of action are still under investigation. Although clinical requirements and experiences are already advanced compared with experimental findings, basic biochemical pathways for CyA still need to be determined. Problems of absorption and metabolism occur after homografting as described by Wonigeit et al. (1983). Orally applied CyA can normally be absorbed from the intestine and secreted unchanged or biochemically modified by the liver. Possible enterohepatic recirculation of the unmodified drug is also assumed. Disturbances of bile production or intestinal function can reduce the absorption rate to an unknown degree. This can be overcome by intravenous treatment until the underlying disturbances stop. Impaired graft function due to ischemia, or toxic or immunological damage can reduce drug metabolism and ex-

Course after liver transplantation in a tumor patient with cirrhosis (R.K. 50 ♂)

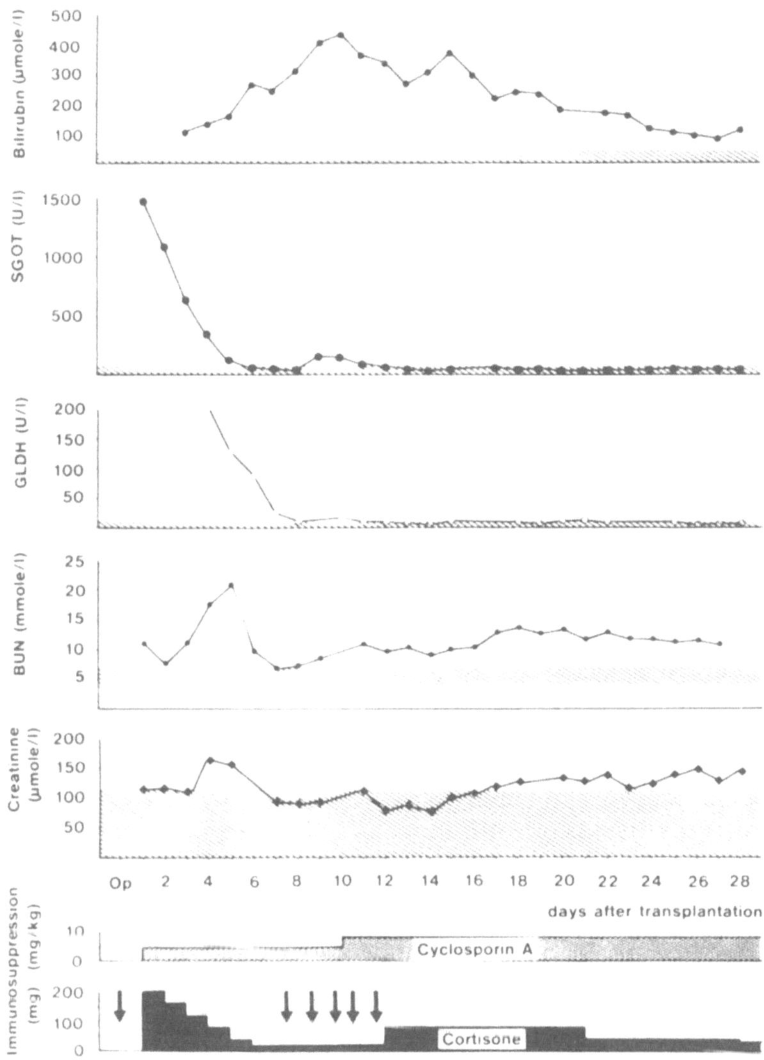

Fig. 3. Postoperative course with nephrotoxicity and acute controllable rejection. Following transplantation and immunosuppressive therapy with 5 mg/kg body weight i.v. CyA in combination with cortisone, nephrotoxicity occurred within 3 days, diagnosed by marked elevation of serum creatinine and blood urea nitrogen (*BUN*). Appropriate dosage adjustment led to reversibility of renal functional impaiment. Acute controllable rejection occurred on day 7 with consecutive treatment by steroid bolus. CyA was administered orally after day 10 at a dosage of 10 mg/kg body weight. *SGoT*, serum glutamic oxaloacetic transaminase

cretion leading to excessive drug accumulation and intoxication. In addition, disurbances of various types are not mutually exclusive. Combinations or rapid changes from one to another may render the choice of the correct CyA dosage very difficult.

Fortunately, toxic effects of high CyA administration are fully reversible as soon as the dosage is correctly adjusted. Thus far, CyA has been convincingly effective and results of clinical liver transplantation are steadily improving. Since in liver transplantation clinical experience and experience with large animals usually precede findings in the microsurgical animal model, there will be a progressive need for immunological research, which – in many ways – can be best carried out in the rat animal model. Finally, after technical and physiological problems were overcome, the rat model turned out to be of appropriate assistance in clinical liver grafting.

References

Andrzejewski W, Brölsch CE (1982) Postoperative reactions of rats of orthotopic liver transplantation: a model for human response. Eur Surg Res 14:428–439

Ascher N, Najarian J (1984) Hepatic transplantation and biliary atresia: early experience in eight patients. World J Surg 8:57–63

Balner H (1980) The DR system of rhesus monkeys: a brief review of serology, genetics and relevance to transplantation. Transplant Proc 12:502–508

Bockhorn H, Lauchart W, Ringe B, et al. (1974) Die Immunreaktion nach allogener Schweinelebertransplantation. Langenbecks Arch Chir [Suppl Chir Forum] 135–138

Brölsch CE, Neuhaus P, Pichlmayr R (1982) Gegenwärtiger Stand der Lebertransplantation. Z Gastroenterol 17:117–122

Brölsch CE, Neuhaus P, Burdelski M, et al. (1984) Aktueller Stand der Lebertransplantation. Diagnostik 17:12–16

Calne RY (1969) Surgical aspects of clinical liver transplantation. Br J Surg 56:729–735

Calne RY, Sells RA, Pena JR, et al. (1969) Induction of immunological tolerance by porcine liver allografts. Nature 223:472–476

Calne RY, Davies DR, Pena JR, et al. (1970) Hepatic allografts and xenografts in primates. Lancet 1:103–106

Calne RY, McMaster P, Portmann P, et al. (1977) Observations on preservation, bile drainage and rejection in 64 human orthotopic liver allografts. Ann Surg 186:282–290

Calne RY, McMaster P, Smith D, et al. (1979a) Use of partial cardiopulmonary bypass during the anhepatic phase of orthotopic liver grafting. Lancet 2:612–614

Calne RY, Rolles K, White DJ, et al. (1979b) Cyclosporine A initially as the only immunosuppressant in 34 recipients of cadaveric organs: 32 kidneys, 2 pancreases and 2 livers. Lancet 2:1033–1036

Calne RY, Williams R, Lindop M, et al. (1981) Improved survival after orthotopic liver grafting. Br Med J 283:115–118

Engemann R, Ulrichs K, Thiede H, et al. (1982) Value of physiological liver transplant model in rats. Transplantation 33:566–568

Fortner JG, Beattie EJ, Shiu MH, et al. (1970) Orthotopic and heterotopic liver homografts in man. Ann Surg 172:23–32

Gütgemann A, Lie TS, Esser G, et al. (1971) Operative Aspekte bei der orthotopen, humanen Lebertransplantation: Erfahrungen bei drei Lebertransplantationen. Chirurg 42:167–172

Houssin D, Branco D, Perthelot P, et al. (1980) Heterotopic liver transplantation in end stage HbsAG positive cirrhosis. Lancet 1:990–992

Iwatsuki S, Shaw BW, Starzl TE (1984) Liver transplantation for biliary atresia. World J Surg 8:51–56

Jaffe WP, Symes MO, Terblanche J (1967) Observations on the immunological reaction of pigs. Colston Papers 19:351–357
Kamada N, Davies HS, Roser BJ (1981) Fully allogeneic liver grafting and the induction of donor-specific unreactivity. Transplant Proc 13:837–841
Krom RAF, Gips CH, Newton D, et al. (1983) A successful start of a liver transplantation program. Transplant Proc 15:1276–1281
Lee S, Charters AC, Chandler JG, et al. (1973) A technique for orthotopic liver transplantation in the rat. Transplantation 16:664–669
Müller GH, Hopt UT, Bockhorn H (1983) Are liver transplants less immunogenic than heart or kidney transplants? Chir 83:173–176
Pichlmayr R, Brölsch CE (1983) Lebertransplantation: Entwicklungsstand und Tendenzen. Med Klin 78:744–752
Pichlmayr R, Brölsch CE, Tidow G, et al. (1981) Indikation zur Lebertransplantation bei Lebertumoren. In: Zelder, Röher, Fischer, Bode (eds) Experimentelle und klinische Hepatologie. Schattauer, Stuttgart, pp 53–57
Pichlmayr R, Brölsch CH, Neuhaus P, et al. (1983) Report on 68 human orthotopic liver transplantations with special reference to rejection phenomena. Transplant Proc 15:1279–1283
Starzl TE, Kaupp HA, Brock DR, et al. (1961) Studies on rejection of the transplanted homologous dog liver. Surg Gynecol Obstet 112:135–144
Starzl TE, Marchioro TL, Faris TD, et al. (1966) Avenues of future research in homotransplantation of the liver. Am J Surg 112:391–400
Starzl TE, Marchioro TL, Porter KA, et al. (1967) The use of heterologous antilymphoid agents in canine renal and liver homotransplantation and in human renal homotransplantation. Surg Gynecol Obstet 124:301–318
Starzl TE, Porter KA, Putnam CW, et al. (1976) Orthotopic liver transplantation in minety-three patients. Surg Gynecol Obstet 142:487–505
Starzl TE, Iwatsuki S, van Thiel D, et al. (1982) Evolution of liver transplantation. Hepatology 2:614–636
Van Es AA, Balner H (1979) Influence of blood transfusions and matching for DR antigens on kidney allograft survival in unrelated rhesus monkeys. Transplantation 28:135–139
Van Thiel DH, Schade RR, Starzl TE, et al. (1982) Liver transplantation in adults. Hepatology 2:637–640
Wonigeit K, Bockhorn H, Pichlmayr R (1979) Posttransplant changes in specific precursor T-cell reactivity. Comparison between liver and kidney allograft recipients. Transplant Proc 11:1250–1255
Wonigeit K, Brölsch CE, Neuhaus P, et al. (1983) Social aspects of immunosuppression with cyclosporine in liver transplantation. Transplant Proc [Suppl 1] 15:2586–2591
Wolff H, Otto E (1981) Neue Erkenntnisse auf dem Gebiet der klinischen Lebertransplantation. Zbl Chir 106:1448–1451
Zimmermann F, Knoll P, Davies H, et al. (1983) The fate of orthotopic liver allografts in different rat strain combinations. Transplant Proc 15:1272–1275

Present Status of Clinical Kidney Transplantation and Possible Developments with Respect to Experimental Microsurgery Models

F. W. Eigler

Kidney transplantation may be looked at as an established therapeutic method in patients with terminal chronic renal failure. But there are still several problems to be solved. The most important of them are

1. Organ procurement
2. Side effects of maintenance and/or antirejection therapy
3. Problems of long follow-up (chronic rejection, hypertension, development of malignant tumors)

The problem of organ procurement is apparent in most centers, especially in the Federal Republic of Germany, which has a mean waiting time of approximately 2 years for a recipient (Pichlmayr 1983). While the main aim must still be to improve the cooperation between transplantation centers and hospitals with potential donors, some progress could be achieved with better organ preservation methods.

The gap between transplanted and needed organs could also be closed in part, if the results of transplantation in kidney function or survival time were to improve: But while in the past 10 years a stepwise improvement has been noted with respect to recipient survival, results of organ function time have shown nearly no change. This development can be demonstrated in the center of Essen, which has performed over 500 transplantations.

In Fig. 1 four groups of patients in different periods, each group with about or more than 100 patients, are compared. The differences in the treatments used in the four groups are listed in Table 1. Each group showed a better survival rate of patients than the previous one, ending with a figure of more than 95% in the 1st year after transplantation.

The lowered letality seems to be mainly due to the overall reduction in the dosages of corticosteroids and more careful indications in antirejection therapy. On the other hand, as shown in Fig. 2, in the same groups the organ survival time did not change over the years: all attempts with better tissue typing, use of antilymphocyte globulin, or change in blood transfusion policy did not improve the overall results with respect to the transplanted kidneys.

Looking back to the many hopeful results in small subgroups a word of caution seems appropriate: Table 2 summarizes some of the known factors which may in-

Microsurgical Models in Rats
for Transplantation Research
Eds. A. Thiede et al.

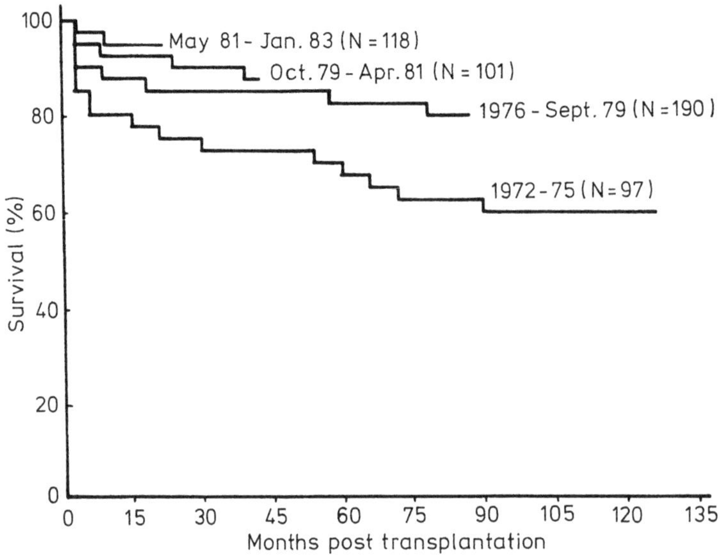

Fig. 1. Actuarial patient survival in four periods with different immunosuppressive regimens (Dept. of General Surgery, University of Essen, 1972–1983)

Table 1. Immunosuppression following renal transplantation (Dept. of General Surgery, University of Essen, 1972–1983)

	Period	Basic therapy	Antirejection therapy
1.	7.72–12.75	High-dose steroids (appr. 3.9 g/1 month) Azathioprine (preop., 4 mg/kg body wt.; postop., 1–2 mg/kg body wt.	High-Dose steroids (up to 25 g/ months) Graft irradiation Actinomycin D Cyclophosphamide
2.	1.76– 9.79	High-dose steroids Azathioprine	Reduction of steroids (max. 10 g/3 months) No other treatment
3.	10.79– 4.81	High-dose steroids Azathioprine FAC. ATG (2 mg/kg body wt./ 3 weeks)	Reduced steroids FAC. ATG
4.	5.81– 1.83	Low-dose steroids (appr. 1.9 g/1 month) Azathioprine ATG (2 mg/kg body wt./3 weeks)	Reduced steroids FAC. ATG

ATG, antithymocyte globulin

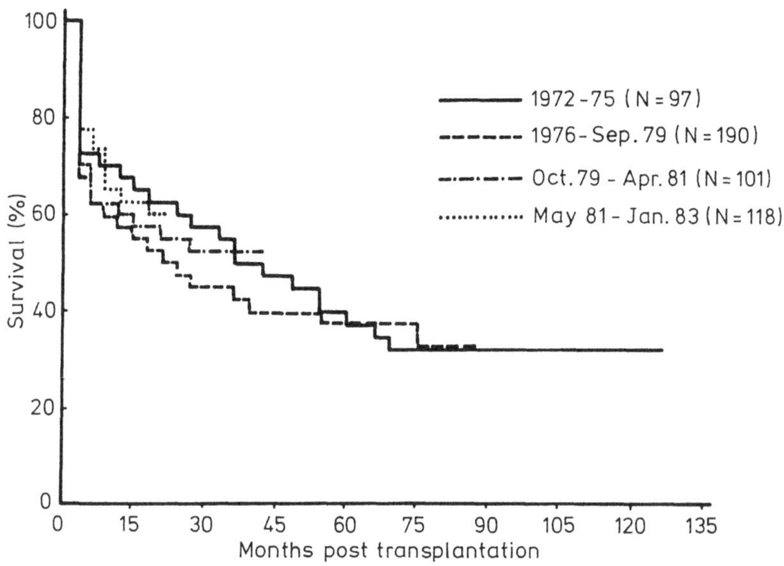

Fig. 2. Actuarial graft survival in four periods with different immunosuppressive regimens (Dept. of General Surgery, University of Essen, 1972–1983)

Table 2. Factors known to influence the outcome of kidney transplantation

1. Responder status
2. Blood transfusions
3. HLA-DR system
4. HLA-AB-EM system
5. ABO-Lewis system
6. Race
7. Sex
8. Age
9. General situation
10. Splenectomy
11. Compliance

fluence the outcome of kidney transplantation (Eigler et al. 1981). Therefore it appears to be difficult to draw conclusions from small investigated groups, because other factors may eventually influence the results more than the modification under investigation.

The introduction of the new drug cyclosporin A promises a real step forward despite all the reservations. So far reports in the literature (Beveridge et al. 1982; Starzl et al. 1982) show a marked improvement of survival time of the transplanted kidney, around 80% after 1 year compared with 60% in conventional therapy. Importantly, these seem to be overall results and not from special subgroups, though one large multicenter study excluded patients with primary acute renal failure after transplantation (Beveridge et al. 1982).

Table 3. Immunological problems

Basic problems (for example, role of different cells in acute and chronic rejection processes, enhancement)
Influence of pretransplant blood transfusions
Biological immunosuppression (for example, spleen: splenectomy, irradiation – blood drainage in portal vein)
Drug testing
New drugs (for example, prostaglandine)
Drug combinations
Combinations of biological and drug immunosuppressives

While the minor side effects of cyclosporin A, namely hirsutism, gingiva hypertrophy, nausea, and tremor, can be avoided by correction of the dosage, nephrotoxicity, hepatoxicity, and tumor induction are still real problems.

In practical kidney transplantation the main problem with the use of cyclosporin A is that of nephrotoxicity, especially if acute tubular necrosis occurs after transplantation. The policy of using kidneys harvested at the same center only seems to be the consequence in some centers. But this cannot be looked at as a good solution with respect to a national and international organ exchange program with all the other advantages. The other problem with cyclosporin A is the possibility of the development of malignant tumors in the long run. Though this problem is also true for all other kinds of immunosuppressive therapy it should be followed very carefully.

In summary then, kidney transplantation has become safer with time but several aspects need to be investigated clinically and experimentally.

In the present situation of renal transplantation the following research projects, especially in microsurgery models, are suggested though species differences should not be neglected (Collins and Halasz 1982).

It is apparent that immunological investigations as described in this volume are still needed to broaden our knowledge of the rejection process and possible therapeutic measures. But this is true for all organ transplantation. Some suggestions of clinical relevance are listed in Table 3.

The rat model could give more insight for instance into the "biological" possibilities of influencing the antirejection process by removing or damaging the spleen by radiation (Dostal 1978; Jakubowski et al. 1981). The detection of cyclosporin A should not hinder the search for other agents (Strom and Carpenter 1983), the combination of which could further improve the results or lower the side effects (Mahdi et al. 1983).

The special situation of kidney transplantation could be improved by investigations on preservation problems. There are several ideas now on how the results of preservation may be improved (Fischer et al. 1982; Howden et al. 1983; Marshall et al. 1984). Furthermore, more investigations are needed to test the role of ischemia time for the postoperative function. The aim should be to abolish ischemic lesions in the kidney during preservation by one or the other method. Mostly the theoretical expectations with sophisticated investigations of metabolic changes are tested in short-term experiments. Here the transplantation in rats could give some indications for the clinical relevance of these ideas in the long run.

The problem of induction of tumors by cyclosporin A should be followed again in the rat model, though the conditions might be difficult to compare with the human being. But in the case of positive results caution would be mandatory.

As in the tumor problem there are several other aspects of long-term side effects with kidney transplantation and immunosuppressive therapy. The most important example in this respect can be seen in blood pressure behavior. Hypertension may occur after transplantation or may be present long before. The special conditions should be followed again in the rat model.

The innovation of experimental microsurgery promises further advantages in the important field of kidney transplantation. It is therefore hoped that corresponding research programs will be encouraged by this symposion.

References

Beveridge T, Maurer W, Poole TW, Wood AJ, Wiskott E (1982) Steroidfreie Behandlung nierentransplantierter Patienten mit Cyclosporin A. Eine europäische multizentrische Studie. Klin Wochenschr 60:1137–1142

Collins GM, Halasz NA (1982) A species difference in the efficacy of two intracellular flush solutions. Transplantation 33:324

Dostal G (1977) Der Einfluß des Splenektomie-Zeitpunktes auf die Überlebenszeitverlängerung nach Nierentransplantation bei der Ratte. Langenbecks Arch Chir [Suppl] 250:250–253

Eigler FW, Jakubowski HD, Letsch R (1981) Immunsuppressive Therapie – ist weniger mehr oder anderes besser? Mitt Arbeitsgemeinschaft Klin Nephrol 10:81–87

Fischer JH, Reiferscheidt G, Fuhs M, Wenzel M, Isselhard W (1982) Deuterium oxide (D_2O) for organ preservation. In: Pegg DE, Jacobson IA, Halsz NA (eds) Organ preservation – basic and applied aspects. MTP Press, Lancaster, p 199

Howden B, Rae D, Jablonski P, Marshall VC, Tange J (1983) Studies of renal preservation using a rat kidney transplant model (evaluation of citrate flushing). Transplantation 35:311–314

Jakubowski HD, Dostal G, Eigler FW (1981) Verlängerung der Überlebenszeit von Nierentransplantaten bei der Ratte nach selektiver Röntgenbestrahlung der Milz. Langenbecks Arch [Suppl 1981]:191–194

Mahdi NA, Hutchinson IV, Brent L (1983) Specific unresponsiveness to fully allogenic kidney allografts in rats induced by procarbazine hydrochloride and antilymphocyte serum. Transplantation 36:480–485

Marshall V, Jablonski P, Howden B, Rae D, Rigol G, Scott DF, Tange J (1984) Citrate magnesium solutions improve both flush and pump preservations of the rat kidney. Transplant Proc 16:170–172

Pichlmayr R (1983) Transplantation. Langenbecks Arch Chir 361:65–70

Starzl TE, Hakala TR, Rosenthal JT, Iwatsuki S, Shaw BW Jr (1982) Variable convalescence and therapy after cadaveric renal transplantation under cyclosporin A and steroids. Surg Gynecol Obstet 154:819–825

Terry B, Strom TB, Carpenter CB (1983) Prostaglandin as an effective antirejection therapie in rat renal allograft recipients. Transplantation 35:279–281

Conclusion

A. Thiede

This book is the outcome of an international microsurgical workshop. The workshop was held with two particular goals in mind. One was to discuss the relevance of the current microsurgical models in the species rat for transplantation research in general. The other was to introduce young transplantation surgeons to the problems and implications of certain experimental transplantation models. In general, the question should be raised whether a given experimental model is of practical value, i.e., whether it is a preclinical model that can contribute to the solution of clinical problems or whether it is of purely theoretical value. The present volume addresses the questions raised by clinicians as to the practical applicability of preclinical models. Therefore, it attempts to analyze the preclinical relevance of microsurgical models in rats.

If we proceed from the assumption, which is probably correct, that immunological reactions in the species rat follow a course that is similar or analogous to that in humans, we can, in principle, apply the results gained with rats to humans, given analogous models. To ensure that experimental models are as relevant, preclinically, as possible, they should differ from approaches used in human transplantation in only one parameter, namely, the different species.

So far this is clearly the case only in vessel and kidney transplantation and in orthotopic liver transplantation with arterialization. Thus the results of rat liver transplantation in the fully physiological arterialized model, which correspond to human liver transplantation, are particularly relevant, clinically, in contrast to nonarterialized or auxiliary models. For this reason only this model should be used in experiments.

In pancreas transplantation in rats, approaches that are analogous to human transplantation methods are also possible: pancreas transplantation with duct occlusion. By comparing models with duct occlusion (unphysiological models) and models with duct drainage (physiological models), it can be clearly demonstrated experimentally that the physiological models are both short-term, from an immunological point of view, and long-term, from a morphological and functional point of view, more satisfactory than models with duct occlusion. Thus experimental data in rats appear to justify the assumption that duct occlusion models in human pancreas transplantation are, in the long run, a step in the wrong direction that needs to be corrected.

Microsurgical Models in Rats
for Transplantation Research
Eds. A. Thiede et al.

The experimental results of heart and small-intestine transplantation in rats thus far are based on bypass models. They have provided interesting immunological data of theoretical value. Before, however, preclinical results can be achieved that are of practical relevance for clinical transplantation, physiological models must be established. Great efforts are being made to do just this in small-intestine transplantation.

In summary, we can say that the preclinical applicability of a model is greatest from an immunological, morphological, and functional point of view when it is fully physiological and/or compatible with human transplantation. In future preclinical transplantation research this should be heeded more than it has been up till now, to increase the relevance of such research for the solution of clinical problems.

Subject Index

Main references in *italics*

MIX
Papier aus verantwortungsvollen Quellen
Paper from responsible sources
FSC® C105338

If you have any concerns about our products,
you can contact us on
ProductSafety@springernature.com

In case Publisher is established outside the EU,
the EU authorized representative is:
Springer Nature Customer Service Center GmbH
Europaplatz 3, 69115 Heidelberg, Germany

Printed by Libri Plureos GmbH
in Hamburg, Germany